Novel
Cosmetic
Delivery
Systems

COSMETIC SCIENCE AND TECHNOLOGY

Series Editor

ERIC JUNGERMANN

Jungermann Associates, Inc.
Phoenix, Arizona

Novel Cosmetic Delivery Systems

edited by
Shlomo Magdassi
Elka Touitou

The Hebrew University of Jerusalem
Jerusalem, Israel

CRC Press
Taylor & Francis Group
Boca Raton London New York

CRC Press is an imprint of the
Taylor & Francis Group, an **informa** business

CRC Press
Taylor & Francis Group
6000 Broken Sound Parkway NW, Suite 300
Boca Raton, FL 33487-2742

First issued in paperback 2019

© 1999 by Taylor & Francis Group, LLC
CRC Press is an imprint of Taylor & Francis Group, an Informa business

No claim to original U.S. Government works

ISBN-13: 978-0-8247-1703-2 (hbk)
ISBN-13: 978-0-367-40022-4 (pbk)

**Visit the Taylor & Francis Web site at
http://www.taylorandfrancis.com**

**and the CRC Press Web site at
http://www.crcpress.com**

About the Series

The Cosmetic Science and Technology series was conceived to permit discussion of a broad range of current knowledge and theories of cosmetic science and technology. The series is made up of books either written by a single author or edited with a number of contributors. Authorities from industry, academia, and the government are participating in writing these books.

The aim of this series is to cover the many facets of cosmetic science and technology. Topics are drawn from a wide spectrum of disciplines ranging from chemistry, physics, biochemistry, and analytical and consumer evaluations to safety, toxicity, and regulatory questions. Organic, inorganic, physical, and polymer chemistry, emulsion technology, microbiology, dermatology, and toxicology all play a role in cosmetic science.

There is little commonality in the scientific methods, processes, or formulations required for the wide variety of cosmetics and toiletries manufactured. Products range from hair care, oral care, and skin care preparations to lipsticks, nail polishes and extenders, deodorants, body powders and aerosols, to over-the-counter products, such as antiperspirants, dandruff treatments, antimicrobial soaps, and acne and sunscreen products.

Cosmetics and toiletries represent a highly diversified field. Even in these days of high technology, art and intuition continue to play an important part in the development and formulations, their evaluation, and the selection of raw materials, but there is a move toward more sophisticated scientific methodologies in the fields of claim substantiation, safety testing, product evaluation, and chemical analyses, as well as a better understanding of the properties of skin and hair.

Emphasis in the Cosmetic Science and Technology series is placed on reporting the current status of cosmetic technology and science in addition to historical reviews. The series has grown to nineteen books dealing with the constantly changing technologies and trends in the cosmetic industry, including globalization. Several of the books have been translated into Japanese and Chinese. Contributions range from highly sophisticated and scientific treatises, to primers, practical applications, and pragmatic presentations. Authors are encouraged to present their own concepts, as well as established theories. Contributors have been asked not to shy away from fields that are still in a state of transition, nor to hesitate to present detailed discussions of their own work. Altogether, we intend to develop in this series a collection of critical surveys and ideas covering diverse phases of the cosmetic industry.

Novel Cosmetic Delivery Systems is the nineteenth book published in the series. It covers new techniques developed in recent years for delivering active performance chemicals to the skin and hair. In addition to providing cosmetic effects, modern cosmetics must have health functionality providing skin or hair treatment and making positive contributions to the condition of the skin. Alpha hydroxy acids, vitamins, and liposomes are recent examples of ingredients in skin care products that demonstrate such positive effects. These new ingredients often require or benefit from the new systems described in this book.

I want to thank all the contributors for taking part in this project and particularly the editors, Professor Shlomo Magdassi and Professor Elka Touitou from The Hebrew University of Jerusalem, for developing the concept of this book and contributing several chapters. Special recognition is also due to Sandra Beberman and the editorial staff at Marcel Dekker, Inc. In addition, I would like to thank my wife, Eva, without whose constant support and editorial help I would never have undertaken this project.

Eric Jungermann, Ph.D.

Preface

Common cosmetic vehicles are usually composed of classical, disperse systems, which include emulsions and suspensions. More sophisticated disperse systems, such as liposomes and microemulsions, although known for many years, were only recently introduced in the development of cosmetic products. Although many books in cosmetic science deal with classical formulations, only a few publications describe the novel cosmetic delivery systems.

The objective of this book is to offer a comprehensive presentation of scientific principles and methodology, as well as guidelines for the design of new cosmetic delivery systems. The book provides a useful source of scientific and technical data for scientists of various backgrounds, whose interests lie in the design, development, and use of cosmetic delivery systems with controlled characteristics.

The book can be helpful to formulators, in that it provides the basic concepts and methodology required to achieve a high success rate in compiling their systems. The information presented in the various chapters is a valuable resource and tool for scientists in the cosmetic and pharmaceutical industry as well as research institutions and academies that deal with the design or use of novel cosmetic and pharmaceutical delivery systems.

The topics presented in the book were divided into two sections and are presented in seventeen chapters. The first section of the book (Parts I and II) covers the biological aspects related to the skin and the cosmetic products, while the second section deals with specific cosmetic delivery systems.

Following an introductory chapter on the classical cosmetic formulations and on the state of the art of present cosmetic delivery systems related to cosmeceutics, Chapters 2–4 provide a concise review of the structure and functions of the human skin, including routes of penetration, transport through the skin, factors influencing permeability, and skin hydration. Chapter 5 describes the modern methods used to assess the biological activity of cosmetic formulations.

Chapter 6 focuses on a subject that has great significance for all the various systems, namely, stability testing of cosmetic formulations. A comprehensive review of the methods for quantitation of penetrant molecules within the skin is presented in Chapter 7.

The second section of the book (Parts III–V), which describes the various novel delivery systems, includes both systems that are already found in commercial products, and systems that we believe will be used in the future.

Chapters 8–10 describe several unique emulsion systems; Chapter 8 describes the properties and use of multiple emulsions, which may encapsulate an internal phase, such as water-in-oil-in-water emulsions. The formation of unique water-in-oil emulsions (''gel emulsion''), which may contain a very high internal water-phase fraction, is presented in Chapter 9.

Chapter 10 deals with the formation of fluorocarbon emulsions and gels. These have been studied for many years for medical purposes, and have been suggested only recently as oxygen carriers in cosmetic products.

Chapter 11 focuses on amphiphilic association structures that may solubilize oil soluble fragrances and affect the vapor pressure of the perfume.

Liposomes are described in Chapters 12 and 13. These systems have gained importance in recent years, especially when combined with the possibility of enhancing skin penetration of active ingredients, thus leading to ''cosmeceutic'' formulations. Subjects such as entrapment of active components and penetration of liposomes into the skin are raised, as well as descriptions of the evaluation of the overall activity of the various potential systems.

Chapter 14 describes the use of cyclodextrins, which have the ability to encapsulate various compounds, on the molecular level. Chapters 15–17 deal with various particulate systems and cover such topics as formation and properties of microcapsules, a demonstration of the possible use of polyvinyl alcohol microcapsules in cosmetic products, and a demonstration of the properties of microparticulate systems which are based on nylon particles.

The chapters are written by experts from industry and academic institutions, and provide a comprehensive and updated description of the topic. Most of the chapters include examples of commercial products and patents.

Obviously, this book does not cover all possible delivery systems, such as unique nanoparticles (''bio-vectors'') or beads filled with various liquids. We believe that better understanding of the nature of the skin and the requirements set for an effective cosmetic product will lead to the development of new delivery systems in cosmetic science and to more sophisticated cosmetic products.

We hope that this book will stimulate scientists in this field toward formation of more effective and exciting cosmetic products.

Shlomo Magdassi
Elka Touitou

Contents

Contributors

Núria Azemar Departamento de Tecnología de Tensioactivos, Centro de Investigación y Desarollo, Consejo Superior de Investigaciones Científicas (CSIC), Barcelona, Spain

Florence Benech-Kieffer Life Science Research Department, L'Oréal, Aulnay sous Bois, France

Gabriela Calderó Departamento de Tecnología de Tensioactivos, Centro de Investigación y Desarollo, Consejo Superior de Investigaciones Científicas (CSIC), Barcelona, Spain

Li-Lan H. Chen Lavipharm Laboratories, Inc., Piscataway, New Jersey

Yie W. Chien College of Pharmacy, Rutgers University, Piscataway, New Jersey

Dominique Duchêne Laboratoire de Physico-Chimie, Pharmacotechnie, Biopharmacie, Université Paris-Sud, Châtenay Malabry, France

Stig E. Friberg Department of Chemistry, Clarkson University, Potsdam, New York

Maria-José García-Celma* Departamento de Tecnología de Tensioactivos, Centro de Investigación y Desarollo, Consejo Superior de Investigaciones Científicas (CSIC), Barcelona, Spain

Nissim Garti Casali Institute of Applied Chemistry, The Hebrew University of Jerusalem, Jerusalem, Israel

Edward M. Jackson Jackson Research Associates, Inc., Sumner, Washington

Kentaro Kiyama Research & Development, Lion Corporation, Tokyo, Japan

---

* *Current affiliation:* Universidad de Barcelona, Barcelona, Spain.

Marie Pierre Krafft   Chimie des Systèmes Associatifs, Institut Charles Sadron, CNRS, Strasbourg, France

Linda Margaret Lieb   Department of Dermatology, University of Utah Health Sciences Center, Salt Lake City, Utah

Shlomo Magdassi   Casali Institute of Applied Chemistry, The Hebrew University of Jerusalem, Jerusalem, Israel

Victor M. Meidan   Department of Pharmaceutics, School of Pharmacy, Faculty of Medicine, The Hebrew University of Jerusalem, Jerusalem, Israel

Pierfrancesco Morganti   Research & Development, MAVI SUD S.r.l., Aprilia (LT), Italy

V. Parison   Département des Polymeres Fonctionels, Elf Atochem, Paris, France

Marie-Christine Poelman   Département de Dermopharmacie, Université René Descartes, Paris, France

Ramon Pons   Departamento de Tecnología de Tensioactivos, Centro de Investigación y Desarollo, Consejo Superior de Investigaciones Científicas (CSIC), Barcelona, Spain

Thomas E. Redelmeier   Northern Lipids, Inc., Vancouver, British Columbia, Canada

Perry Romanowski   Research & Development, Alberto Culver, Melrose Park, Illinois

Hans Schaefer   Research & Development, L'Oréal, Clichy, France

Randy Schueller   Research & Development, Alberto Culver, Melrose Park, Illinois

Conxita Solans   Departamento de Tecnología de Tensioactivos, Centro de Investigación y Desarollo, Consejo Superior de Investigaciones Científicas (CSIC), Barcelona, Spain

Klaus Stanzl   Cosmetics Division, Dragoco Gerberding & Co. AG, Holzminden, Germany

Elka Touitou   Department of Pharmaceutics, School of Pharmacy, Faculty of Medicine, The Hebrew University of Jerusalem, Jerusalem, Israel

Yelena Vinetsky   Casali Institute of Applied Chemistry, The Hebrew University of Jerusalem, Jerusalem, Israel

Denis Wouessidjewe   Laboratoire de Pharmacie Galénique, Université Joseph Fourier, Meylan-Grenoble, France

1
Cosmeceutics and Delivery Systems

Shlomo Magdassi and Elka Touitou
The Hebrew University of Jerusalem, Jerusalem, Israel

In the formulation of cosmetic products, active ingredients are combined with a variety of other compounds that give the products its physical form and may control the delivery of the active ingredient. The cosmetic industry is constantly seeking new, effective products that will combine both proven biological activity and an efficient delivery system. Although today conventional formulations are the primary cosmetic products seen on the market, numerous advances have been made in the development of new techniques for cosmetic delivery systems. These techniques control the rate of delivery and duration of activity and they target the delivery of an active compound to the tissue.

By far the most conventional and widely used cosmetic delivery system is the traditional oil-water emulsion. Most cosmetic creams and lotions on the market today are emulsions. More sophisticated forms of emulsion—such as microemulsion, multiple, gel, and fluorocarbon emulsions—have been developed to improve appearance, stability, and ease of application and to allow for controlled or sustained release of the active agent.

The cosmetic delivery systems can be divided into three broad types: vesicular (liposomes and niosomes), molecular (cyclodextrins), and particulate (microcapsules and matrix particles) [1–3].

The carrier of the system can affect the delivery of active components by a number of different means, as by interacting with the active agent, controlling the rate of release from the vehicle, altering stratum corneum (SC) resistance, or enhancing stratum corneum hydration [4]. Permeation enhancers may be incorporated in the system to increase the skin delivery of the active agent. They have

1

chemical characteristics that are believed to disturb the packing of the stratum corneum lipid bilayers. Compounds that have been used as penetration enhancers include sulfoxides, alcohols, polyols, alkanes, esters, amines and amides of fatty acids, terpenes, surfactants, and cyclodextrins [5].

As scientific progress deepens our understanding of compound–skin interactions and of skin permeability, it is becoming more and more difficult for products to stay within the definition of cosmetics. Frequently, products that were once considered purely cosmetic preparations have been found to interact with the skin. Moreover, in order to satisfy the demands of today's consumer, who is looking for efficient products, cosmetics are often promoted with strong claims that sometimes border on pharmaceutical benefits. These products, which fall into the gray area between the two categories of cosmetics and pharmaceuticals, are often referred to as *cosmeceutics*.

The Food and Drug Administration (FDA) clearly differentiates between two types of products: cosmetics and pharmaceuticals. Products applied topically that do not have a systemic effect are defined as *cosmetic products*, while compositions that affect the structure or function of the body are *pharmaceuticals* [6]. The categorization of a product has serious implications for marketing, as drugs and cosmetics have very different requirements for testing, labeling, claims, and advertising. The approval and regulation procedure for pharmaceutical products is significantly more expensive and lengthy than that for cosmetic products.

Items such as skin-care products, shampoos, toothpastes, and underarm products can be considered cosmetics or pharmaceuticals, depending on the types of active compounds contained in the product or the effect of the delivery system on the target organ.

Controversial compounds in the cosmetic versus pharmaceutical debate include the alpha hydroxy acids (AHAs), glucans, enzymes, antioxidants, and retinoic acid. AHAs—a group of naturally occurring acids involved in many metabolic processes such as the Krebs cycle, glycolysis, and serine biosynthesis—have been known for years to improve skin characteristics such as texture, wrinkling, and pigmentation [7]. The cosmetic properties of milk, used even in biblical times as a skin-care product, are attributed to lactic acid, an AHA. However, the use of AHAs to improve the skin's appearance has become a complicated issue, as the specific effects of AHAs on the skin that result in cosmetic changes are not clear.

AHAs applied at low concentrations appear to act by causing intraepithelial chemical peeling, brought about by the acidity of the molecule, which results in reduced corneocyte cohesion and desquamation of the stratum corneum. At high concentrations, application of AHAs causes detachment of keratinocytes and epidermolysis. It was also suggested that AHAs may suppress inflammatory reactions in the skin cells [8].

Increasingly vigorous claims are being made for AHA-containing products, sometimes overstepping the boundary between a cosmetic and a pharmaceutical [9]. In addition, the validity of certain claims is still questionable, such as whether any long-term benefits can actually be demonstrated from the use of AHAs. Some of the claims, such as those of exfoliation and hydration, can be measured by a number of different laboratory tests, including ballestometry (skin firmness), impedance measurement (hydration) and ultrasound (clinical assessment of skin thickness) [10].

Two other important points concern the effects of AHAs on the penetration of other ingredients in the formulation and possible effects of AHAs on sensitivity of the skin to UV radiation [11]. These concerns are currently being addressed by the FDA and will contribute to the decision on whether AHAs remain cosmetic ingredients or are transferred to the category of pharmaceuticals, with the accompanying restrictions on testing and labeling.

Another active ingredient used in many cosmeceutic products is β-glucan. The immunostimulatory effects of glucans on the skin are relatively well defined, as is the mechanism by which they carry out this function [12]. β-Glucan is a powerful chemotactic for macrophages, which, when activated, initate immune and reparative functions. Glucans bind to specific receptors on the surface of skin macrophage cells (Langerhans cells), resulting in activation and production of cytokines, epidermal cell growth factor, tumor necrosis factor-a, and angiogenesis factor. β-Glucan is a commonly used component in various cosmetics, such as skin creams, exfoliants, soapless cleanser, moisturizer, sunscreens and suntan oil, and preparations to promote healing of the skin [12]. In addition to immunostimulatory effects, a soluble derivative of β-glucan was shown to have positive effects on skin hydration, rate of cell renewal, and protection against UVA [13].

Enzymes such as papain, bromelin, ficin, lipases, and proteases are also incorporated in cosmeceutical preparations such as skin cleansing or debriding agents [14]. However, the use of enzymes in such skin-cleansing products presents a number of problems related to stability of the enzyme in the presence of surfactants as well as over time. A recent study presented an innovative solution to these problems by modifying the enzymes with glutaraldehyde, polyethylene glycol or succinate, or conjugating the enzymes to dextran [15]. Another creative solution to preservation of enzyme stability involved incorporating enzymes into gelatin microspheres. Results showed that superoxide dismutase in gelatin capsules retained almost 80% of its natural activity but was more stable to a wide variety of destructive conditions [16].

Considerable proteolytic activity has been measured in human skin, and this process may be involved in skin pathology. In another study, UV irritation and dryness of skin were measured in correlation with proteinase activity. While the results were not conclusive, there seemed to be a mild positive correlation in each experiment. Thus, topical application of inhibitors of nonspecific proteinases could be considered as an addition to various skin-care products [17]. Another interesting use for enzymes in cosmetic products—as permeation enhancers—was recently investigated. Pretreatment of the skin with phospholipase C was shown to significantly enhance the permeation of benzoic acid, mannitol, and testosterone.

Externally applied vitamins, whose specific effects are still being studied, are also used in many skin-care preparations. Vitamin E functions as an antioxidant, protecting the cell from photodamage caused by UV, toxic pollutants, environmental assaults, or aging. In its acetate form, vitamin E can penetrate into the skin and provide antioxidant activity similar to that of free vitamin E [18]. In addition, vitamin E linoleate functions as a moisturizer, accumulating in the SC and maintaining the intercellular moisture barrier of the skin [18]. It is absorbed into the skin and may therefore serve as an antiaging constitutent of skin care products [19].

Retinoic acid, a derivative of vitamin A, was also found useful in the treatment of UV-damaged and wrinkled skin. It regulates the growth and activity of epithelial cells, increases the skin's elasticity, and thickens the dermis and epidermis. Retinoic acid exhibits a high potential for producing serious side effects. It is therefore classified as a prescription medical substance [20].

Aside from the issue of classification, validity of the claims is another important concern of regulatory agencies: Are these claims true? Can they be supported and if so, how? With the development of new products as well as new understanding of drug-skin interactions, new ways are needed to determine efficacy and substantiate claims. Noninvasive techniques such as corneometry (skin hydration), sebumetry (total epidermal lipid content for testing), transepidermal water loss (skin barrier function) and potentiometry can be used in the design and evaluation of new products [21]. The amplitude and rate of delivery of the active agent in the various strata of the skin can be measured by novel, sophisticated quantitative methods, as described in chapter 7 in this book.

A recent development in the regulation of cosmetic products, is the FDA's approval, in 1995, of an NDA for a prescription drug, as ''an adjunctive agent for use in the mitigation of time wrinkles, mottled hyperpigmentation, and tactile roughness of facial skin in patients who do not receive such palliation using comprehensive skin care and sun avoidance programs alone.'' This appears to be the first FDA approval of a new drug that can be promoted with FDA-approved antiwrinkle claims, as described by McNamara [6] in a recent publication dealing with the term *cosmeceutics*.

There is a growing belief that many more of the conventional cosmetic preparations we have been using for many decades will gradually be replaced in coming years by delivery systems based on high technology. Growing research in new delivery systems may explain the increase in the number of patents related to cosmetics, shown in Fig. 1. In the United States alone, about 120 patents per

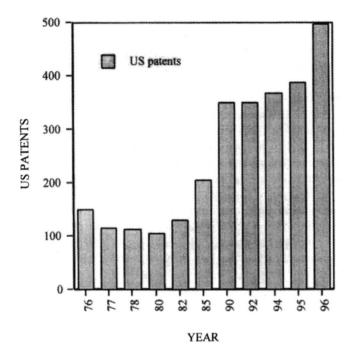

Figure 1 U.S. patents in the field of cosmetics.

year were issued between the years 1976 and 1982. By the mid 1980s, a dramatic increase in the number of cosmetic-related patents was observed; it now reaches about 400 issued patents per year.

The use of new delivery systems is aimed at improving both the performance of presently used active ingredients and the commercial "appeal" of the product. Since there are thousands of cosmetics producers all over the world, the competition is very aggressive. Therefore, the companies also try to be innovative in the appearance of their products (besides the packaging). Such examples are transparent gels that contain dispersed, colored beads (containing an imiscible oil), translucent microemulsions, two-phase systems, etc.

As seen in the market today, the leading cosmetic formulations contain a large number of ingredients, both active ones and excipients. Obviously, a suitable delivery system should be developed for each "multiple active component" formulation; the simple, conventional cosmetic delivery systems are no longer suitable. Having said that, it must be emphasized that the majority of the cosmetic products available in the market today are still based on the well-known delivery systems such as oil-in-water and water-in-oil emulsions (creams and lotions).

Therefore, cosmetic formulators perform extensive research to improve the "feel" of their formulations by changing the rheological properties, or the composition ("oil-free"), and by formulating products while aiming that each product will treat a specific problem. This later approach often solves the problem of combining components that require different conditions: for example, skin-peeling products that contain AHA require low pH, but this pH may not allow the activity of enzymes present in the formulation.

The following chapters describe various types of delivery systems that we believe will be widely applied in cosmetic products in the near future. These systems include unique emulsions and gels, vesicular systems, and particulate systems such as microcapsules and microspheres. Such delivery systems can be applied as components in a complex cosmetic product—e.g., nylon microspheres in a conventional oil-in-water emulsion or as the only component, such as liposomes in aqueous solution. However, owing to the complexity of simultaneous biological activities required, it seems most likely that future cosmetic products will combine components of various delivery systems in one formulation.

REFERENCES

1. Nacht S. Encapsulation and other topical delivery systems. Cosmet Toilet 1995; 110: 25.
2. Fox C. Advances in cosmetic science and technology: IV. Cosmetic vehicles. Cosmet Toilet 1995; 110:59.
3. Magdassi S. Delivery systems in cosmetics. In: Colloids and Surfaces. In press.
4. Zatz JL. Enhancing skin penetration of actives with the vehicle. Cosmet Toilet 1994; 109:27.
5. Smith EW, Maibach HI, eds. Percutaneous Penetration Enhancers. Boca Raton, FL: CRC Press. 1995.
6. McNamara SH. FDA regulation of cosmeceuticals. Cosmet Toilet 1997; 112:41.
7. Berardesca E, Maibach HI. AHA mechanisms of action. Cosmet Toilet 1995; 110: 30.
8. Morganti P. Alpha hydroxy acids in cosmetic dermatology. J. Appl. Cosmetol. 1996; 14:35.
9. Calogero C. Regulatory review. Cosmet Toilet 1996; 111:16.
10. Smith WP. Hydroxy acids and skin aging. Cosmet Toilet 1994; 109:41.
11. Price SNC. FDA brings AHA safety drug-claim concerns to CIR. Cosmet Toilet 1996; 111:9.
12. Mansell PWA. Polysaccharides in skin care. Cosmet Toilet 1994; 109:67.
13. Zulli F, Suter F, Biltz H, et al. Carboxymethylated b-(1-3)-glucan. Cosmet Toilet 1996; 111:91.
14. Fox C. Skin and skin care. Cosmet Toilet 1996; 111:21.
15. Ohta M, Goto A, Mori K, et al. Cosmet Toilet 1996; 111:79.

16. Fox C. Enzymes. Cosmet Toilet 1997; 12:23.
17. Voegeli R, Meier J, Doppler S, et al. Elastase and tryptase determination on human skin surface. Cosmet Toilet 1996; 111:51.
18. Djerassi D. The role of vitamins in aged skin. J Appl Cosmetol 1993; 11:29.
19. Wester RC, Maibach HI. Absorption of tocopherol into and through human skin. Cosmet Toilet 1997; 112:53.
20. Umbach W. Cosmeceuticals—the future of cosmetics? Cosmet Toilet 1995; 110: 33.
21. Rodrigues L, Jaco I, Melo M, et al. About claims substantiation for topical formulations: an objective approach to skin care product's biological efficacy. J Appl Cosmetol 1996; 14:93.

2
The Skin and Its Permeability

Hans Schaefer
L'Oréal, Clichy, France

Thomas E. Redelmeier
Northern Lipids, Inc., Vancouver, British Columbia, Canada

Florence Benech-Kieffer
L'Oréal, Aulnay sous Bois, France

I. INTRODUCTION

The skin is not a uniform surface [1–3]. The surface area of the exposed skin is increased by approximately 30% by the fine wrinkles as compared with a flat surface. A consequence of the wrinkles is that topical applications do not uniformly cover the skin surface [4]. There are significant regional differences in the topography, since these fine wrinkles are not found on either the soles of the feet or the palms of the hands.

A mature human weighing 65 kg will have approximately 18,000 cm^2 of skin surface area. The superficial region, termed the *stratum corneum* or *horny layer*, is between 10 and 20 μm thick. Underlying this region is the viable epidermis (50–100 μm), dermis (1–2 mm) and hypodermis (1–2 mm). Because of the large surface area as well as the volume of the compartments, the skin is the body's largest organ, weighing approximately 7 kg and representing more than 10% of the total body mass. However, though the skin comprises a very large volume, the barrier to percutaneous absorption lies within the stratum corneum, the thinnest and smallest compartment. The properties of the skin vary according to the anatomical locations (also termed *regional variation*) [5]. For example, there are many hair follicles in the scalp region and none on the palm of the hand. Regional variations in the thickness [6] and composition of the stratum corneum [7,8] are also clearly seen, largely related to the activity of the underlying viable epidermis and to a lesser extent to the density and activity of sebaceous and sweat glands.

II. THE STRATUM CORNEUM

A. Composition and Structure

The stratum corneum consists of corneocytes or horny cells, which are flat, polyhedral, nonnucleated cells approximately 40 μm long and 0.5 μm in diameter [6]. The size of the corneocytes is not completely fixed and varies with anatomical location [6], age [9], and conditions that increase epidermal proliferation. The corneocytes are cell remnants of the terminally differentiated keratinocytes found in the viable epidermis (Fig. 1). Their cellular organelles and cytoplasm have disappeared during the process of cornification. In turn, this is accompanied by a remodeling of the remaining protein constituents to form the corneocytes. They are composed primarily of insoluble bundled keratins [10,11] surrounded by a cell envelope stabilized by cross-linked proteins [12] and covalently bound lipid [13,14]. Interconnecting the corneocytes of the stratum corneum are polar structures such as corneodesmosomes [15,16], which ascertain the cohesion of the stratum corneum.

Intercellular lipid is generated primarily from the exocytosis of lamellar bodies during the terminal differentation of the keratinocytes [17–19] and, less importantly, from sebaceous secretions [20], which are predominantly deposited in the upper layers of the stratum disjunctum [21]. The intercellular lipid is pivotal for a competent skin barrier and forms the only continuous domain in the stratum corneum [22]. It follows a tortuous path within the stratum corneum, a structural feature that may account in part for the barrier properties of the skin [23].

The stratum corneum comprises approximately 15 layers, though at sites of increased pressure—such as the soles of the feet—this number is significantly (5- to 10-fold) increased. The upper layer, termed the *stratum disjunctum*, contains approximately 3–5 layers and is constantly undergoing desquamation. The stratum compactum (lower three layers) is thicker, more densely packed, more regular, and contains structures that more closely reflect the underlying epidermis [16,24]. The lower stratum compactum has more water associated with it (30% by weight) as compared with the stratum disjunctum (15% by weight), though both are considerably less hydrated than the viable dermis (70% by weight) [25,26]. These differences correlate with the amino acid [27,28] and lipid [29] content of the layers. Further, differences are observed for the rigidity of the cellular membranes, perhaps reflecting the maturation process of the corneocyte cell envelope during the passage from the epidermis to the surface of the skin and the final shedding (desquamation) [12]. Finally, the stratum compactum has a higher density of corneodesmosomes [30], suggesting that their proteolysis is

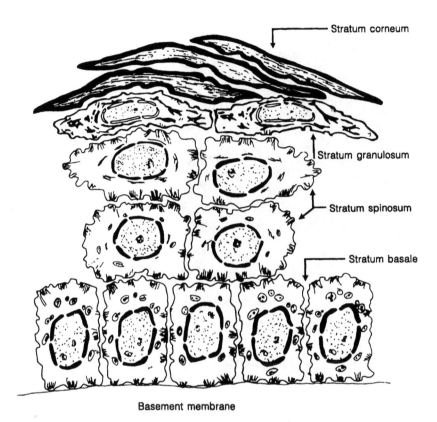

Stratum corneum

Stratum granulosum

Stratum spinosum

Stratum basale

Basement membrane

Figure 1 Diagrammatic representation of the viable epidermis indicating three layers: stratum basale, spinosum, and granulosum. The basal layer contains stem cells and transit-amplifying cells. This layer interacts with the basement membrane, which exerts an influence on the rate of proliferation. The suprabasal layers contain keratinocytes in the process of terminal differentiation. Nutrients and growth factors are obtained by diffusion from below. Migration from the basal to granular layers takes 2–4 weeks, during which the precursors to the stratum corneum are synthesized.

required for the separation of mature corneocytes [31,32] (Fig. 2). Taken together, this indicates that the stratum corneum is not uniformly homogeneous, that it continuously evolves from below to the skin surface, and that the layers represent various stages of corneocyte and intercellular lipid maturation.

B. Formation of the Skin Barrier

Thus the barrier function does not simply arise from the dying, degeneration, and compaction of the underlying epidermis. Rather, the processes of cornifica-

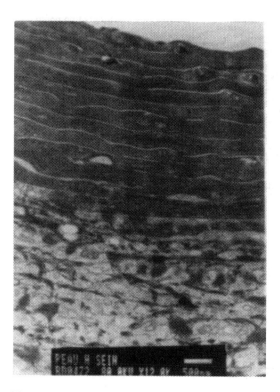

Figure 2 Electron micrograph of stratum corneum, granular, and cornified layers. Keratohyalin granules, lamellar bodies, and desmosomes accumulate in the keratinocytes present in the suprabasal layers. ×12,000; bar = 0.5 μm. The arrangement of corneodesmosomes is particularly striking in the transition region between the viable and nonviable layers. (Courtesy of F. Bernerd.)

tion and desquamation are intimately linked; synthesis of the stratum corneum occurs at the same rate as loss. There is substantial evidence that the formation of the skin barrier is under homeostatic control. This is illustrated by the epidermal response to barrier perturbation by skin stripping or solvent extraction of lipids.

The stratum corneum layers are formed at the rate of one layer per 5 h in the perinatal rat, which decreases to one layer per 8 h postnatally [33]. To support this level of synthesis, keratinocytes must mount from the basal to the suprabasal layer at approximately three cells per square centimeter. The transit time across the epidermis is approximately 1–2 weeks. For adult human skin, the rate of proliferation is slower. Keratinocytes take approximately 2–4 weeks to migrate from the stratum basale to the stratum granulosum and 2 weeks to transit the stratum corneum. Since there are approximately 15 layers in the stratum corneum, this means that it takes about 24 h to form one layer and that the stratum corneum is completely renewed every 15 days.

C. Epidermal Response to Perturbation of the Skin Barrier

Current understanding of the formation of the stratum corneum comes from studies of the epidermal responses to perturbation of the skin barrier. There are several experimental procedures to disrupt the skin barrier: (1) extraction of skin lipids with apolar solvents, (2) physical stripping of the stratum corneum using adhesive tape, and (3) chemically or physically induced irritation.

The most studied experimental system is the treatment of mouse skin with acetone. This leads to a marked and immediate increase in transepidermal water loss (TEWL), indicating a decrease in skin barrier function [34]. Since acetone treatment selectively removes glycerolipids and sterols from the skin, this suggests that these lipids are necessary though perhaps not sufficient in themselves for a barrier function.

A decrease in the ratios of free fatty acids to cholesterol and free fatty acids to ceramides after three and five strippings, respectively, confirms the importance of this level of stratum corneum lipids in skin barrier properties [35].

The ability of cholesteryl derivatives to promote the recovery effect in skin damaged by sodium lauryl sulfate was shown by their effects on water-holding capacity and transepidermal water loss in vivo [36].

The repair process subsequent to tape stripping or acetone treatment can be roughly classified into early (0–30 min), intermediate (3–6 h), and late (>6 h) periods, though the absolute kinetics of the response depend upon the extent of barrier perturbation and the animal species. In mice, the early stage correlates with little or no improvement as measured by the decrease in the rate of TEWL (<10%), whereas a 40–60% reduction has been observed by the intermediate period and an 85–95% reduction by 24 h [19]. In humans, the return to normal barrier function also appears biphasic, though the recovery is distinctly slower; 50–60% of barrier recovery is typically seen by 6 h, but complete normalization of barrier function requires 5–6 days [37]. Thus, the skin has an extraordinary capacity to react to a loss in barrier function by presumably synthesizing the (precursor) molecules required for barrier recovery and subsequently assembling them to form the barrier.

After barrier perturbation, the early stage of repair is characterized by lamellar body excretion [19] and an increase in fatty acid and cholesterol biosynthesis [38]. This is consistent with an increase in the activity of the enzymes responsible for these biosynthetic steps [39,40]. The intermediate stage is characterized by an increase in the activity of enzymes involved in ceramide [41] and glucosylceramide biosynthesis [42,43]. At the late stage, increased DNA synthesis but not

SG-SC
interface   Lower SC   Mid- to
                                                                outer SC

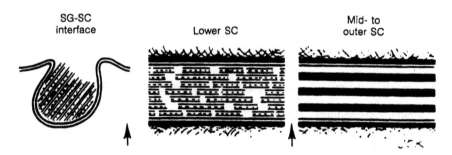

Figure 3 Several of the steps required for the synthesis of the unit membrane found in the intercellular space of the stratum corneum (SC). Lipid precursors are synthesized by the keratinocytes in the upper layers of the stratum spinous layer and packaged into lamellar bodies. Secretion of the lamellar bodies is followed by fusion to form broad lipid sheets. These are subsequently processed to form the unit membrane structure. The location of the arrows indicates that secretory phospholipase A2 and cerobrosidase may mediate these transitions. SG = Stratum granulosum. (Courtesy of P. Elias.)

bulk protein formation is observed [44]. Thus, the recovery of the skin barrier function runs parallel to the synthesis of the primary precursors to the intercellular lipid layers.

D. Assembly of the Intercellular Lipid Layers

Figure 3 summarizes the processes required for the formation of the structures present in the intercellular lipid spaces: (1) synthesis of precursor lipids in the suprabasal keratinocyte, (2) assembly of these lipids into lamellar bodies, (3) exocytosis of the lamellar bodies, and (4) subsequent processing of the precursor lipids to form the lamellar structures characteristic of the intercellular spaces. The intercellular lipid domain consists of roughly equimolar concentrations of free fatty acids, cholesterol, and ceramides [8,13].

III.   THE BARRIER TO PERCUTANEOUS ABSORPTION

Several lines of evidence indicate that the skin barrier lies within the stratum corneum. Perhaps the most convincing proof is provided by the demonstration that physically stripping off the outermost layers of the skin results in a dramatic increase in the permeability to water [45] and other compounds [46]. The results in Fig. 4 present the relation between the removal of the stratum corneum and loss of barrier function as measured by TEWL. A sequential increase in transepidermal water loss is observed as the stratum corneum is progressively removed, indicating that the entire stratum corneum provides the barrier to diffusion [47].

However, it should not be inferred from this discussion that stripping of the stratum corneum or solvent extraction of the intercellular lipids completely removes the protection of the body by a skin barrier. In general, these perturbations increase water or solute permeability by approximately 10- to 20-fold. This is equivalent to the maximum decrease in barrier function that is seen for some pathological conditions. However, the water permeability is approximately 100-fold less than that observed for a stratified layer of keratinocytes. Thus, a substantial barrier remains across the skin even after extensive barrier perturbation. There are several possible reasons for this. First, these perturbations may not completely remove the stratum corneum [48]. Approximately 10–30% of the stratum corneum remains after skin stripping, though this will depend upon the protocol and species. Second, homeostatic responses to barrier perturbation are rapid, and it is difficult to measure absorption rapidly in the complete absence of these responses. Third, secretion and coagulation of interstitial fluid may occur rapidly after barrier perturbation, leading to a reduction in transepidermal water loss. Finally, it is

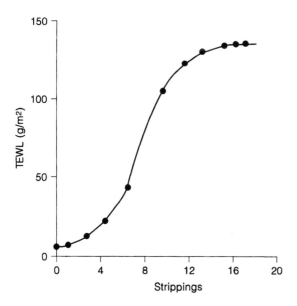

Figure 4 Removal of the stratum corneum by tape stripping reveals the location of the skin barrier as indicated by TEWL measurements. (From Ref. 47.)

possible that lipid derived from lamellar body excretion in the stratum spinosum and granulosum reduces transepidermal water loss.

A. Structural Basis for Percutaneous Absorption Pathways

These routes are referred to as (1) appendiceal, (2) transcellular, and (3) intercellular (Fig. 5). The relevance of these routes to percutaneous absorption of a compound depends upon their number per surface area and path length as well as the diffusivity and solubility of the compound in each domain. These pathways should not be treated as mutually exclusive. Hair follicles are the most important appendages in terms of surface area. The possibility that they form a pathway for percutaneous absorption has been elegantly demonstrated using fluorescence microscopy [49–51]. This is consistent with previous investigations relying upon autoradiography, which have demonstrated penetration of hair follicles by several substances [52].

B. Transcellular vs. Intercellular Pathways

The rate-limiting step for permeation includes a hydrophobic barrier—i.e., the intercellular lipid. Available evidence suggests that the only continuous domain within the stratum corneum is formed by the intercellular lipid space [53,54]. This suggests that the majority of compounds penetrating the stratum corneum must pass through intercellular lipid, though it does not exclude the possibility that compounds can also enter into the inner lumens of corneocytes (Fig. 6).

There is additional evidence that other compounds can and do penetrate the corneocytes. It is well established, for example, that occlusion or immersion of skin in a bath leads to swelling of the corneocytes, consistent with the entry of water.

Low-molecular-weight moisturizers like glycerol are likely to partition into the corneocytes and alter their water-binding capacity [55]. Thus, the penetration of compounds into corneocytes cannot be excluded from the consideration of percutaneous absorption pathways.

IV. CUTANEOUS METABOLISM

The skin contains a wide range of enzyme activities, including phase I oxidative, reductive, and hydrolytic reactions; phase II conjugative reactions; and a full complement of drug-metabolizing enzymes [56,57].

Metabolism of compounds in the stratum corneum will directly reduce the percutaneous absorption of compounds, though it is accepted that this is not gen-

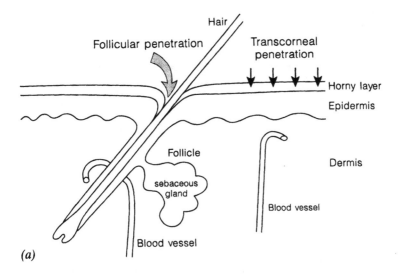

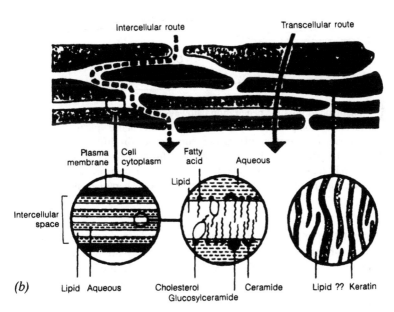

Figure 5 Model of penetration pathways. (a) Penetration occurs via appendages that exhibit a reduced barrier to diffusion but occupy a relatively small surface area. (b) Permeation through the stratum corneum (transcorneal permeation) may be considered to occur through the intercellular lipid domain or through the corneocytes (transcellular route). (Courtesy of P. Elias.)

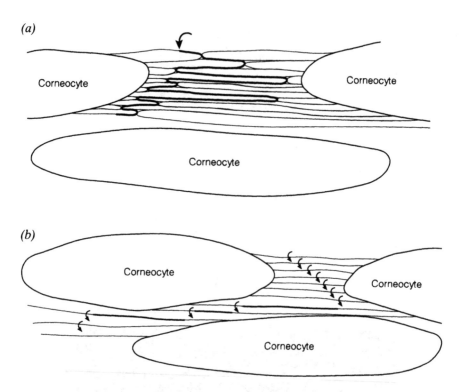

Figure 6 Schematic of possible penetration pathways through the intercellular lipid domain: (a) diffusion of compounds may occur along lipid lamellae (pictured as a single line), which occasionally penetrate the stratum corneum, or (b) diffusion occurs across the lamellae in a mechanism which is analogous to diffusion across lipid bilayers. In (a) the pathway is indicated by a heavy line whereas in (b) the pathway is denoted by an arrow to indicate translamellar diffusion and lines to denote lateral lamellar diffusion.

erally of quantitative importance. Exceptions to this rule may include compounds that are natural substrates for enzymes involved in the formation of the intercellular lipid domains or in desquamation [58,59]. For example, the level of phospholipase, A_2 or cerebrosidase activity may metabolize exogenously added phospholipids or sphingomyelin. In contrast, metabolism in viable tissues does not directly reduce the level of percutaneous absorption, though it is relevant to consideration of the bioavailability of compounds and the bioequivalence of different formulations or devices.

Investigations of cutaneous metabolism in vivo cannot, in general, distinguish cutaneous from systemic metabolism. The capillaries of the microvasculature reach high up in the skin, as close as 50–100 μm from the surface. Thus the actual site of metabolism is very small, as is the amount of compound transformed under even the most optimal conditions. This presents a formidable technical challenge to the estimation of cutaneous metabolism—a task that is further complicated by the fact that the clearance of compounds from the viable tissues competes with the level of metabolism [60].

Metabolism can be studied in vitro in cell homogenates [61] or in tissue culture cells [62]. Though useful for establishing potential metabolic pathways, these approaches cannot quantitatively predict cutaneous metabolism in vivo. Investigations using Franz cells have been adapted to assess the level of cutaneous metabolism [63–65]. Fresh preparations of epidermal sheets are recommended in these investigations in order to maintain tissue and enzyme viability. In addition, a "physiological" receptor phase (tissue culture medium with 10% fetal calf serum for hydrophobic compounds) should be chosen. The viability of the epidermal cells can be monitored by following the consumption of glucose [66] or by using the MTT assay, which measures the mitochondrial activity [67]. Viability is maintained for up to 24 h in the presence of a physiological receptor phase, whereas cells lose viability after 6 h if this is replaced with phosphate-buffered saline and after 1 h for distilled water [68]. Freezing or improper storage of epidermal sheets may further reduce enzyme activity as well as result in the release of enzymes from the epidermis and dermis. Control experiments should demonstrate that metabolism is located in the skin sheets and not due to enzymes released into the receptor phase. The effects of freezing and azide treatment of in vitro human skin on the flux and metabolism of 8-methoxypsoralen lowered the skin barrier properties to its transport through the skin [69].

A recent investigation provides an example of how the viability of the skin can dramatically influence the extent of skin metabolism [68]. In vitro percutaneous absorption of benzoic acid, p-aminobenzoic acid, and benzocaine in the hairless guinea pig is only modestly influenced by the viability of the epidermal sheets (Table 1), though the distribution of p-aminobenzoic acid between the receptor fluid and the viable tissues is significantly altered. Benzoic acid undergoes modest but significant metabolism, with less than 18% of the absorbed dose converted to other products. A similar level of metabolism of p-aminobenzoic acid is observed; however, note that the metabolic product is preferentially distributed into the receptor fluid. This accounts for the observation that decreased viability of the epidermal sheets increased the concentration of radioactivity in the receptor fluid. A similar behavior has been reported for benzo[a]pyrene metabolism in in vitro experiments [70]. This may influence the calculation of percutaneous absorption if the levels found in viable tissues are not incorporated into these estimations. Further, it will result in an overestimation of metabolism if the ratio of product:substrate is radically different for the receptor fluid relative to viable tissues. In general, it is recommended that the level of metabolism should be calculated as the amount of converted compound expressed as a frac-

Table 1 Metabolism in the Skin During Penetration

Compound	Receptor fluid		Skin	
	Viable	Nonviable	Viable	Nonviable
Benzoic acid				
Total[a]	47.3	59.1	2.2	1.0
Benzoic acids	73.8	92.3	83.2	84.2
Hippuric acid	6.9	0.1	0.6	0.0
Polar	11.66	2.5	3.8	8.1
PABA				
Total	5.0	18.7	20.7	14.7
PABA	24.6	93.0	86.0	83.9
Acetyl-PABA	60.7	1.6	6.7	6.2
Benzocaine				
Total	75.1	76.0	1.5	1.8
Benzocaine	4.4	66.8	25.8	30.0
N-Acetylbenzocaine	82.4	7.2	31.7	20.5
PABA	0.2	5.4	5.5	12.7
Acetyl-PABA	6.2	7.7	9.7	8.1
Polar	2.5	6.0	14.6	16.6

[a] The data for the total is expressed as the percentage of the compound recovered in the respective compartment. The data for the metabolic products refers to the % of the metabolite recovered in a particular compartment.

tion of the total compound recovered in both the receptor fluid and the viable tissues.

In contrast to that of the other two compounds, benzocaine metabolism is very significant, resulting in acetylation of approximately 82% of the absorbed compound (Table 1). Note that this conversion is largely sensitive to the viability of the cells, a characteristic of enzymes that require a cellular energy sources to carry out their activity. Nevertheless, total benzocaine absorption was not significantly different between the viable skin and the skin treated with the enzyme inhibitor. In fact, the stratum corneum is probably the primary barrier to the absorption of benzocaine, with metabolism in the viable epidermis occurring after the rate-determining step in the absorption process [71].

For other enzymes such as esterases, tissue viability in not likely to be as important, since the acetylsalicylic acid–cleaving activity on mouse and rat skin was maintained following repeated freezing and thawing pretreatment [72].

The experiments outlined here demonstrate that cutaneous metabolism may play a role in the bioavailability of compounds in target tissues other than the skin.

Moreover, since metabolism of compounds is likely to be influenced by the concentration of compound in the viable tissues, the level of metabolism will be influenced by choice of formulation, skin barrier activity, and the presence of enhancers. For example about 18% of topically applied retinyl palmitate were absorbed from acetone vehicle by human skin and 44% of the absorbed compound was hydrolized to retinol through viable skin [73]. When the same compound was incorporated into a O/W cosmetic emulsion, only 4% of the applied dose was absorbed and no endogeneous metabolism was observed probably due to its weak cutaneous bioavailability in this type of vehicle [74].

In vitro investigations, however, may overestimate the level of cutaneous metabolism, since the clearance of a compound from viable tissues may be reduced due to the shorter dermal path length for permeants in vivo [75].

In order to circumvent this, cutaneous metabolism has been studied in vitro in perfused porcine skin [76–79]. These model systems have the distinctive advantage of not only maintaining the epidermis and dermis in a viable state but also allowing the the cutaneous microvasculature to remain functional. The level of resorption for certain compounds may be controlled by alterations in the rate of perfusion and the protein composition of the perfusate. Investigations using this approach have demonstrated that the level of propoxur metabolism is lower in perfused porcine skin than in a skin explant [78]. Moreover, changing the rate of resorption by altering the flow rate or protein concentration of the perfusate results in corresponding changes in the level of cutaneous metabolism.

This observation suggests that the level of metabolism measured in a Franz cell may be overestimated because the level of clearance is reduced. This is also consistent with parathion and carbaryl metabolism in the isolated perfused porcine skin, where the level of metabolism is significantly increased by recycling the perfusate [80]. Taken together, these investigations outline the technical problems encountered in estimating the level of first-pass cutaneous metabolism.

V. PREDICTION AND MEASUREMENT OF PERCUTANEOUS ABSORPTION

A. Definition of Terms

The following terms are most commonly used in discussing issues of passage of substances into and through the skin [46] (Fig. 7):

Adsorption indicates the reversible, noncovalent interaction of compounds with structures such as the binding of drugs to keratin filaments. It is used to describe a state and not the process; it should be differentiated from the term *substantivity*, which refers to irreversible binding (for example of suncreen compounds to the upper layers of the stratum corneum). The term *absorption* is used to describe the process of intake of substances, as by an organism. *Percutaneous*

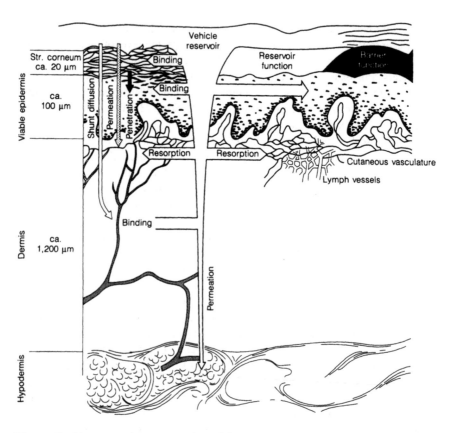

Figure 7 Diagrammatic representation of four compartments of the skin: stratum corneum, viable epidermis, dermis, and hypodermis. The skin barrier is located within the outermost layer, the stratum corneum. Appendages originating from the dermis permeate this layer infrequently and represent sites for penetration. The skin vascular network does not reach into the epidermal layer. Note the relative thickness of the compartments: dermis ($\cong$ 1 mm) and hypodermis ($\cong$ 1 mm); viable epidermis ($\cong$ 0.1 mm) and stratum corneum ($\cong$ 0.01 mm).

absorption is thus a global term describing the passage of compounds across the skin, though it does not necessarily indicate their eventual fate. This process can be subdivided into three steps. *Penetration* is the entry of a substance into a particular layer or structure, such as the entrance of a compound into the stratum corneum. It is to be differentiated from the term *permeation*, which indicates that the compound has diffused from one layer to another distinct layer. Finally, *resorption* is defined as the uptake of substances through the vascular system into the central or inner compartment. Thus, we do not consider that compounds which have penetrated into the stratum corneum should be considered to be absorbed into the body, since in vivo this amount may be reduced by several factors such as the physiological desquamation and/or binding effect; rather, this penetration contributes to the stratum corneum *reservoir function.*

B. In Vivo Percutaneous Absorption Measurements

In vivo percutaneous absorption remains the goal to which all measurements should be compared. Procedures for measuring percutaneous absorption in vivo have recently been reviewed [81,82]. Investigations are performed to evaluate or establish (1) the systemic bioavailability of a drug to be delivered transdermally, (2) the bioavailability of a compound in a skin compartment, (3) the bioequivalence of different topical formulations, (4) the safety assessment of cosmetics, and (5) systemic toxicological risk assessment of drugs and environmental chemicals.

The extrapolation of percutaneous absorption studies from animal species to human should be done with caution. Of the common laboratory animals, rats provide the most reasonable alternative to humans. Percutaneous absorption is typically 2- to 5-fold higher in rats than in humans. Other acceptable alternatives include weanling pigs [83] and the rhesus monkey [84].

1. Studies in Human Volunteers

Percutaneous absorption of cosmetic ingredients in humans in vivo may be followed by measuring the secretion of compounds in the urine and excreta. These investigations are considered to be the most relevant for the estimation of systemic exposure provided that mass balance is adequate (90 ± 5%). However, in practice, this approach is significantly limited by the sensitivity of detection, ethical considerations, the regulatory limitations for the use of radioactive materials in humans, and the efficiency of excretion of compounds and their metabolic products. Serum levels can also, in theory, provide an indication of the level of percutaneous absorption, though this approach is not practical for most cosmetic ingredients.

2. Stripping Method

By far the most convenient in vivo procedure for predicting percutaneous absorption in humans involves measuring the level of compound that has penetrated the stratum corneum by physically stripping off the individual layers of the stratum corneum (Table 2). This is an in vivo approach in humans which can be considered virtually noninvasive.

If this is done before significant levels of compound have penetrated the body, then the amount corresponds to that which will eventually be absorbed.

Table 2 Stripping Procedure

1	Choose an appropriate hairless area
2	Clean the area gently with tissue paper and water and subsequently dry
3	Stick a cell (typically 1 cm^2) over the application area using silicone glue
4	Apply the formulation for 30 min (typically 20 μL total volume)
5	Wash the application area with an appropriate medium and dry with a cotton swab; repeat this step once
6	Rinse with 300 μL distilled water and gently dry with a cotton swab; repeat this step once
7	Remove the cell
8	Apply tape (micropore 3M) gently and remove it with a single fluid motion; discard the first strip
9	Strip application area 15 times for human skin (6 times for hairless rat skin)
10	For radioactive monitoring, the strips may be dissolved by incubating for 2 days in 15 mL soluene (United Technologies Packard, Chicago, Ill., USA); subsequently, 2 mL of the sample is incubated with 15 mL of Hionic Fluor (United Technologies Packard) for 24 h to reduce chemiluminescence and subsequently counted using standard techniques for scintillation counting
11	For nonradioactive analysis, the compound is solubilized using appropriate solvents from the strips and analyzed by established procedures: control experiments measure the efficiency of compound recovery from the strips

Skin stripping is done by firmly pressing an adhesive tape to the skin surface and tearing it off. The material recovered is then quantified by appropriate analytical procedures [85,86]. The application site should be clean, free of cosmetics or other extraneous agents, and dry. The removal of free residual formulation at the end of the penetration period and before stripping is particularly important, and care should be taken to ensure that the choice of medium (mild detergent or solvent) does not cause subsequent redistribution in the layers underneath the surface. The yield of the washing step is another important point, as one must be sure that all material applied to the skin surface is correctly removed during this step. Ideally, this amount should be quantified. The first or even second tape strip should be discarded, because these contain superficial formulation residue.

The results presented in Fig. 8a demonstrate the amount 4-chlortestosterone found in human stratum corneum as a function of the number of tape strips. The exponential decline reflects, as a first approximation, the concentration gradient across the stratum corneum. In order to reduce interindividual variability in the amount of stratum corneum harvested as a function of depth, data can be expressed as the amount of compound per milligram or harvested stratum corneum as determined by weighing of the strips [87]. Surprisingly, there may be considerable interindividual variability in the material recovered in the strips (Fig. 8b).

The accuracy and reliability of the spectroscopic examination of the tape

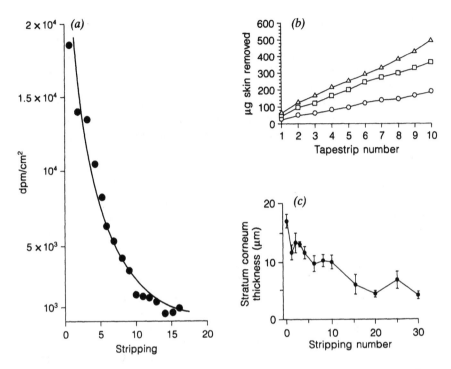

Figure 8 (a) Distribution of radioactivity in the stratum corneum after topical application of 4-chlortestosterone as determined by tape stripping. (b) Cumulative stratum corneum weight removed from human inner forearm skin for three individuals following a 24-h exposure to 0.05% betamethasone dipropionate cream. (From Ref. 90.) (c) Confocal microscopy determination of residual stratum corneum following the indicated number of tape strippings. (From Ref. 48.)

were compared with those of weighing [88]. The light absorption by the proteins on the tape was correlated to the weight of the stratum corneum material. Compared with weighing, however, this method was less reliable because the light scattering of the stratum corneum on the tape largely overshadowed the absorption of the proteins. Nevertheless, with direct spectroscopic measurements, a tape strip can be examined laterally to determine its homogeneity.

Moreover, the yield of stratum corneum may be influenced by the length of contact time as well as by some [89] though not all formulations [90]. Changes in stripping properties occurred in response to the influences of vehicle, which may be interpreted as changes in the mechanical strength of the intercellular spaces and other junctional regions [89]. Thus, it is likely that the yield of stratum corneum will vary for different protocols as well as laboratories.

Another consideration is that tape stripping of skin does not result in the complete removal of all of the stratum corneum. It is, for example, well established that tape stripping does not result in the quantitative removal of hair follicles [91]. In addition, as demonstrated in a recent study (Fig. 8c), extensive tape stripping (30 strippings with Blenderm adhesive tape) removed stratum corneum corresponding to only 12 of 17 μm [48]. This will vary among experimental protocols. Thus, the material recovered in tape strips should be treated as a significant fraction of that present in the stratum corneum but not as 100%.

The analysis of the specific binding of molecules to stratum corneum proteins, especially for highly lipophilic compounds, can be suitable, as the ratio bound fraction:unbound fraction affects the dynamic pathways of epidermal-dermal uptake and ultimate absorption by the body fluids [92].

However, given these caveats, there is an extremely good correlation between the amount of material recovered by tape stripping 30 min after application and the percutaneous absorption of a compound as detected by whole-body analysis after 4 days [86]. Importantly, this correlation holds for compounds which differ widely in physicochemical properties and which exhibit a range in percutaneous absorption values of 500-fold. The correlation in Fig. 9a indicates that the total absorption of a compound after 4 days corresponds to 1.8 times the amount found in the strips after a 30-min application. In addition to being applicable to a variety of compounds, this procedure has been shown to effectively predict the influence of (1) concentration of the compound at the site of application (Fig. 9b), (2) vehicle composition (Fig. 9c), and (3) anatomical location (Fig. 9d) on the level of percutaneous absorption [86]. In addition, in situations where there is a linear relationship between percutaneous absorption and exposure to a particular compound, this approach can be used to extrapolate to different time periods.

Thus, tape stripping can be used to predict the percutaneous absorption of compounds after a relatively short-term application. It has been applied to the safety assessment of cosmetic ingredients and in the bioequivalence assessment of several topical corticosteroid formulations [93]. It is particularly attractive since it is relatively easy, accurate, and rapid and allows the use of analytical procedures that circumvent or limit the use of radioisotopes. Most importantly, this in vivo approach in humans can be considered virtually noninvasive. However, though it has found extensive use in several laboratories, it cannot yet claim to be validated in a wide variety of laboratories.

C. In Vitro Approaches to Predicting Percutaneous Absorption

In vitro measurements on skin specimen mounted in a diffusion cell are the most established procedure for predicting percutaneous absorption [94–100]. It has been shown to be particularly applicable to the development of transdermal drug

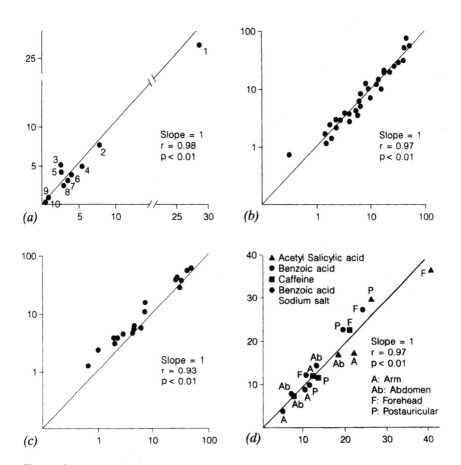

Figure 9 The correlation between penetration as determined by stripping and in vivo percutaneous absorption in hairless rats [86]. Where drawn, the slope of the line is 1. The stripping procedure successfully predicts the variation in percutaneous absorption observed for (a) different compounds: benzoic acid (1), acetyl salicylic acid (2), dehydroepiandrosterone (3), sodium salicylate (4), testosterone (5), caffeine (6), thiourea (7), mannitol (8), hydrocortisone (9), and dexamethasone (10); (b) four different compounds at five different concentrations; (c) several different mixed alcohol vehicles, and (d) several different anatomical locations.

delivery and the safety assessment of cosmetics [101] as well as the risk assessment of potentially toxic compounds [102].

1. Skin Models

Where possible, human skin is used for in vitro studies. If human skin is not available, then pig [83,103] or rat skin [103,104] can be considered. The results presented in Fig. 10 demonstrate a comparison of an in vitro percutaneous absorption study for two sunscreens and a reference compound with high skin permeability, salicylic acid, using split-thickness human and pig skin [105]. The qualitative agreement between human and pig skin is very good; the rank order of percutaneous absorption is maintained for both species. The amount recovered from epidermis, dermis, and receptor fluid is similar for the two sunscreens. The amounts of [^{14}C]-salicylic acid in the human epidermal layer are larger than in pig skin, while the amounts in the dermis and receptor fluid are similar.

However, if percutaneous absorption is calculated as the amount of compound recovered in viable tissues as well as receptor fluid, then the pig skin provides a very good substitute for human skin.

Reconstituted skin—i.e., cultures of skin equivalents—has been investigated with a view to generating in vitro systems for predicting percutaneous absorption and the irritation potential of compounds [106–108].

In these cultures, keratins 1 and 10 [109] and the chymotryptic enzyme, a protease involved in desquamation [110], are correctly expressed in the stratified epithelial layers. However, involucrin, transglutaminase, and filaggrin are ex-

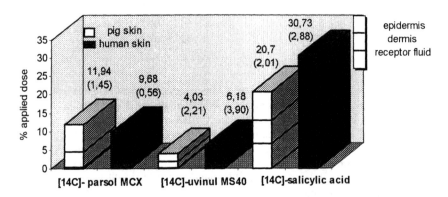

Figure 10 In vitro percutaneous absorption of two sunscreens (Parsol MCX, Uvinul MS40) and salicylic acid using split-thickness human and pig skin. (From Ref. 105.) Good quantitative agreement is obtained for the two models if percutaneous absorption is calculated as the amount of compound found in the viable tissues as well as the receptor fluid.

pressed in the lower layers of the stratum spinosum, indicating that epidermal differentiation does not completely reflect the in vivo tissue distribution [109]. More importantly, the lipid composition of the stratum corneum does not reflect that found in vivo.

The most distinct difference between the lipid composition of these cultures and normal skin is the three- to fourfold increase in the level of lanosterol, a relatively minor lipid component of the stratum corneum. In addition, though the absolute ceramide content of the cultures is similar to that of normal skin, the amount of the more apolar fractions (1 and 2) is increased, whereas the level of polar fractions is significantly decreased (4, 5, and 6). Taken together, these results suggest that further improvements in cultured skin equivalents are ''just around the corner'' and that these will soon be considered to be morphologically and physically similar to human skin.

In vitro percutaneous absorption experiments may use whole-thickness or split-thickness skin or epidermal sheets from which the hypodermis and dermis have been removed. The dermis retains substances that, because of its relative size and properties, would normally be expected to enter the receptor fluid. Several approaches have been used to separate epidermis from dermis: proteases, incubation in concentrated solutions [111], heat-soaking techniques [112], and the dermatome [113]. A standard approach for separation involves heating the epidermal sheet at 60°C for 45 s, followed by a subsequent peeling of the epidermis from the dermis. Currently, the dermatome is widely used for the preparation of epidermal sheets of defined thickness [113]. The often pronounced concern that these separation techniques might leave holes at the sites of the orifices of the hair follicles and sweat glands appears to be unsubstantiated. Isolated epidermis contracts to about 70% of its original surface area (Juhlin, personal communication). Apparently, the orifices become closed by this contraction.

The results presented in Fig. 11a indicate the percutaneous absorption of two sunscreens, parsol MCX and uvinul MS40, as well as salicylic acid across whole- and split-thickness pig skin [105]. The qualitative comparison between these two models for all three compounds is very good and the rank order of compounds is maintained. However, the presence of the dermis reduces the amount of compound that appears in the receptor fluid during the experiment (Fig. 11b). Importantly, these results confirm that measurement of compounds in the receptor fluid alone may be insufficient, especially for highly lipophilic compounds that penetrate the skin slowly. In other words, the amounts situated in the residual dermis should be taken into account for safety evaluation when dermatomed skin or whole skin specimens are investigated.

2. Diffusion Cell

The design of diffusion cells (Franz cells) has recently been reviewed [114]. In vitro percutaneous absorption measurements can be made with one- or two-

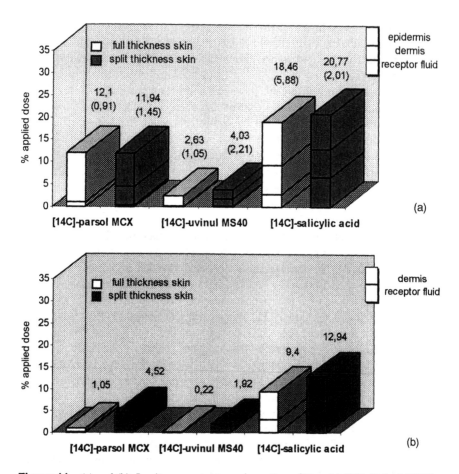

Figure 11 (a) and (b). In vitro percutaneous absorption of Parsol MCX, Uvinul MS40, and salicylic acid across full- and split-thickness pig skin. (From Ref. 105.) Good quantitative agreement is obtained for the two methods if percutaneous absorption is calculated as the amount of compound found in the viable tissues as well as the receptor fluid. This is especially important for relatively hydrophobic compounds.

chamber diffusion cells. One-chamber cells are topologically equivalent to application of a compound on the skin surface in vivo—i.e., with a finite applied dose. For two-chamber cells, the skin flap is placed between the two chambers and the compound in question diffuses from a (aqueous) donor phase through the skin into an (aqueous) acceptor phase. This is referred to as *infinite dose conditions*, where the concentration of compound in the donor compartment is not significantly diminished during the study. This is a suitable design for structure-activity analysis. However, this approach results in excessive hydration of the stratum corneum and subsequently a diminished skin barrier [115]. This is especially a concern with mouse skin, where increased percutaneous absorption is observed as early as 4 h after the initiation of the experiment (116,117). This design is generally not directly applicable to drug development or safety assessment of cosmetics; consequently one-chamber devices are more commonly used.

3. Assessment of Barrier Integrity of Skin Samples

Assessment of the barrier integrity of skin samples used in percutaneous absorption experiments is particularly important. Each sample must be checked for integrity before application of the test compound, and the test used must not alter the parameter it is intended to measure. The integrity can be evaluated by measuring permeability to tritiated water or reference compounds [118], but these methods take several hours, add to the length of the experiment, and, depending on the compound, can modify the stratum corneum itself.

As, in vivo, there is a well-known linear relationship between TEWL and penetration [119], TEWL measurement seems to be a suitable alternative method to use in vitro [120]. TEWL measurements in vitro were performed on dermatomed human skin (Fig. 12a), full-thickness human skin (Fig. 12b), and corresponding damaged samples (prepared by removing part of the stratum corneum). There was no significant increase in TEWL with the exposure time (24 h), confirming that the maintenance of stratum corneum integrity and mean TEWL from stripped skin samples were significantly higher than for the corresponding undamaged samples [120]. As observed in vivo [121], significant degrees of intra- and intervariability were observed. This must be taken into account in establishing criteria for the rejection of data.

4. Receptor Fluid

The selection of an appropriate receptor fluid for use in an in vitro percutaneous absorption study is crucial to obtaining meaningful data and properly interpreting them [122]. It must preserve the integrity of the skin, and the solubility of the test compound in the medium must be guaranteed.

For the measurement of lipophilic compounds, a receptor fluid into which the test compound freely partitions must be used. Surfactants additive (PEG 20 oleyl ether) or inclusion of bovine serum albumin (BSA) may be effective in increasing partitioning of lipophilic compound [122].

5. Pharmacokinetic Parameters

An "ideal" curve of an in vitro percutaneous absorption experiment using a finite dose would indicate that, in contrast to infinite-dose conditions, the cumulative

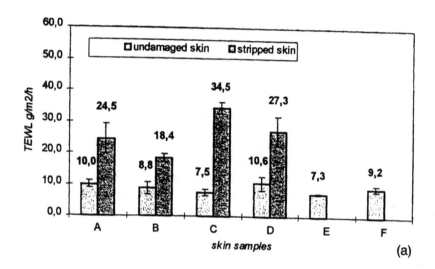

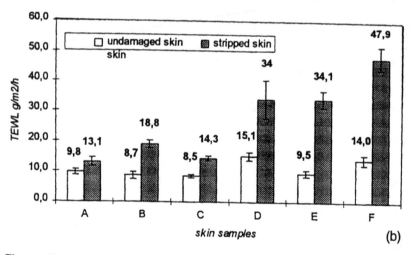

Figure 12 (a) and (b). TEWL measurements in vitro. Intraindividual variability and stripping effect on human dermatomed skin (a) and full-thickness skin (b). Averaged values for all exposure times (1, 2, 3, 5, and 24 h) pooled together, mean ± SD, n = 6. (From Ref. 120.)

amount appearing in the compartment is limited. However, the form of the curve would be similar in all other respects. This behavior is most frequently observed for in vitro percutaneous absorption investigations of transdermal formulations where the device is prepared such that a steady state is achieved rapidly and is maintained for the duration of the experiment.

However, in practice, for most cosmetic compounds and dermatological formulations, percutaneous absorption is not described by such straightforward pharmokinetics. Steady state is either nor achieved during the appropriate exposure period (reflecting in-use conditions) or is difficult to demonstrate. The rate of percutaneous absorption reflects several processes, all of which may be occurring simultaneously: (1) diffusion within the vehicle, (2) depletion of compound from the vehicle, (3) changes in the thermodynamic activity at the skin surface, (4) changes in the hydration status of the skin, (5) diffusion through the barrier, and (6) breakdown in the integrity of the skin in the diffusion cell. For most applications, the principal parameters that are relevant to the prediction of percutaneous absorption are (1) the cumulative absorbed dose, equal to the area under the curve ($\mu g/cm^2$), (2) the time to achieve maximum flux (h), and (3) the maximum flux ($\mu g/cm^2/h$).

VI. SAFETY ASSESSMENT OF A COSMETIC INGREDIENT

A. Initial Safety Estimate

The strategy outline in Table 3 describes a step-by-step approach for deciding if and when percutaneous absorption experiments are required for the safety assessment of an ingredient in a cosmetic formulation. The steps can be broken down

Table 3 Safety Estimate

1	Compound under consideration
2	Systemic toxicology (calculate NOAEL)
3	In-use conditions
4	Preliminary MoS calculation assuming 100% percutaneous absorption
5a	If this MoS < 1 reconsider ingredient
5b	If this MoS > 100 accept ingredient
5c	If this MoS is between 1 and 100 proceed to 6
6	Synthesis and formulate into prototype (is purity/stability suitable?)
7	In vitro investigation using in-use conditions
8	Revise MoS calculation using bioavailability and reject or accept ingredient as indicated
9	In vivo investigation (if necessary)

Table 4 Safety Calculation

	Example
Maximum applied amount of active ingredient (I)	1 mg
Typical body weight (adult human)	60 kg
Maximum absorption[a] (in vitro/in vivo investigation)	10%
Dermal absorption per application I (mg) $\times$ A%	0.10 mg
Systemic exposure dose (SED) I(mg) $\times$ A%/60 kg BW	0.002 mg/kg BW
NOAEL	0.2 mg/kg BW
Margin of safety = NOAEL/SED = 0.002 mg/kg BW/0.2 mg/kg BW MoS = 100	

[a] In an initial calculation, MoS will be assumed to be equal to 100% of the applied dose. If the initial calculated MoS is between 1 and 100 then further percutaneous absorption investigations are recommended.

into (1) identification of an ingredient, (2) evaluation of systemic toxicity in an animal model, (3) description of in-use conditions, (4) assessment of potential toxicity assuming a worst-case scenario, and (5) measurement of percutaneous absorption if necessary. For the vast majority of compounds, the procedure ends at step (4) where it is concluded that at the level of exposure indicated, the ingredient is not a cause for concern even if 100% absorption is assumed.

In a first approach, a margin of safety (MoS) for a cosmetic ingredient can be calculated from the ratio of the no-adverse-effect level (NOAEL) to the exposure (Table 4). This is a worst-case scenario assuming 100% penetration. The acceptable MoS for a particular cosmetic ingredient is judged on a case-by-case basis by the appropriate regulatory authorities. This value will vary for different ingredients depending upon exposure conditions, though a figure of 100 is generally considered reasonable [123]. We will use this value for the purposes of our discussion. If preliminary calculations indicate that the exposure (assuming 100% bioavailability) leads to a MoS of greater than 100, then no further studies should be required and the ingredient can be considered safe. In contrast, if the MoS is less than 0.1, then the compound may generally be excluded from consideration as a potential ingredient, since it has too much potential for toxicity. However, there may be extenuating circumstances for certain compounds. In vitro investigations of percutaneous absorption are generally required if the calculated MoS is between 1 and 100; i.e., this is the ''window'' within which further analysis is required.

For example, assuming a worst-case scenario (100% absorption of applied dose), the application of a cream formulation containing 1% of the compound under consideration would result in the absorption of only 10 $\mu g/cm^2$ of this compound per day (180 mg/day if one assumes a whole-body application). Complete dilution into the body (60 kg) would result in a serum concentration of 3 $\mu g/mL$ or a concentration of 10 M (for a compound with a molecular mass of 300 Da), if one assumes no elimination or metabolism. In practice, the systemic concentrations are more than 1000-fold lower than this value under even the most optimal conditions for absorption.

B. In-Use Conditions

The safety assessment of cosmetics must be made on the basis of their degree of exposure. The following need to be addressed: (1) duration of exposure, (2) surface area of application, (3) site of contact, (4) dose, (5) formulation effects, (6) frequency of application, (7) nature of consumer, and (8) potential for misuse. Table 5 indicates the levels to which the skin surface is exposed by cosmetic products.

Leave-on products such as sunscreens and skin-care products are generally left on the skin for 16 h (day) or 6–8 h (night). In contrast, the period of exposure for rinse-off products is less than 1 h, typically less than 10 min. Extrapolation of data from studies that do not reflect the appropriate exposure conditions should be avoided.

Often the question is raised whether the absorption is influenced by the quantity of a cosmetic product applied. For cosmetics, more than 2 mg of semi-solid product will overload the skin; i.e., a homogeneous contact is not obtained. Application of a greater amount does not result in a proportionate increase in percutaneous absorption. For hair dyes, the large surface area of hair (20- to 100-fold) compared with that of the scalp competes for the formulation and reduces the amount of hair product that comes into contact with the skin. This is consistent with the observation that removal of surface hair from rats increases percutaneous absorption of 2-nitre-p-phenylenediamine twofold. Finally, in some cases it may be more appropriate to rely upon estimates of consumer use to judge an appropriate dose.

For example, consumers apply maximally 10 g of sunscreen per use or approximately 0.5 mg/cm^2, assuming a whole-body application. This value may be a more appropriate dose that 2 mg/cm^2, for safety assessment.

It is far harder to provide general rules concerning how to treat frequent applications of a cosmetic. In general, if sufficient time is allowed between applications, the percutaneous absorption can be calculated as if the applications were independent [124]. However, if the applications are repeated over a relatively

Table 5 Consumption of Cosmetics

Product type	Typical quantity per application, g	Frequency applications/ day	Exposure levels, g/day	
			Normal use	Extensive use
Mucous membrane contact				
Toothpaste[a]	1.4	1–2	0.24	0.48
Mouthwash[b]	10	1–5	1	5
Lipstick	0.01	2–6	0.02	0.06
Eye makeup				
Powder	0.01	1–3	0.01	0.03
Mascara	0.025	1	0.025	0.025
Eye liner	0.005	1	0.005	0.005
Leave-on products				
Face cream	0.8	1	0.8	0.8
General-purpose cream	1 mg/cm² skin	1–2	1.0	2.0
Body lotion	8	1–2	8	16
Antiperspirant roll-on	0.5	1	0.5	0.5
Hair styling products[d]	5	1–2	0.5	1
Rinse-off products[c]				
Make-up remover	2.5	1–2	2.5	5
Shower gel	5	1–2	5	10
Shampoo[d]	12	2–7/week	0.34	1.2
Hair conditioner[d]	14	1–2/week	0.2	0.4

[a] 0.24 g is assumed to be ingested per application.
[b] Only 10% is ingested.
[c] Multiply final exposure levels by 0.1 since 10% rinse-off coefficient is assumed.
[d] A partition coefficient of 10% is applied on the assumption that only 10% of product remains on scalp.

short period of time, it is likely that saturation of the stratum corneum reservoir will reduce the subsequent absorption of the following application.

C. Exposure Limits

As opposed to systemic dosage, where multiples of a standard dose can be administered, the amount of a drug or xenobiotic that can enter into an effective contact with the skin surface is limited. Thus, a patient or consumer applies no more than 0.5–2 mg/cm² of a semisolid (drug or cosmetic) preparation; more is felt as unpleasant and consciously or unconsciously rubbed off or spread to a larger surface.

For example, if a cream containing 1% of an active compound is applied to the face, chest, or scalp (with a surface of about 1000 cm^2) no more than 10 μg/cm^2 or 10 mg/1000 cm^2 of the ingredient can be brought into efficient contact with the skin. The same limitation holds for solid material, the amount of which is restricted by adherence to the surface and subsequent dissolution in stratum corneum lipids and its aqueous phase. It also holds—for obvious reasons—for liquids. An increase of concentration of a given compound in a formulation is equally limited by the changes this imposes on a vehicle in terms of stability, of viscosity, for the incorporation of other ingredients, and finally in terms of cosmeticity or acceptance by the consumer or patient.

On the other hand, it is physically impossible to apply less than 0.1 mg of any material evenly to the skin surface. It follows that there is a natural "window of dosage" between a minimum and a maximum topical dosage, which should be estimated before developing a topical drug or investigating the percutaneous absorption of a xenobiotic.

D. Absorbed Dose

The safety assessment of the compound is based upon cumulative dose absorbed (μg/cm^2) after a suitable period of absorption (normally 24 h). Though the percutaneous absorption can also be expressed as a percentage of the dose applied, this invites extrapolation to different doses, which is generally not warranted. The dose absorbed should include not only the compound in the receptor fluid but also an estimate of the amount in "viable tissues"—i.e. the epidermis and dermis. This is especially important for investigations on whole-thickness skin samples.

There is currently some debate as to whether the material that has penetrated the stratum corneum and epidermis but has not entered into the systemic compartment should be included in a safety assessment. Provided that the completeness of the diffusion process from the reservoir into the receptor fluid has been assessed, this material, when recovered from the horny layer and epidermis at the end of the experiment, cannot be considered to have penetrated the barrier during this time, nor would it significantly contribute to the overall process after prolonged exposure. In fact, it is the amount entering the systemic compartment of the body which is relevant for the comparison with a systemic NOAEL and a subsequent calculation of the safety margin of cosmetic ingredients. This corresponds to the quantity which, at the end on an appropriate exposure period in vitro, has permeated the stratum corneum barrier and the epidermis and has entered into the dermis and the receptor fluid—that is, into the compartment corresponding to the vascularized tissue in vivo.

E. Standard Protocols and Good Laboratory Practices

It is of paramount importance that experiments for safety assessment be performed using good laboratory practices (GLP) or quality control close to GLP. This entails (1) establishing standard operating protocols, (2) training staff, and (3) performing routine quality control analysis.

With in vitro methods, strictly standardized protocols are crucial to obtaining reproducible data.

An interlaboratory comparison of percutaneous absorption of sunscreens ([^{14}C]-octyl methoxycinnamate and [^{14}C]-benzophenone-4) and a reference compound with high skin permeability ([^{14}C]-salicylic acid) on pig skin in vitro showed that the data obtained in two different laboratories using the same protocol were of the same order [125]. Detailed analysis of skin distribution showed some differences between the two laboratories, especially with the highly permeating compound; these were probably due to differences in the main steps of protocol—i.e., washing and stripping steps, which are subject to technical variations.

F. Examples of Safety Assessment of Cosmetic Formulations

1. Ethyl Alcohol

Concern has been expressed as to the potential risk of exposure to cosmetics containing significant quantities of ethyl alcohol. Such risks would be of special concern to children. However, a simple calculation can be used to indicate that the flux of ethanol from a topical application is not sufficient to cause concern. The permeability coefficient of ethanol across human skin is 8×10^{-8} mol/cm^2 from aqueous solutions [126]. Given the volatility of ethanol, the maximum exposure period after topical application would be significantly less than 1 h. Moreover, the application area for these formulations is generally restricted to less than 1000 cm^2. Thus, the percutaneous absorption from 70% ethanol solution would be approximately 100 mg. This corresponds to the amount of alcohol in 1.5 mL of wine (10% alcohol by volume), suggesting that skin exposure to ethanol in cosmetics is not a safety concern.

2. Hair Dyes

The use of synthetic hair dyes to color hair for cosmetic purposes has become widespread. They can be classified into temporary, semipermanent, and permanent dyes. The temporary hair dyes are high-molecular-weight compounds that are deposited on the hair fiber but do not penetrate it. Their high molecular weight precludes significant percutaneous absorption. In contrast, the semipermanent hair dyes use dyes of relatively low molecular weight that can penetrate the hair cortex and bind keratin. A formulation typically contains two or more dyes. These products are more resistant to washing, though they can be leached off the hair. The permanent hair dyes rely upon the permeation of low-molecular-weight precursor molecules into the hair fiber (classically an amine and a coupler) followed by the formation of high-molecular-weight dyes via oxidative polymerization, which effectively traps the polymers in the hair fiber.

Maibach and coworkers [127,128] studied the percutaneous absorption of seven hair dyes applied under in-use conditions. The results, shown in Table 6, indicate that only a very small percentage of the application can be detected in the urine (<0.1%), but the protocol that was followed did not allow the measurement of mass balance. Percutaneous absorption of p-phenylenediamine has also been followed using an independent assay for detection of metabolites in the urine [129]. Several points in this study are of general interest. During the coloring process, consumption of the precursors on and within the hair reduces the availability of these compounds for percutaneous absorption. Moreover, the relative surface area of the hair is greater than that of the scalp (75- to 100-fold), which further reduces the percutaneous absorption. Finally, the amount of material applied to the hair saturates the available scalp surface area. Thus, the percutaneous absorption of hair dyes is unlikely to be sensitive to the volume of formulation used. Together, these factors reduce the absolute amount of compound absorbed to levels corresponding to less than 1% of the formulation (i.e., in the range of milligrams or less).

Table 6   Percutaneous Absorption of Hair Dyes

	Dye, % (w/w)	Dose, g	Absorbed %	Absorbed mg
Permanent				
PPD	2.7	1.43	0.19	2.7
Resorcinol	1.225	0.65	0.076	0.5
4-amino-2-hydroxytoluene	0.69	0.37	0.200	0.2
DAA	1.74	0.92	0.022	0.2
Semipermanent				
2-Nitro-PPd	1.36	1.09	0.143	1.6
4-Amino-2-nitrophenol	0.433	0.23	0.235	0.5
HC blue 1	1.48	1.18	0.151	1.8

3. Sunscreens

Sunscreens are applied to large areas of the body for extended periods of time. Their use is expected to offer considerable benefit in protection against the deleterious effects of exposure to UV radiation. Recent in vitro investigations of the percutaneous absorption of several benzophenone derivatives indicate that absorption was relatively modest. Though the maximum flux of the compounds with the highest absorption was 4.8 ng/cm^2/h, the reported lag time was 4–6 h [130].

Figure 10 demonstrate the results of an in vitro investigation of the percutaneous absorption of parsol MCX [105]. The experiment was designed to reflect in-use conditions. The compound was radiolabeled (specific activity = 30μCi/mg) and judged to be of greater than 96% purity. The parsol MCX was incorporated into a semisolid commercial formulation at 5% (w/w). It was applied to skin samples using a spatula, which was subsequently weighed to measure the exact amount of compounds applied (approximately 2 mg/cm^2). Mass balance indicated that >95% of the radioactivity was recovered following completion of the experiment.

The results obtained on human skin indicate that for all conditions, most of the material was recovered from the skin surface, approximately 88% of the applied dose. The cutaneous distribution showed that the majority of the absorbed dose is recovered from the epidermis and especially from the stratum corneum—i.e., 8% of the applied dose. This indicates that the systemic exposure to Parsol MCX is low, about 2% of the applied dose. Thus, the percutaneous absorption of Parsol MCX incorporated at 5% (w/w) in this vehicle, assuming a whole-body application at 0.5 mg/cm^2, could be estimated at 9 mg.

VII. SUMMARY

The skin is particularly effective as a selective barrier to the penetration of a diverse range of compounds. The principal barrier function of the skin resides in the stratum corneum. It consists of two major components: corneocytes or the keratinocyte cell remnants surrounded by intercellular lipid. The lipid located in the intercellular spaces plays a key role in limiting the diffusion of compounds through the stratum corneum.

Percutaneous absorption involves a series of individual transport processes occurring in sequence. First, molecules must be adsorbed at the surface of the stratum corneum; then they must diffuse through it and penetrate into the viable epidermis and dermis until they reach the systemic blood flow.

In general, the percutaneous absorption of a cosmetic ingredient is small, and this precludes in vivo monitoring except for compounds that permeate very readily. An alternative in vivo approach uses skin stripping to predict percutaneous absorption. Though not currently validated in a large number of laboratories, this approach is likely to be more accepted in the future. However, in general, there is an increasing reliance on in vitro experiments that use either split-thickness human or pig skin to predict systemic exposure. The predicted systemic exposure can be calculated as the amount of ingredient found in viable tissues, including perfusate, 24 h following the application of the cosmetic. The available evidence indicates that this approach is warranted provided that good laboratory practices are followed.

The safety assessment of cosmetics and their ingredients requires that suitable semichronic toxicity tests be conducted to estimate the NOAEL (no-adverse-effect level). A MoS (margin of safety) can be calculated from the systemic exposure. Most regulatory agencies consider that a MoS of greater than 100, assuming 100% cutaneous bioavailability, indicates that the ingredient is safe. Percutaneous absorption experiments are required for more accurate estimates of systemic exposure only if the calculated MoS is less than 100; i.e., investigations are carried out if the MoS is between 10 and 100. It is especially important to ensure that the in use conditions (dose and length of exposure) are duplicated in these investigations. Thus, cosmetics that come into contact with the skin for relatively short periods of exposure (shampoos, hair dyes, and skin cleansers) should be differentiated from those that are left on after application (sunscreens, makeup, skin-care products).

REFERENCES

1. Barton SP, Black DR. Surface contour: Variability, significance and measurements. In: Marks R, Barton SP, Edwards C, eds. The Physical Nature of the Skin. Lancaster, England: MTP Press, 1998:23–30.
2. Kligman AM, Zheng P, Lavker RM. The anatomy and pathogenesis of wrinkles. Br J Dermatol 1985; 113:37–42.
3. Fiedler M, Meier W-D, Hoppe U. Texture analysis of the surface of the human skin. Skin Pharmacol 1995; 8:252–265.
4. Brown S, Diffey BL. The effect of applied thickness on sunscreen protection: In vivo and in vitro studies. Photochem Photobiol 1986; 44:509–513.
5. Wester RC, Maibach HI. Regional variations in percutaneous absorption. In: Bronaugh RL, Maibach HI, eds. Percutaneous Absorption. Basel: Decker, 1989:111–120.
6. Marks R, Barton SP. The significance of size and shape of corneocytes. In: Marks R, Plewig G, eds. Stratum Corneum. Berlin: Springer-Verlag, 1983:175–180.
7. Green RS, Downing DT, Pochi PE, Strauss JS. Anatomical variation in the amount and composition of human skin surface. J Invest Dermatol 1970; 54:240–247.

8. Lampe MA, Burlingame AL, Whitney J, et al. Human stratum corneum lipids: Characterization and regional differences. J Lipid Res 1983; 24:120–130.
9. Grove GL, Lavker RM, Holze E, Kligman AM. The use of nonintrusive methods to monitor age-associated changes in human skin. J Soc Cosmet Chem 1981; 32: 15–26.
10. Eckert RL. Structure, function and differentiation of the keratinocyte. Physiol Rev 1989; 69:1316–1345.
11. Steinert PM. Structure, function, and dynamics of keratin intermediate filaments. J Invest Dermatol 1993; 100:729–733.
12. Richert U, Michels, Schmidt R. The cornified envelope: a key structure of terminally differentiating keratinocytes. In: Darmon M, Blumenberg M, eds. The Keratinocytes. San Diego, CA: Academic Press, 1993:107–150.
13. Wertz PW, Madison KC, Downing DT. Covalently bound lipids of the stratum corneum. J Invest Dermatol 1989; 92:109–111.
14. Wertz PW, Swartzendruber DC, Kitko DJ, et al. The role of the corneocyte lipid envelopes in cohesion of the stratum corneum. J Invest Dermatol 1989; 93:169–172.
15. White FH, Gohari K. Some aspects of desmosomal morphology during differentiation of hamster cheek pouch. J Submicrosc Cytol 1984; 16:407–422.
16. Fartasch M. The epidermal barrier in disorders of the skin. Microsc Res Techniq 1997; 38:361–372.
17. Grayson S, Johnson-Winegar AG, Wintraub BU, et al. Lamellar body-enriched fractions from neonatal mice: preparative techniques and partial characterization. J Invest Dermatol 1985; 85:285–289.
18. Landmann L. Epidermal permeability barrier: transformation of lamellar-granule disks into intercellular sheets by a membrane-fusion process: a freeze-fracture study. J Invest Dermatol 1986; 87:202–209.
19. Menon GK, Feingold KR, Elias PM. The lamellar secretory response to barrier disruption. J Invest Dermatol 1992; 98:279–289.
20. Elias PM, Menon GK. Structural and lipid biochemical correlates of the epidermal permeability barrier. Adv Lipid Res 1991; 24:1–26.
21. Bommannan D, Potts RO, Guy RH. Examination of the stratum corneum barrier function in vivo by infrared spectroscopy. J Invest Dermatol 1990; 95:403–408.
22. Nemaniak MK, Elias PM. In situ precipitation: a novel cytochemical technique for visualization of permeability pathways in mammalian stratum corneum. J Histol Cytochem 1980; 28:573–578.
23. Potts RO, Francoeur ML. The influence of stratum corneum morphology on water permeability. J Invest Dermatol 1991; 96:495–499.
24. Hou SYE, Mitra AK, White SH, et al. Membrane structures in normal and essential fatty acid–deficient stratum corneum: characterization by ruthenium tetroxide staining and x-ray diffraction. J Invest Dermatol 1991; 96:215–223.
25. Warner RR, Myers MC, Taylor DA. Electron probe analysis of human skin: determination of the water concentration profile. J Invest Dermatol 1988; 90:218–224.
26. von Zglinicki T, Lindberg M, Roomans GM, Forslind B. Water and ion distribution profiles in human skin. Acta Derm Venereol (Stockh) 1993; 73:340–343.

27. Scott IR, Harding CR, Barrett JG. Histidine-rich protein of the keratohyalin granules: source of free amino acids, urocanic acid and pyrolidone carboxylic acid in the stratum corneum. Biochim Biophys Acta 1982; 719:100–117.

28. Horii I, Nakayama Y, Obata M, Tagami H. Stratum corneum hydration and amino acid content in xerotic skin. Br J Dermatol Res 1989; 121:588–592.

29. Yamamura T, Tezuka T. The water-holding capacity of the stratum corneum as measured by ^{1}H-NMR. J Invest Dermatol 1989; 93:342.

30. Mils V, Vincent C, Croute F, Serre G. The expression of desmosomal and corneo-desmosomal anti-gens shows specific variations during terminal differentiation of epidermis and hair follicle epithelia. J Histochem Cytochem 1992; 40:1329–1337.

31. Lundstrom A, Egelrud T. Cell shedding from human plantar skin in vitro: evidence of its dependence on endogenous proteolysis. J Invest Dermatol 1988; 91:340–343.

32. Imokawa G, Abe A, Jin K, et al. Decreased levels of ceramides in stratum corneum of atopic dermatitis: an etiologic factor in atopic dry skin? J Invest Dermatol 1991; 96:523–526.

33. Hoath SB, Tanaka R, Boyce ST. Rate of stratum corneum formation in the perinatal rat. J Invest Dermatol 1993; 100:400–406.

34. Grubauer G, Elias PM, Feingold KR. Lipid content and lipid type as determinants of the epidermal permeabilities barrier. J Lipid Res 1989; 30:89–96.

35. Bonté F, Saunois A, Pinguet P, Meybeck A. Existence of a lipid gradient in the upper stratum corneum and its possible biological significance. Arch Dermatol Res 1997; 289:78–82.

36. Ishii H, Mikami N, Sakamoto K. Lipo-amino acid cholesteryl derivatives promote recovery effect for damaged skin. J Soc Cosmet Chem 1996; 47:351–362.

37. Ghadially R, Williams ML, Hou SY, Elias PM. Membrane structure abnormalities in the stratum corneum of the autosomal recessive ichthyosis. J Invest Dermatol 1992; 99:755–763.

38. Feingold KR. The regulation and role of epidermal lipid synthesis. Adv Lipid Res 1991; 24:57–82.

39. Proksch E, Elias PM, Feingold KR. Regulation of 3-hydroxy-3-methylglutaryl-coenzyme A reductase in murine epidermis: modulation of enzyme content and activitation state by barrier requirements. Br J Dermatol 1990; 85:874–882.

40. Mao-Qiang M, Elias P, Feingold KR. Fatty acids are required for epidermal permeability barrier homeostasis. J Clin Invest 1993; 92:791–798.

41. Holleran WM, Gao WN, Feingold KR, Elias PM. Localization of epidermal sphingolipid synthesis and serine palmitoyl transferase activity: alterations imposed by permeability barrier requirements. Arch Dermatol Res 1995; 287:254–258.

42. Holleran WM, Feingold KR, Man M-Q, et al. Regulation of epidermal sphingolipid synthesis by permeability barrier function. J Lipid Res 1991; 32:1151–1158.

43. Holleran WM, Mao-Qiang M, Gao WN, et al. Sphingolipids are required for mammalian barrier function: inhibition of sphingolipid synthesis delays barrier recovery after acurate perturbation. J Clin Invest 1991; 88:1338–1345.

44. Proksch E, Holleran WM, Menon GK, et al. Barrier function regulates epidermal lipid and DNA synthesis. Br J Dermatol 1993; 128:473–482.

45. Scheuplein RJ, Blank IH. Permeability of the skin. Physiol Rev 1971; 51:702–747.

46. Schaefer H, Zesch A, Stuttgen G. Skin Permeability. Berlin: Springer-Verlag, 1982.

47. Bommannan D, Potts RO, Guy RH. Examination of the stratum corneum barrier function in vivo by infrared spectroscopy. J Invest Derm 1990; 95:403–408.

48. Corcuff P, Bertrand C, Leveque JL. Morphometry of human epidermis in vivo by real-time confocal microscopy. Arch Dermatol Res 1993; 285;475–481.

49. Rolland A, Wagner N, Chatelus A, et al. Site-specific drug delivery to pilosebaceous structures using polymeric microspheres. Pharmacol Res 1993; 10:1738–1744.

50. Rolland A, Wagner N, Chatelus A, et al. Polymeric microspheres, as a novel topical site specific drug delivery system for targeting a naphtoic acid derivative, adapalene to the pilo-sebaceous unit. J Invest Dermatol 1993; 100:218.

51. Rolland A. Particulate carriers in dermal and transdermal drug delivery: myth or reality. In: Walters KA, Hadgraft J, eds. Pharmaceutical Particulate Carriers: Therapeutic Applications. New York: Marcel Dekker, 1993:367–421.

52. Illel B, Schaefer H, Wepierre J, Doucet O. Follicles play an important role in percutaneous absorption. J Pharm Sci 1991; 80:424–427.

53. Middleton JD. The mechanism of water binding in stratum corneum. Br J Dermatol 1986; 80:437–450.

54. Fartasch M, Bassukas ID, Dipegen TL. Structural relationship between epidermal lipid lamellae, lamellar bodies and desmosomes in human epidermis: an ultrastructural study. Br J Dermatol 1993; 128:1–9.

55. Batt MD, Davis WB, Fairhust E, et al. Changes in the physical properties of the stratum corneum following treatment with glycerol. J Soc Cosmet Chem 1988; 39: 367–381.

56. Kao J, Carver MP. Cutaneous metabolism of xenobiotics. Drug Metab Rev 1990; 22:363–410.

57. Mukhtar H, Khan WA. Cutaneous cytochrome P-450. Drug Metab Rev 1989; 20: 657–673.

58. Mao-Qiang M, Feingold KR, Jain M, Elias PM. Extracellular processing of phospholipids is required for permeability barrier homeostasis. J Lipid Res 1995; 36: 1925–1935.

59. Mao-Qiang M, Jain M, Feingold KR, Elias PM. Phospholipase A_2 activity is required for permeability barrier homeostasis. J Invest Dermatol 1996; 106:57–63.

60. Potts RO, McNeill SC, Desbonnet CR, Wakshull E. Transdermal drug transport and metabolism: II. The role of competing events. Pharm Res 1989; 6:119–124.

61. Andersson P, Edsbacker S, Ryrfeldt A, von Bahr C. In vitro transformation of glucocorticoids in liver and skin homogenate fraction from man, rat and hairless mouse. J Steroid Biochem 1982; 16:787.

62. Coomes MW, Norling AH, Pohl RJ, et al. Foreign compound metabolism by isolated skin cells from the hairless mouse. J Exp Pharmacol Ther 1983; 225:770.

63. Higo H, Ninz RS, Lau DTW, et al. Cutaneous metabolism of nitroglycerin: II. Effect of skin conditions and penetration enhancement. Pharm Res 1992; 9:299–302.

64. Collier SW, Sheikh NM, Sahr A, et al. Maintenance of skin viability during in vitro percutaneous absorption/metabolism studies. Toxicol Appl Pharmacol 1989; 99:522–533.

65. Bronaugh RL. Methods for in vitro skin metabolism studies. Toxicol Meth 1995; 5:275–281.

66. Bronaugh RL, Collier SW, Storm JA, Stewart RF. In vitro evaluation of skin absorption and metabolism. J Toxicol Cutan Ocular Toxicol 1990; 8:453–467.

67. Jewel C, Clowes HM, Heylings JR, et al. Absorption and metabolism of dinitrochlorobenzene through mouse skin in vitro. In: Brain KR, James VJ, Walters KA, eds. Prediction of Percutaneous Penetration, vol 4b. Cardiff, Wales: STS, 1996: 218–221.

68. Nathan D, Sakr A, Lichtin JL, Bronaugh RL. In vitro skin absorption and metabolism of benzoic acid p-aminobenzoic acid, and benzocaine in the hairless guinea pig. Pharm Res 1990; 7:114.

69. Shaikh NA, Ademola JI, Maibach HI. Effects of freezing and azide treatment of in vitro human skin on the flux and metabolism of 8-methoxypsoralen. Skin Pharmacol 1996; 9:274–280.

70. Storm JE, Collier SW, Stewart RF, Bronaugh RL. Metabolism of xenobiotics during percutaneous penetration: role of absorption rat and cutaneous enzyme activity. Fund Appl Toxicol 1990; 15:132–141.

71. Kraeling MEK, Lipicky RJ, Bronaugh RL. Metabolism of benzocaine during percutaneous absorption in the hairless guinea pig: acetylbenzocaine formation and activity. Skin Pharmacol 1996; 9:221–230.

72. Frantz SW, Dittenber DA, Eisenbrandt DL, Watanabe PG. Evaluation of a flow-through in vitro penetration chamber method using acetone-deposited organic solids. J Toxicol Cutan Ocular Toxicol 1990; 9:277–299.

73. Boehnlein J, Sakr A, Lichtin JL, Bronaugh R. Characterization of esterase and alcohol dehydrogenase activity in skin: metabolism of retinyl palmitate to retinol (vitamin A) during percutaneous absorption. Pharm Res 1994; 11:1155–1159.

74. Leclerc C, Wegrich P, Fouchard F, et al. Cutaneous bioavailability and metabolism or retinyl palmitate in human skin in vitro: role of an exogenous enzyme. In: Brain KR, James VJ, Walters KA, eds. Perspectives in Percutaneous Penetration, vol 5a. Cardiff, Wales: STS, 1997:122.

75. Liu P, Higushi WI, Ghanem AH, Good WR. Transport of β-estradiol in freshly excised human skin in vitro: diffusion and metabolism in each skin layer. Pharm Res 1994; 11:1777–1784.

76. Carver MP, Williams PL, Riviere JE. The isolated perfused porcine skin flap: III. Percutaneous absorption pharmacokinetics of organophosphates, steroids, benzoic acid, and caffeine. Toxicol Appl Pharmacol 1989; 97:324–327.

77. Bikle DD, Halloran BP, Riviere JE. Production of 1,25-dyhydroxyvitamin D_3 by perfused pig skin. J Invest Dermatol 1994; 102:796–798.

78. Elliot GR, de Lange J, van de Sandt JJM, et al. A comparison of in vitro and in vivo methods for determining the transdermal penetration, permeation and metabolism of propoxur. In: Brain KR, James VJ, Walters KA, eds. Prediction of Percutaneous Penetration, vol 4b. Cardiff, Wales: STS Publishing, 1995. In press.

79. Elliott GR, de Lange J, Poot CAJ, et al. The "with blood perfused pig ear": a model for detecting irritant chemicals. In: Brain KR, James VJ, Walters KA, eds. Predictions of Percutaneous Penetration, vol 4b. Cardiff, Wales: STS Publishing, 1995. In press.

80. Chang SK, Williams PL, Dauterman WC, Riviere JE. Percutaneous absorption: dermatolapharmokinetics and related bio-transformation studies of carbaryl, lindane, malathion, and parathion in isolated porcine skin. Toxicology 1994; 91:269–280.

81. Shah VP, Flynn GL, Guy RH, et al. In vivo percutaneous penetration/absorption, Washington, May 1989. Pharm Res 1991; 8:1071.

82. Franz T, Lehman P, McGuire E. In vivo methods for assessment of percutaneous absorption in man. In: Zatz JL, ed. Skin Permeation: Fundamentals and Application. Allured Publishing Corp., Wheaton, IL, 1993:73–92.

83. Reifenrath WG, Chellquist EM, Shipwash EA, Jedeberg WW. Evaluation of animal models for predicting percutaneous penetration in man. Fundam Appl Toxicol 1984; 4:S224–S230.

84. Wester RC, Maibach HI. Animal models for percutaneous absorption. In: Shaw VP, Maibach HI, eds. Topical Drug Bioavailability, Bioequivalence and Penetration. New York: Plenum Press, 1993:333–350.

85. Rougier A, Dupuis D, Lotte C, Maibach HI. Stripping method for measuring percutaneous absorption in vivo. In: Bronaugh RL, Maibach HI, eds. Percutaneous Absorption Mechanism, Methodology, Drug Delivery. Basel: Marcel Dekker, 1989: 415–434.

86. Rougier A, Lotte C. Predictive approaches. I. The stripping technique. In: Shaw VP, Maibach HI, eds. Topical Drug Bioavailability, Bioequivalence and Penetration. New York, Plenum Press, 1993:163–182.

87. Pershing LK, Parry GE, Bunge A, et al. Assessment of topical corticosteroid treatment in vivo and in vitro. In: Shah VP, Maibach HI, eds. Topical Drug Bioavailability: Bioequivalence and Penetration. New York: Plenum Press, 1993:351–391.

88. Marttin E, Neelissen-Subnel MTA, de Haan FHN, Boddé HE. A critical comparison of methods to quantify stratum corneum removed by tape stripping. Skin Pharmacol 1996; 9:69–77.

89. Tsai JC, Cappel MJ, Weiner ND, et al. Solvent effects on the harvesting of stratum corneum from hairless mouse skin through adhesive tape stripping in vitro. Int J Pharm 1991; 68:127–133.

90. Pershing L, Lambert L, Shah VP, Lam SY. Variability and correlation of chromameter and tape stripping methods with the visual skin blanching assay in the quantitative assessment of topical 0.050% betamethasone dipropionate bioavailability in humans. Int J Pharm 1992; 86:201–210.

91. Finlay A, Marks R. Determination of corticosteroid concentration profiles in the stratum corneum using the skin surface biopsy technique. Br J Dermatol 1982; 107: 33.

92. Miselnicky SC, Lichtin JL, Sakr A, Bronaugh RL. The influence of solubility, protein binding, and percutaneous absorption on skin reservoir formation. J Soc Cosmet Chem 1988; 39:169–177.

93. Pershing L, Lambert L, Wright ED, et al. Topical 0.050% betamethasone dipropionate: pharmacokinetic and pharmodynamic dose-response studies in humans. Arch Dermatol 1994; 130:740–747.

94. Poulson BJ, Flynn GL. In vitro methods used to study transdermal delivery and percutaneous absorption. In: Bronaugh RL, Maibach HI, eds. Percutaneous Absorp-

tion: Mechanisms-Methodology-Drug Delivery. New York: Marcel Dekker, 1985: 431–459.

95. Bronaugh R, Collier S. In vitro methods for measuring skin permeation. In: Zatz JL, ed. Skin Permeation: Fundamentals and Application. Allured Publishing Corp., Wheaton, IL, 1993:93–112.

96. Scott RC, Carmichael NG, Huckle KR, et al. Methods for measuring transdermal penetration of pesticides. Food Chem Toxicol 1993; 31:523–529.

97. Ainsworth M. Methods for measuring percutaneous absorption. J Soc Cosmet Chem 1969; 11:69–74.

98. Franz TJ. Percutaneous absorption: on the relevance of in vitro data. J Invest Dermatol 1975; 64:190–195.

99. Franz TJ. The finite dose technique as a valid in vitro model for the study of percutaneous absorption in man. In: Simon GA, Paster N, Klinberg MA, Kaye M, eds. Current Problems in Dermatology, vol 7. Basel: Karger, 1978:58–68.

100. Scott RC. Percutaneous absorption in vivo: in vitro comparisons. In: Shroot B, Schaefer H, eds. Skin Pharmacokinetics, vol 1. Basel: Karger, 1987:103–110.

101. Beck H, Bracher M, Faller C, Hofer H. Comparison of in vitro and in vivo skin permeation of hair dyes. In: Scott RC, Guy RH, Hadgraft J, Boddé HE, eds. Prediction of Percutaneous Penetration: Methods, Measurements and Modelling, vol 2. London: IBC Technical Services Ltd, 1991:441–450.

102. Scott RC. Biological monitoring for pesticide exposure: measurement, estimation and risk reduction. In: Wang RGM, Franklin CA, Honeycutt RC, Reinert JC, eds. ACS Symposium Series 382. Washington, DC: American Chemical Society, 1989: 158–168.

103. Bronaugh RL, Stewart RF, Congdon ER, Giles AL Jr. Methods for in vitro percutaneous absorption studies: 1. Comparison with in vitro results. Toxicol Appl Pharmacol 1982; 62:474–480.

104. Tregear RT. The permeability of skin to molecules of widely different properties. In: Rook VA, ed. Progress in Biological Science in Relationship to Dermatology. London: Cambridge University Press, 1964.

105. Benech-Kieffer F, Berthelot B, Wegrich P, et al. A comparative study of in vitro skin models: Percutaneous absorption of sunscreens on pig skin and human skin. In: Brain KR, James VJ, Walters KA, eds. Prediction of Percutaneous Penetration, vol 4a. Cardiff, Wales: STS Publishing, 1995: 68.

106. Ponec M, Haverkorft, M, Soei YL, et al. Use of human keratinocytes and fibroblast cultures for toxicity studies of topically applied compounds. J Pharm Sci 1990; 79: 312–316.

107. Régnier M, Asselineau D, Lenoir MC. Human epidermal skin reconstructed on dermal substrates in vitro: an alternative to animals in skin pharmacology. Skin Pharmacol 1990; 3:7–85.

108. Contard P, Bartel RL, Jacobs L II, et al. Culturing keratinocytes and fibroblasts in a three-dimensional mesh results in epidermal differentiation and formation of a basal lamina-anchoring zone. J Invest Dermatol 1993; 100:35–39.

109. Ponec M. Reconstitution of human epidermis on de-epidermized dermis: expression of differentiation-specific protein markers and lipid composition. Toxicol In Vitro 1991; 5:597–606.

110. Egelrud T, Regnier M, Sondell B, et al. Expression of stratum corneum chymotryptic enzyme in reconstructed human epidermis and its suppression by retinoic acid. Acta Derm Venereol 1993; 73:181–184.

111. Scott RC, Ramsey JD. Comparison of the in vivo and in vitro percutaneous absorption of a lipophilic molecule (cypermethrin, a pyrethroid insecticide). J Invest Dermatol 1987; 89:142–146.

112. Scott RC, Walker M, Dugard PH. In vitro percutaneous absorption experiments: a technique for the production of intact epidermal membranes from rat skin. J Soc Cosmet Chem 1986; 37:35–41.

113. Clowes HM, Smith FM, Scott RC. Preparation of intact epidermal membranes from whole skin: an ''in-depth'' assessment of the dermatome. In: Brain KR, James VJ, Walters KA, eds. Predictions of Percutaneous Penetration, vol 3b. Cardiff, Wales: STS Publishing, 1993:123–127.

114. Bronaugh RL. Diffusion cell design. In: Shaw VP, Maibach HI, eds. Topical Drug Bioavailability, Bioequivalence and Penetration. New York: Plenum Press, 1993: 117–125.

115. Blank IH. The effect of hydration on the permeability of the skin. In: Bronaugh RL, Maibach HI, eds. Percutaneous Absorption. New York: Marcel Dekker, 1985: 97–105.

116. Bond JR, Barry BW. Hairless mouse skin is limited as a model for assessing the effects of penetration enhancers in human skin. J Invest Dermatol 1988; 90:810–813.

117. Walker M, Dugard PH, Scott RC. In vitro percutaneous absorption studies: a comparison of human and laboratory species. Hum Toxicol 1982; 2:561.

118. Bronaugh RL, Stewart RF, Simon M. Methods for in vitro percutaneous absorption studies: VII. Use of excised human skin. J Pharm Sci 1986; 75:1094–1097.

119. Lotte C, Rougier A, Wilson DR, Maibach HI. In vivo relationship between transepidermal water loss and percutaneous absorption of some organic compounds in man: effect of anatomic site. Arch Dermatol Res 1987; 279:351–356.

120. Benech-Kieffer F, Wegrich P, Schaefer H. Transepidermal water loss as an integrity test for skin barrier function in vitro: assay standardization. In: Brain KR, James VJ, Walters KA, eds. Perspectives in Percutaneous Penetration, vol 5a. Cardiff, Wales: STS Publishing, 1997:56.

121. Pinnagoda J, Tupker RA, Smit JA, et al. The intra- and interindividual variability and reliability of transepidermal water loss measurements. Contact Dermatitis 1989; 21:255–259.

122. Bronaugh RL. Methods for in vitro percutaneous absorption. Toxicol Methods 1995; 4:265–273.

123. Johannsen FR. Risk assessment of carcinogen and non-carcinogen chemicals. Crit Rev Toxicol 1990; 20:341–367.

124. Wester RC, Noonan PK, Maibach HI. The effect of frequency of application on percutaneous absorption of hydrocortisone. Clin Res 1977; 25:102A.

125. Benech-Kieffer F, Berthelot B, Wegrich P, et al. An interlaboratory comparison of percutaneous absorption of sunscreens on pig skin in vitro. In: Lisansky S, Macmillan R, Dupuis J, eds. Alternatives to Animal Testing. CPL press, Newbury, UK, 1996:246–250.

126. Scheuplein RJ, Blank IH. Mechanisms in percutaneous absorption: IV. Penetration of nonelectrolytes (alcohols) from aqueous solutions and from pure liquids. J Invest Dermatol 1973; 60:286–296.

127. Maibach HI, Wolfram LJ. Percutaneous penetration of hair dyes. J Soc Cosmet Chem 1981; 32:223–229.

128. Wolfram LJ, Maibach HI. Percutaneous penetration of hair dyes. Arch Dermatol Res 1985; 277:235–241.

129. Goetz N, Lasserre P, Boré P, Kalopissis G. Percutaneous absorption of p-phenylene diamine during an actual hair dyeing procedure. Int J Cosmet Sci 1988; 10:63–73.

130. Monti D, Saettone MF, Centini M, Anselmi C. Substantivity of sunscreens: in vitro evaluation of the transdermal permeation characteristics of some benzophenone derivatives. Int J Cosmet Sci 1993; 15:45–52.

3
Enhancement of Skin Permeation

Li-Lan H. Chen
Lavipharm Laboratories, Inc., Piscataway, New Jersey

Yie W. Chien
Rutgers University, Piscataway, New Jersey

I. INTRODUCTION

The skin is the largest organ of human body. It has long been considered a prime site for drug administration for diseases of the skin or other organs. In recent decades, transdermal drug delivery has been an active field of biomedical research that has advanced rapidly [1]. The advantages associated with transdermal drug delivery are well documented [2,3]. However, the transdermal route of administration is not appropriate for all drugs, because the skin is a very efficient barrier. This allows only a small amount of drug to be absorbed through the skin and reach the systemic circulation. In general, transdermal drug delivery is effective only in highly potent drugs whose daily dose is on the order of a few milligrams [3] for topical application. However, the transdermal delivery of even such highly potent drugs still requires skin permeation enhancement.

The epidermal barrier of the body is known to be a complex, dynamic, biochemical environment that responds to ambient conditions to maximize its protective function. Diffusional resistance is known to reside in the stratum corneum and consists of a complex interaction between lipid and proteinaceous components that create very distinct hydrophilic and lipophilic transport pathways [4]. Recent research has led to a greater understanding of the function of the stratum corneum, which, in turn, has resulted in the formulation of a diverse range of compounds and techniques that have been tested for their ability to facilitate skin permeation.

It is known that the biochemical order of the intercellular lipid matrices of the stratum corneum must be altered to increase the penetration of compounds

51

[5]. However, an ideal skin permeation enhancer should specifically promote penetration of drugs across the skin barrier without permanently changing its properties.

II. CONCEPTUALIZATION: TO ENHANCE OR NOT TO ENHANCE

Development of transdermal drug delivery systems for any topical or systemic medications requires incorporation of penetration-enhancing techniques to increase the local concentration or systemic plasma level of transdermally administrated medication. The skin is a complex organ that can exhibit a variety of pharmacological, pathological, and toxicological responses toward medications, delivery systems, added permeation enhancers, or incorporated physical enhancing forces. Drug and delivery system suitability and feasibility should be carefully examined before any attempt is made to develop transdermal or topical formulations for cosmetic or pharmaceutical purposes. The subsequent decision of whether skin permeation enhancement should be employed is totally dependent upon the purposes of the medications (topical or systemic), the physicochemical properties of drugs (partition coefficient, diffusivity, stability, solubility, and permeability), chemical and physical interactions (among the diffusant, enhancer, vehicle, and skin) and toxicological responses (localized irritant, allergic cutaneous reactions, and systemic toxicity) [6]. According to Katz and Poulsen [7], an ideal skin permeation enhancing technique should:

1. Be pharmacologically inert and possess no action of itself at receptor sites in the skin or in the body generally.
2. Be nontoxic, nonirritating, and nonallergenic.
3. Produce immediate onset of penetration-enhancing action with predictable and suitable duration of effect.
4. Allow the skin to immediately and fully recover its normal barrier properties after the material is removed.
5. Reduce the barrier function of the skin in one direction only, so as to promote penetration into the skin. Body fluids, electrolytes, or other endogenous materials should not be lost to the atmosphere.
6. Be chemically and physically compatible with a wide range of drugs and pharmaceutical adjuvants.
7. Be an excellent solvent for drugs.
8. Spread well on the skin and possess a suitable skin ''feel.''
9. Allow formulation into any cosmetic and/or pharmaceutical dosage form, such as lotions, suspensions, ointments, creams, gels, aerosols, and/or patches.

10. Be inexpensive, odorless, tasteless, and colorless, so as to be cosmetically and pharmaceutically acceptable.

It is unlikely that any material could be found that possesses all of these desirable properties. However, some enhancing techniques or materials may possess several of these attributes. Our goal is to find the optimal skin permeation–enhancing technique with minimal or no side effects.

III. CLASSIFICATION OF SKIN PERMEATION ENHANCEMENT

This section describes the three most common approaches for enhancing skin permeation. The first approach utilizes physical forces, such as electricity and ultrasound, to alter skin permeability or increase thermodynamic activity to enhance skin absorption. The second approach incorporates chemical enhancers into delivery systems to reduce skin resistance or modify drug molecules to improve the partition coefficient and/or diffusivity within the stratum corneum. The third approach integrates drug molecules into vehicle carriers, such as lipid micelles, liposomes, and Transfersomes to facilitate deposition of the drugs into the skin.

IV. PHYSICAL ENHANCEMENT

Two physical forces have been investigated for their potential to enhance skin permeation: electricity and ultrasound. There are two techniques that utilize electricity to enhance skin penetration: iontophoresis, which uses a physiologically acceptable electric current to enhance the skin permeation of charged molecules, and electroporation, which uses a high-voltage current for a very short period of time to reduce skin resistance. The use of ultrasound as an enhancer for transdermal drug delivery is referred to as *phonophoresis* or *sonophoresis*.

A. Iontophoresis: Mechanisms and Applications

Iontophoresis shares the advantages of transdermal drug delivery, which include bypass of gastrointestinal vagaries and hepatic "first-pass" effect, controlled plasma level of potent drugs with short biological half-lives, increased patient compliance, and ease of drug delivery termination. This noninvasive drug delivery system also minimizes trauma, risk of infection, and damage to existing wounds, and it is an important alternative to parenteral injection [8]. Its advantage in treating local conditions lies in the reduced incidence of systemic side effects resulting from minimal systemic uptake of drugs and high local drug concentrations [9,10].

1. Theory and Mechanisms

The term *iontophoresis* appears in the medical literature and is defined as the facilitated penetration of ions into surface tissues such as skin, oral mucosae, and other epithelia under an externally applied potential difference. The rate of skin permeation of drugs can be enhanced by means of an external energy source. Iontophoresis is a modern technique to provide this external energy source to drug ions in their movements in the biomembrane. It is based on the general principle of electricity. Migration of ions through the skin can be expressed as an ionic current [11]. Under the influence of an electrical field, a positively or negatively charged molecule (D+ or D−) is repelled by the same-polarity electrode and moves toward the skin. At equilibrium, the quantity of cations entering equals that of anions departing. Figure 1 illustrates a simplified scheme for transdermal iontophoresis by either anodic or cathodic delivery under a constant current or constant voltage [12]. Anodic or cathodic delivery is determined by molecular charge, which is dependent on the medium's pH, and on the molecules' pK_a

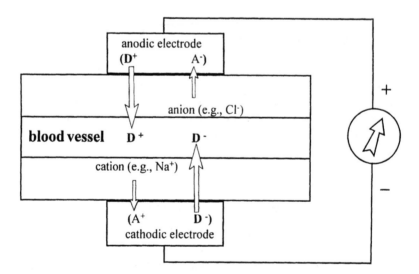

Figure 1 A schematic illustration of transdermal iontophoresis by either anodic or cathodic delivery under a constant current or constant voltage. (From Ref. 12.) The charged molecules (D⁺ or D⁻) are repelled from the same polarity electrode [anode (+) or cathode (−)] and move toward the skin.

or P_{iso}. Iontophoresis has also been observed to enhance permeation of neutral molecules by electroosmosis or convective flow [13–16].

Increased penetration of an ionic species under an applied electric field can be due to (1) the electrochemical potential gradient across the skin, (2) increased skin permeability under the applied electrical field, and/or (3) current-induced water-transport effect (such as electroosmosis or convective transport) [17]. Therefore, the transdermal iontophoretic flux of ion i can be described by the following equation:

$$J_i = J_{pi} + J_{ci} + J_{ci}$$

where J_{pi} is the passive permeation flux, given by:

$$J_{pi} = DK \frac{dC}{dh}$$

J_{ci} is the flux due to electric current facilitation, given by

$$J_{ci} = \frac{D_i C_i Z_i F}{d\Psi \ RT} \frac{d\Psi}{dh}$$

J_{ci} is the flux due to convective flow, given by

$$J_{ci} = CV_c$$

where D is the diffusivity across the skin, K is the partition coefficient between donor medium and stratum corneum, C is the concentration in the skin tissue, dC/dh is the concentration gradient across the skin. D_i is the diffusivity of ion i across the skin, C_i is the concentration of the ionic species i, Z_i is the electric valence of ionic species i, F is the Faraday constant. R is the gas constant, T is the absolute temperature, $d\Psi/dh$ is the electric potential gradient across the skin, and V_c is the velocity of convective flow which is proportional to the applied current density.

The major factors affecting skin permeation enhancement by iontophoresis include physiochemical properties of drugs (molecular size, pK_a/P_{iso}, charge, concentration, and partition coefficient), formulation factors (type of vehicles/ buffers, pH, viscosity, ionic strength, competing ions, and buffer capability), electrode characteristics (type and size), biological variations (gender, race, age, administration site, disease, and regional blood flow), and electrical parameters (type of current, waveforms, frequency, on/off ratio, and duration).

2. Applications of Iontophoresis

Iontophoresis is a noninvasive therapy gaining considerable attention among anesthesiologists, pediatricians, general and orthopedic surgeons, burn specialists, and the dental community [18]. Iontophoresis applications in transdermal medication, dermatology, and other fields have been reviewed by Costello et al. [19], Sloan et al. [20], and Singh et al. [8]. Dermatologists have reportedly used iontophoresis in the treatment of plantar warts, lichen planus, scleroderma, infected burn wounds, and for inducing local skin anesthesia [9]. Physical therapists have treated bursitis and other musculoskeletal disorders by corticosteroid iontophoresis [21]. Local deep tissue penetration of corticosteroids and local anesthetics is also facilitated by iontophoresis [22,23]. Local anesthesia has been successfully achieved for myringotomy procedures by iontophoresis of lidocaine [21].

In addition to the topical applications of iontophoresis mentioned above, systemic applications of iontophoresis are receiving increased attention because they enable noninvasive and controlled administration of drugs for their systemic effects. Transdermal iontophoresis is capable of delivering drugs that are ionized at the physiological pH, have large molecular size, and are hydrophilic in nature. Proteins and peptides belong to this class of compounds. The importance of peptides and proteins in clinical therapy has gained growing recognition in the past few years owing to the development of recombinant DNA technology—e.g., insulin in the treatment of diabetes mellitus, vasopressin for diabetes insipidus, and luteinizing hormone–releasing hormone (LHRH) for reproductive cancers. Transdermal iontophoresis has also shown promise in delivering charged polypeptides across the skin [17,24,25].

B. Electroporation

1. Mechanisms

The principle of electroporation is based on the observation that short electric impulses above a certain field strength can make biomembranes transiently more permeable without damaging the membrane [26]. This technique has been used to introduce foreign genes into biological cells [27]. A biological membrane is a cooperatively stabilized organization of lipids and proteins that contains dynamic and structural defects. These local defects are well suited for the onset of further perturbations that can be induced by an electric field, leading to formation of permeation sites for materials exchange. As shown in Fig. 2, when a cell is exposed to a high electric field between a pair of parallel electrodes, the ions inside the cell will move along the field until they reach and are held back by the cell membrane. The conductivity of which is several orders of magnitude less than that of the intracellular fluid. The accumulation of ions on the surface of the membrane will generate a large transmembrane potential, which is maximal around the two loci opposing the electrodes. Opposite electrical charges on the membrane attract each other, exerting pressure on the membrane, which, in turn, causes thinning of the membrane. Generally speaking, if a cell membrane is polarized very rapidly (<10 to $100 \, \mu s$) to a very high voltage, an electrical breakdown

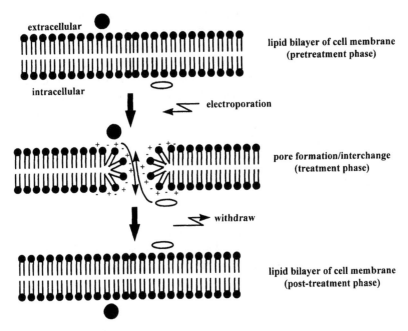

extracellular

intracellular

lipid bilayer of cell membrane
(pretreatment phase)

electroporation

pore formation/interchange
(treatment phase)

withdraw

lipid bilayer of cell membrane
(post-treatment phase)

Figure 2 A schematic illustration of pore formation under electroporation. (Modified from Ref. 26.)

of the membrane is observed, which is associated with a dramatic, reversible increase in membrane conductivity and permeability. The membrane voltage drops to a very low value because of the high conductance state of the membrane. This phenomenon of electrical breakdown is different from the mechanical breakdown, which is defined as an irreversible destruction of the membrane [26]. The effects induced in a membrane by electrical breakdown are completely reversible and, after a certain time interval, the original membrane resistance and permeability are restored. If the field strength exceeds the critical field (the breakdown voltage) or if the exposure time of the membrane to the field is too long, the electrical breakdown phenomenon is accompanied by a mechanical and irreversible breakdown [28]. The creation of pores causes this increase in membrane permeability, allowing exchange of intracellular and extracellular components by diffusion [27,28].

2. Applications

The most common application of electroporation is for induction of DNA into cells [29,31]. The first application of electroporation in transdermal delivery was in 1992 [30]. The mechanisms by which electroporation of the stratum corneum's multilamellar intercellular lipid bilayers enhances transdermal drug delivery may be similar to those reported for other bilayer lipid membranes or cells: namely, the reversible permeation of the lipid bilayers involving the creation of transient aqueous "pores" by the application of an electric pulse [27] and the electrophoretic drift of the drug by electric current [31,32]. So far, no clinical applications of transdermal electroporation have been reported. More research is needed to investigate the safety and feasibility of transdermal electroporation for topical or systemic application of medication.

C. Sonophoresis

1. Mechanisms

Sonophoresis is defined as the migration of the drug molecules, contained in a coupling contact agent, through intact skin into soft tissue under the influence of ultrasonic perturbations. As shown in Fig. 3, ultrasonic waves can penetrate

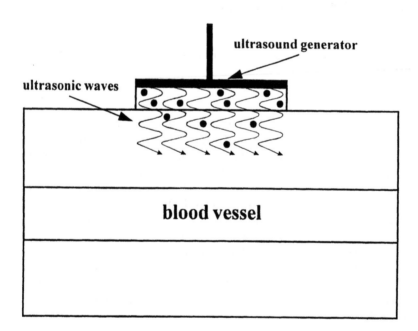

Figure 3 A schematic illustration of transdermal sonophoretic delivery. (From Ref. 12.) The ultrasonic waves can penetrate skin to a depth of 5 cm. (From Ref. 33.) Both the active drug (spheres) and contact agent (in rectangle) should be able to transmit ultrasound. (From Ref. 55.)

skin to a depth of 5 cm [33]. Sonophoresis was first used in transdermal delivery of hydrocortisone from an ointment formulation for treatment of polyarthritis by Fellinger as early as 1954 [34]. The exact mechanisms by which sonophoresis enhances the skin permeation of drugs are still unknown despite the frequency of sonophoresis use in physical therapy clinics [35]. Sonophoretic enhancement is thought to result from the thermal, mechanical, and chemical alterations of biological tissues induced by ultrasonic waves [33,36].

The thermal effect is a result of ultrasound's mechanical disturbance of absorbing media or interfaces and the continuous conversion of mechanical energy associated with the sound field into heat [33,36–38]. Ultrasound can cause deep penetrating hyperthermia, which could increase the diffusivity and solubility of penetrants: it can also lead to vasodilatation and increased blood flow to facilitate vascular skin permeation. These effects could be attributable to ultrasonic "stirring" or pore size–enlarging activity associated with cavitation, the formation of small gaseous inclusions [36]. Cavitation may cause mechanical stress, temperature elevation, or enhancement of chemical reactivity; hence it may facilitate drug transport. Hill et al. [39] and Mitragotri et al. [40] have suggested that the major factor affecting interaction of ultrasonic waves with living cells is the occurrence of "stable" or "resonant" bubble-type cavitation in the cellular medium under ultrasound irradiation. Mechanical shear resulting from this cavitation leads to disruption of the cell membrane. Ultrasound does appear to cause rapidly reversible cell damage [41,42]. It has been speculated that ultrasound affects the organization of lipid structure in the stratum corneum [43,44]. Chemical effects of ultrasound exposure include oxidation, hydrolysis, and depolymerization [36].

The efficiency of sonophoresis depends on several parameters, such as frequency (20 kHz ∼ 10 MHz), intensity (<3 W/cm²), exposure time (<5 min), coupling agents, and contact time. Most studies documented that sonophoresis was associated with an increase in drug diffusion at the subcutaneous level.

2. Applications

Major medical applications of sonophoresis aim to enhance delivery of steroidal anti-inflammatory drugs [45–48], such as hydrocortisone, for topical treatments. Sonophoresis has been also investigated for enhancing the skin permeation of anesthetics [49–52], such as lidocaine, and proteins/peptides, such as insulin [53,54].

To maximize the effectiveness of sonophoresis treatments, scientists and clinicians should (1) select only topical agents that transmit ultrasound, (2) check the skin carefully for moistness and hydration, (3) pretreat the skin with ultrasound, (4) position the patient to maximize circulation to the area being treated, (5) use an intensity in the thermal range (1.5 W/cm² or higher), and (6) leave the drug on the skin with an occlusive dressing after treatment [55].

The use of ultrasound as an enhancer for transdermal drug delivery is still a relatively new concept requiring more research. Transitivity of the medication is a critical variable to consider when selecting ultrasound as an enhancer. Both the active drug and the contact agent should transmit ultrasound [55].

V. CHEMICAL ENHANCEMENT

Several chemical skin permeation–enhancing techniques have been integrated into some transcutaneous delivery systems. These include chemical modifications (prodrugs or analogues) to improve stability, solubility, lipophilicity, diffusivity, and permeability and the addition of chemical enhancers to increase thermodynamic activity or alter skin barrier properties. In most cases, addition of chemical enhancers may still be required for the prodrugs or analogues.

A. Mechanisms of Chemical Enhancement

Penetration enhancers may promote drug permeation across the skin by a variety of mechanisms, and the molecular basis for their actions is becoming clearer. Long-term tissue hydration may form polar channels for drug diffusion [56]. There is no doubt the occlusive nature of transdermal systems will result in hydration of the underlying skin, particularly the stratum corneum. This can have significant consequences for the penetration rate of therapeutic agents. Some lower alcohols have been shown to enhance skin permeation. The enhancing ability of these alcohols appears to be related to their ability to extract lipids from the stratum corneum, increase the vehicle solubility, and/or improve keratin solvation. Another class of solvent-type chemical enhancers [e.g., azone and dimethyl sulfoxide (DMSO)] can cause lipid bilayer disruption, lipid extraction, displacement of bound water, loosening of horny cells, and delamination of stratum corneum.

Surfactants are major components of pharmaceutical, cosmetic, and food formulations. Increased skin permeation resulting from the addition of surfactants in transdermal delivery systems at low concentration is normally attributed to the ability of the surfactant molecule to penetrate and eventually disrupt the cell's membrane structure. Anionic surfactants can penetrate and interact strongly with skin [57,58]. Most anionic surfactants can induce swelling of the stratum corneum and the viable epidermis [59,60]. It has been speculated that the hydrophobic interaction of the alkyl chains with the substrate leaves the negative end group of the surfactant exposed, creating additional anionic sites of the membrane [59]. This results in the development of repulsive forces that separate the protein matrix, uncoil the filaments, and expose more water-binding sites, possibly increasing the hydration level of the tissue. Separation of the protein matrix would also result in disruption of the long-range order within the keratinocyte, possibly lead-ing to increased intracellular diffusivity. Cationic surfactants are more irritating than the anionics, therefore they have not been widely studied as skin penetration enhancers. Of the three major classes of surfactants, nonionics have long been recognized as those with the least irritancy potential. The mode of action of these surfactants appears to be linked to their ability to partition into the intercellular lipid phases of the stratum corneum. This results in increased fluidity in this region, which presumably reduces diffusional resistance. At high concentration, some lipid extraction will occur and result in reducing diffusional resistance. Another possible mode of action involves penetration of the surfactants into the intracellular matrix, followed by interaction and binding with the keratin fila-ments.

Fatty acids and alcohols have been shown to be very effective skin perme-ation enhancers [61,62]. They undoubtedly act upon the lipids of the stratum corneum, and their relative effectiveness is related to the ease with which they partition into the stratum corneum.

Figure 4 shows the proposed action sites for enhancers in the intercellular space of the horny layer as presented by Barry [63]. Table 1 summarizes the types of skin permeation enhancers and possible penetration-enhancing mechanisms.

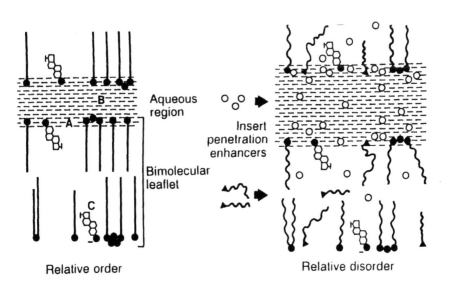

Figure 4 Proposed sites of action for chemical enhancers in the intercellular space of the horny layer by Barry. (From Ref. 63.) Linear chains represent azone, bent chains correspond to *cis*-unsaturated enhancers and the circles stand for small polar solvent such as DMSO and its analogues, the pyrrolidones and propylene glycol.

Table 1 Skin Penetration Enhancers and Possible Permeation-Enhancing Mechanisms

Specific enhancers	Mechanism of action
1. Solvents	
Water	Hydration, association with lipid head groups, keratin swelling
Alcohol	
Short-chain (C_2—C_5)	Lipid extraction (or swelling)
Long-chain ($>C_6$)	Lipid bilayer disintegration
Azone	Lipid bilayer disruption
Amide enhancers	Lipid interaction
Dimethylsulfoxide (DMSO)	Lipid extraction, delamination, horny cell loosening
2. Surfactants	
Anionic	Protein interaction
Cationic	Epidermal proteins binding
Nonionic	Increased membrane fluidity, solubilization, lipid extraction
3. Fatty acids and esters	
Oleic acid	Lipid interaction
Isopropyl myristate	Lipid interaction
4. Biodegradable enhancers	
Alkyl esters	Increased lipid fluidity
Unsaturated cyclic ureas	Possible stratum corneum lipids interaction
Alkyl N,N-dialkyl-aminoacetates	Lipid interaction
N-(2-hydroxyethyl)-2-pyrrolidones	Possible skin lipid interaction
ε-Aminocaproic acids	Possible skin lipid interaction
5. Other	
Amino acids	Keratin loosening
Terpenes	Lipid domain disruption, increased drug partitioning
Pyrrolidones	Skin hydration, protein interaction
Urea compounds	Stratum corneum hydration, keratolysis
Macrocyclic enhancers	Increased drug solubility
Phospholipids	Surfactant properties

B. Types of Chemical Enhancers

Based on the various penetration mechanisms previously mentioned, skin penetration enhancers are categorized into five groups, as summarized in Table 1. The first category is solvents, which includes water, alcohols, alkyl methyl sulfoxides, pyrrolidones, and polyols, which increase solubility, improve partitioning coefficient, hydrate the stratum corneum, extract lipids, and loosen horny cells. The second group is surfactants, which interact with proteins, swell the stratum corneum, and increase fluidity in the intercellular lipid phases. The third group is long-chain fatty acids, esters and alcohols, which disrupt the lipids in the stratum corneum. The fourth group is biodegradable enhancers, which degrade after permeating through the skin and whose mode of action for enhancing skin permeation is to increase lipid fluidity through possible interaction with skin lipids. The remaining skin penetration enhancers are classified as ''other,'' a group which includes amino acids, which loosen keratin; terpenes, which increase drug partitioning; urea compounds, which hydrate the stratum corneum; macrocyclic enhancers, which increase solubility; and phospholipids, which have surfactant properties.

C. Applications in Pharmaceutics and Cosmetics

The goal of chemical enhancers is to ensure that compounds are delivered to the systemic circulation or targeted topical site. It is apparent that if diverse classes of drugs are to be delivered via the transdermal route, most will require substantial permeation enhancement to overcome the skin's naturally low permeability. The role of chemical enhancers in topical formulations has therefore been firmly established.

VI. VESICLE CARRIERS

Several vesicle carriers are being studied for their ability to increase skin deposition, such as lipid micelles, liposomes, and Transfersomes.

A. Lipid Micelles

Lipid micelles are most often spherical, disk-shaped, or block-like, but they may have other shapes. Their typical dimension is rather small, comparable to the lipid bilayer thickness [64]. Micelles are appreciably smaller than most other types of lipid aggregates and provide a suitable ''small supramolecular permeant'' control [65].

B. Liposomes

In the past decade, liposomal formulations have been extensively employed to enhance the efficiency of drug delivery via several routes of administration. Liposomes have become increasingly important as vehicles for controlled delivery of cosmetics and pharmaceutics. The major advantages of topical liposomal drug formulations include (1) reduction of serious side effects and incompatibilities that may arise from undesirably high systemic absorption of drug; (2) significant enhancement of drug accumulation at the site of administration as a result of the high substantivity of liposomes with biological membranes; and (3) ready incorporation into a wide variety of hydrophilic and hydrophobic drugs [66]. Additionally, liposomes are nontoxic and biodegradable and can be prepared on a large scale.

Liposomes consist of amphiphilic lipids, typically phospholipids; they have a polar head group and two long fatty acid chains. They are spherical vesicles, about 30–100 nm in diameter, formed by one or several concentric lipid bilayers about 4–5 nm thick, enclosing an aqueous core [67]. Table 2 summarizes the main liposome types and their features [68]. Drug molecules can either be encapsulated in the aqueous space or intercalated into the lipid bilayer, the exact location of a drug in the liposome depending upon its physicochemical characteristics and the composition of the lipids [69,70]. Such encapsulation offers improved uptake, adhesion, and persistence of active ingredients in skin-surface and hair products. Deposition of liposomes into the stratum corneum would create a ''drug reservoir'' for topical delivery.

C. Transfersomes

Transfersomes, which were created and introduced in 1992 by Cevc [71], are defined as special lipid aggregates that can penetrate efficiently even through pores or constrictions that would be confining for other particles of comparable size. Cevc [72] claimed that this capability is due to the self-adaptable and extremely high deformability of the Transfersome membrane.

D. Applications in Pharmaceutics and Cosmetics

Generally speaking, there are at least three groups of substances considered to be candidates for liposomal topical application: (1) drugs that are known to be

Table 2 Types and Features of Liposomes

Type of vesicle	Size (mm)	Features of liposomes
Multilamellar vesicle	0.05–10	Large retention volume, good stability
Small unilamellar vesicle	0.025–0.05	Uniform size and shape, small retention volume, easy refusion
Large unilamellar vesicle	>0.1	Large retention volume, not uniform size

Source: From Ref. 68.

effective when applied systematically but not after topical administration—e.g., interferon [73]; (2) drugs that currently show insufficient effects after topical application—e.g., herbal drugs, such as hammamelis distillate [74]; and (3) drugs that show no optimal benefit/risk ratio by the conventional, topical method of administration—e.g., topical glucocorticoids, such as betamethasone dipropionate [75].

Recently, a great deal of interest in the use of liposomes in skin gels or skin creams has been generated in the field of cosmetics. Vegetable phospholipids are widely used for topical applications in cosmetics and dermatology, since they have a high content of esterified essential fatty acids, especially linoleic acid, which is believed to increase the barrier function of skin and decrease water loss within a short time after application [76,77]. Soya phospholipids or other vegetable phospholipids, owing to their surface activity and their ability to form liposomes, are also an ideal source for possible transport of linoleic acid into the skin [78,79].

VII. FUTURE PROSPECTS

The ultimate goal of skin permeation enhancement is to ensure that the compounds are delivered, preferably at a specific rate, to remain at the topical site or to reach the systemic circulation. The customization of drug vehicles and enhancers may be the most promising future direction for transdermal enhancement and drug delivery.

REFERENCES

1. Xu P, Chien YW. Enhanced skin permeability for transdermal drug delivery: physiopathological and physicochemical considerations. Crit Rev Ther Drug Carriers Syst 1991; 8:211.
2. Barry BW. Structure, function, diseases and topical treatment of human skin, Dermatological Formulations: Percutaneous Absorption. New York: Marcel Dekker, 1993: 21.
3. Guy RH, Hadgraft J. Transdermal drug delivery: the ground rules are emerging. Pharm Int 1985; 6:112.
4. Walker RB, Smith EW. The role of percutaneous penetration enhancers. Adv Drug Del Rev 1996; 18:295.
5. Harada K, Murakami T, Yata N, Yamamoto S. Role of intercellular lipids in stratum corneum in the percutaneous permeation of drugs. J Invest Dermatol 1992; 99:278.
6. Schmidt RJ. Cutaneous side effects in transdermal drug delivery: avoidance strategies. In: Hadgraft J, Guy RH, eds. Transdermal Drug Delivery. New York: Marcel Dekker, 1989:83.

7. Katz M, Poulsen BJ. In: Brodie BB, Gillete J, eds. Handbook of Experimental Pharmacology, vol 28. New York: Springer-Verlag, 1971:103.

8. Singh P, Maibach HI. Iontophoresis in drug delivery: basic principles and applications. Crit Rev Ther Drug Carrier Syst 1994; 12(2&3):161.

9. Sloan JD, Soltani K. Iontophoresis in dermatology. J Am Acad Dermatol 1986; 15: 671.

10. Kark MR, Gangarosa LP. Iontophoresis: an effective modality for the treatment of inflammatory disorders of the temporomandibular joint and myofascial pain. J Craniomandib Pract 1990; 8:108.

11. Singh J, Maibach HI. Topical iontophoretic drug delivery in vivo: historical development, devices and future perspectives. Dermatology 1993; 187:235.

12. Chen LH, Lin S, Chien YW. Transdermal route of peptides and proteins drug delivery. In Lee VHL, ed. Peptide and Protein Drug Delivery, 2nd ed. New York: Marcel Dekker, 1998. In press.

13. Pikal MJ. Transport mechanisms in iontophoresis: I. A theoretical model for the effect of electroosmotic flow on flux enhancement in transdermal iontophoresis. Pharm Res 1990; 7:118.

14. Pikal MJ, Shah S. Transport mechanisms in iontophoresis II. Electroosmotic flow and transference number measurement for hairless mouse skin. Pharm Res 1990; 7: 213.

15. Wearley LL, Chien YW. Enhancement of the in vitro skin permeability of AZT via iontophoresis and chemical enhancers. Pharm Res 1990; 1:34.

16. Sims SM, Higuchi WI, Sirnivasan V, Skin alteration and convective flow effects during iontophoresis: I. Neutral solute transport across human skin, Int J Pharm 1991; 69:109.

17. Chien YW, Lelawongs P, Siddigui O, et al. Facilitated transdermal delivery of therapeutic peptides and proteins by iontophoretic delivery devices. J Contr Rel 1990; 13:263.

18. Fay MF. Indications and applications for iontophoresis. Todays OR Nurse 1989; 11: 10.

19. Costello CT, Jeske AH. Iontophoresis: applications in transdermal medication delivery. Phy Therp 1995; 75:554.

20. Sloan JB, Slotani K. Iontophoresis in dermatology: a review. J Am Acad Dermatol 1986; 4:671.

21. Comeau M, Brummett R, Vernon J. Local anesthesia of the ear by iontophoresis of lidocaine. Arch Otolaryngol 1975; 101:418.

22. Singh P, Roberts MS. Iontophoretic transdermal delivery of salicylic acid and lidocaine to local subcutaneous structures. J Pharm Sci 1993; 82:127.

23. Glass JM, Stephen RL, Jacobsen SC. The quantity and distribution of radiolabeled dexamethasone delivered to tissue by iontophoresis. Int J Dermatol 1980; 19: 519.

24. Meyer BR, Kreis W, Eschbach J, et al. Transdermal versus subcutaneous leuprolide: a comparison of acute pharmacodynamic effect. Clin Pharmacol Ther 1990; 48: 340.

25. Siddqui O, Sun Y, Liu JC, Chien YW. Facilitated transport of insulin. J Pharm Sci 1987; 76:341.

26. Weaver JC, Powell KT. Theory of electroporation. In: Neumann E, Sowers AE, Jordan CA, eds. Electroporation and Electrofusion in Cell Biology. New York: Plenum Press, 1989:111.

27. Chang D, Chassy D, Saunders J, Sowers A. Guide to Electroporation and Electrofusion. San Diego, CA: Academic Press, 1992.

28. Tsong TY. Electroporation of cell membranes: mechanisms and applications. In: Neumann E, Sowers AE, Jordan CA, eds. Electroporation and Electrofusion in Cell Biology. New York: Plenum Press, 1989:149.

29. Potter H. Molecular genetic applications of electroporation. In: Neumann E, Sowers AE, Jordan CA, eds. Electroporation and Electrofusion in Cell Biology. New York: Plenum Press, 1989:331.

30. Prausnitz M, Dose V, Langer R, Weaver J. Transdermal drug delivery by electroporation. Proc Int Symp Control Rel Bioact Mater 1992; 19:232.

31. Klenchi VA, Sukharev SI, Serov SM, et al. Electrically induced DNA uptake by cells is a fast process involving DNA electrophoresis. Biophys J 1991; 60:804.

32. Vanberver R, Lecouturier N, Preat V. Transdermal delivery of metoprolol by electroporation, Pharm Res 1994; 11:1657.

33. Tyle P, Agrawala P. Drug delivery by sonophoresis, Pharm Res 1989; 6:355.

34. Fellinger K, Schmid J. Wein Klin Wochenschr 1954; 66:549.

35. Pottenger FJ, Karalfa GL, Utilization of hydrocortisone phonophoresis in United States Army physical therapy clinics. Mil Med 1989; 154:355.

36. Saxena J, Sharma N, Makoid MC, Banakar UV. Ultrasonically mediated drug delivery. J Biomat App 1993; 7:277.

37. Jankowiak J, Majewski C. Electron-microscopic studies of acid phosphatase in neutrophilic granulocytes in the blood of rabbits subjected to ultrasound, Am J Phys Med 1966; 45:1.

38. National Council on Radiation Protection and Measurements. Biological effects of ultrasound: mechanism and clinical implications. NCRP Report No. 74. Bethesda, MD: NCRP, 1983.

39. Hill CR. In: Reid JM, Sikov MR, eds. Proceedings of the Workshop on the Interaction of Ultrasound and Biological Tissues. 1972:57.

40. Mitragotri S, Edwards DA, Blankschtein D, Langer R. A mechanistic study of ultrasonically-enhanced transdermal drug delivery. J Pharm Sci 1995; 84:697.

41. Levy D, Kost J, Meshulam Y, Langer R. Effect of ultrasound on transdermal drug delivery to rats and guinea pigs. J Clin Invest 1989; 83:2074.

42. Kremkau FW. Ultrasonic treatment of experimental animals tumors. Br J Cancer 1982; 45(Suppl V):226.

43. Bommmannan D, Menon GK, Okuyama H, et al. Sonophoresis: II. Examination of the mechanisms of ultrasound-enhanced transdermal drug delivery. Pharm Res 1992; 9:1043.

44. Kost J, Levy D, Langer R. Proceedings of the International Symposium on the Controlled Release of Bioactive Materials. Controlled Release Society, Norfolk, VA. 1986:177.

45. Griffin JE, Echternach JL, Price RE. Patients treated with ultrasonic driven hydrocortisone and with ultrasound alone. Phys Ther 1967; 47:594.

46. Kleinkort JA, Wood AF. Phonophoresis with 1% versus 10% hydrocortisone. Phys Ther 1975; 55:1320.
47. Davick JP, Martin RK, Albright JP. Distribution and deposition on tritiated cortisol using phonophoresis. Phys Ther 1988; 68:1672.
48. Newman JT, Nellermoe MD, Carnett JL. Hydrocortisone phonophoresis: a literature review. J A Podiatr Med Assoc 1992; 82:432.
49. Novak EJ. Experimental transmission of lidocaine through intact skin by ultrasound. Arch Phys Med Rehabil 1964; 45:231.
50. Mollm MM. A new approach to pain: lidocaine and Decadron with ultrasound. USSAF Med Ser Dig 1979; 30:8.
51. Hsu SM, Liu TK, Yu HY. Absorption of lidocaine following topical application in microvascular procedures in rabbits. J Orthop Res 1991; 9:545.
52. Williams AR. Phonophoresis: an in vivo evaluation using three topical anesthetic preparations. Ultrasonics 1990; 28:137.
53. Tachibana K. Transdermal delivery of insulin to alloxan-diabetic rabbits by ultrasound exposure. Pharm Res 1992; 9:952.
54. Langer R. Ultrasound-mediated transdermal protein delivery. Science 1995; 269:850.
55. Byl NN. The use of ultrasound as an enhancer for transcutaneous drug delivery: phonophoresis. Phy Ther 1995; 75:539.
56. Walters KA, Walker M, Olejnik O. Hydration and surfactant effects on methyl nicotinate penetration through hairless mouse skin, J Pharm Pharmacol 1985; 37:81P.
57. Bettley FR. The influence of detergents and surfactants on epidermal permeability. Br J Dermatol 1965; 77:98.
58. Gibson WT, Teall MR. Interaction of C12 surfactants with the skin: studies on enzyme release and percutaneous absorption in vitro. Food Chem Toxicol 1983; 21:581.
59. Rhein LD, Robbins CR, Fernee K, Cantore R. Surfactant structure effects on swelling of isolated human stratum corneum, J Soc Cosmet Chem 1986; 37:125.
60. Blake-Haskins JC, Scala D, Rhein ID, Robbins CR. Predicting surfactant irritation from the swelling response of a collagen film. J Soc Cosmet Chem 1986; 37:199.
61. Cooper ER, Merritt EW, Smith RL. Effect of fatty acids and alcohols on the penetration of acyclovir across human skin in vitro. J Pharm Sci 1985; 74:688.
62. Bennett SL, Barry BW. Effectiveness of skin penetration enhancers propylene glycol, azone, decylmethylsulphoxide and oleic acid with model polar (mannitol) and non polar (hydrocortisone) penetrants, J Pharm Pharmacol 1985; 37:84P.
63. Barry BW. Lipid-protein partitioning theory of skin penetration enhancement. J Controlled Rel 1991; 15:237.
64. Lichtenberg D, Robson RJ, Dennis EA. Solubilization of phospholipids in detergents. Structural and kinetic aspects. Biochim Biophys Acta 1983; 737:285.
65. Fromhera P, Rueppel D. Lipid vesicle formation: the transition from open disks to closed shells, FEBS Lett 1985; 179:155.
66. Egbaria K, Weiner N. Liposomes as a topical drug delivery system. Adv Drug Del Rev 1990; 5:287.
67. Strauss G. Liposomes: from theoretical model to cosmetic tool. J Soc Cosmet Chem 1989; 40:51.

68. Schmid MH, Korting HC. Liposomes: a drug carrier system for topical treatment in dermatology. Crit Rev Ther Drug Car Syst 1994; 11(2&3):97.

69. Gregoriadis G. The carrier potential of liposomes in biology and medicine. N Engl J Med 1976; 295:704.

70. Fendler JH, Romero A. Liposomes as drug carriers. Life Sci 1997; 20:1109.

71. Cevc G. Eur Pat Appl 1992; A 61 k 9/50.

72. Cevc G. Lipid properties as a basis for the modeling and design of liposome membranes. In: Gregoriadis G, ed. Liposome Technology, 2d ed. Boca Raton, FL: CRC Press, 1992:1–36.

73. Du Plessis J, Egbaria K, Ramachansran C, Weiner N. Topical delivery of liposomally encapsulated gamma-interferon. Antivir Res 1992; 18:259.

74. Korting HC, Schafer-Korting M, Hart H, et al. Anti-inflammatory activity of hamamelis distillate applied topically to the skin. Eur J Clin Pharmacol 1993; 44:315.

75. Korting HC, Zienicke H, Schafer-Korting M, Braun-Falco O. Liposome encapsulation improves efficacy of betamethasone dipropionate in atopic eczema but not in psoriasis vulgaris. Eur J Clin Pharmacol 1991; 29:349.

76. Rieger MM. Skin lipids and their importance to cosmetic science. Cosmet Toiletries 1987; 102:36.

77. Prottey C, Hartop PJ, Press M. Correction of the cutaneous manifestations of essential fatty acid deficiency in man by application of sunflower-seed oil to the skin. J Invest Dermatol 1975; 64:228.

78. Lautenschlager H, Roding J, Ghyczy M. The use of liposomes from soya phospholipids in cosmetics. SOFW 1988; 14/88:531.

79. Lautenschlager H. Comments concerning the legal framework for the use of liposomes in cosmetics preparations. SOFW 1988; 18/88:761.

4
Skin Hydration

Pierfrancesco Morganti
MAVI SUD S.r.l., Aprilia (LT), Italy

I. THEORETICAL ASPECTS OF CUTANEOUS HYDRATION

The life of animals and plants is regulated by water, an indispensable component existing in all living organisms. When environmental conditions change or a disorder occurs, the human body reacts through a series of self-defense mechanisms, which replenish the water level if necessary. Unfortunately, these self-regulating mechanisms are extremely sensitive and delicate. They often do not react quickly enough or do not perform their tasks fully, especially if the body is weaker than usual or is no longer young and dynamic. If so, the inability of some biological systems to preserve the balance in the living matter can be treated by supplying such substances as can help the body's defense mechanisms to react more efficiently.

Water is not only a solvent in which biological phenomena occur but also a fundamentally active element in the metabolic functions of cells and tissues, and it helps maintain the physical properties of the stratum corneum [1–5].

Water reaches the epidermis from the dermis, of which it constitutes 70%. In the dermis, water is bound reversibly with glucosoaminoglycans (GAG) and especially with hyaluronic acid, depending on polymerization rate and therefore the hyaluronic acid/hyaluronidase ratio.

When GAG is saturated, water becomes free, and this causes a rise in interstitial edema.

Both intracellular and extracellular water reaches the stratum corneum, exceeds the maximum that can be absorbed, and vaporizes as insensible perspiration.

As a physiological regulator, the skin plays a major role in the general metabolism of water in the body. Therefore the moisture level of the stratum corneum is crucial to keep the cutaneous surface supple and healthy.

Today—thanks to the latest data on the physiology of the cutaneous surface—the stratum corneum, which is directly exposed to the outer environment, is known not to be an inert layer consisting only of dead keratinized cells and sebum. The stratum corneum is able to control its water content through both hygroscopic substances, accounting for 20% of its weight, and intercellular lipids modeled as flattened multilamellar vesicles, controlling transepidermal water loss (TEWL) [6].

The stratum corneum's flexibility and protective function, which are tightly linked with its moisture level (Fig. 1), depend basically on three factors: (1) the rate at which the water in the dermis reaches the stratum corneum; (2) the rate at which water is eliminated by evaporation (TEWL); and (3) the stratum corneum's ability to retain water. This is tightly linked with the role of the surface lipidic film, natural moisturizing factors (NMFs), and polar lipids (glycolipids, phospholipids, and free fatty acids), which make up the well-known "lamellae" in the intercellular spaces of the stratum corneum [6–13].

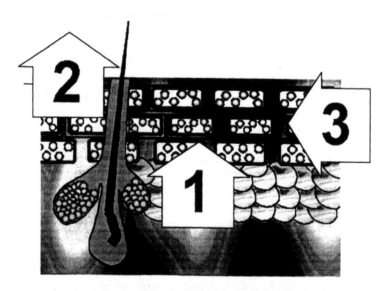

Figure 1   The moisture level in the stratum corneum.

II. CUTANEOUS LIPIDS

Various kinds of lipids can be found in the stratum corneum [14–25] (Fig. 2):

> The lipids from the epidermal cells, which are generally stored in the form
> of lipidic vacuoles in keratinized cells
> Lipidic lamellas, which are found in the intercellular spaces of keratinized
> cells and have been recognized as crucial for water retention

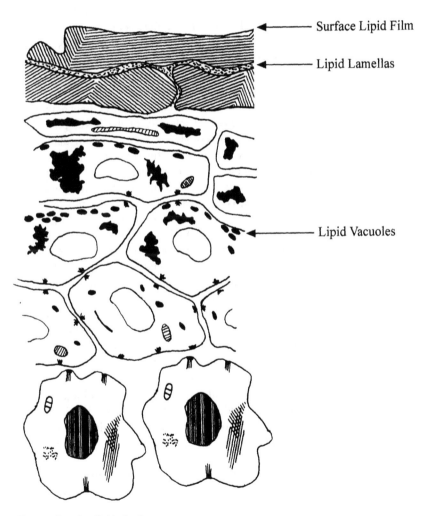

Surface Lipid Film

Lipid Lamellas

Lipid Vacuoles

Figure 2   The lipids in the stratum corneum.

Epidermal lipids, which make up the surface lipidic film (5%)

Sebum lipids, which are produced by sebaceous glands and represent the major component (95%) of the surface lipidic film

Various fatty agents, which are either applied to the skin through cosmetics or come from the outer environment

Lipids can be found in the cellular membrane and the interstices; they change dramatically in quality and quantity as soon as epidermal cells leave the basal layer to turn gradually into desquamating cells. In the basal cells, phospholipids (in bilayer closed vesicles), cholesterol, free fatty acids, triglycerides, and glycolipids prevail, together with a small amount of esterified cholesterol and ceramides. On the other hand, horny cell lipids are made up half of ceramides and half of a mixture of cholesterol and free fatty acids.

Cholesterol esters and glycolipids have been detected immediately before desquamation, showing that these polar lipids, fundamental components of the water barrier, help to maintain the physical integrity of the lower part of the stratum corneum. From a quantitative point of view, it is worth noting that protective lipids increase six times as they reach the stratum corneum from the basal layer, whereas cholesterol and ceramides are, in the main, newly synthetized.

The lipid composition of the keratinocytes changes substantially owing to the increase of free fatty acids released in catabolic processes and mainly to this further lipid synthesis. Specific catabolic enzymes determine the progressive decay and synthesis of products, reflecting highly interesting morphological, structural, and functional conditions in the granular and horny layers. Before the nuclei decay, bilayered liposome-like vesicles (Odland's bodies) arise from the Golgi complex of granular cells and are processed. They function to cement corneocytes and have a water-binding capacity. Passing from the compact to the unbound layer, lamellar bodies change their composition and eventually lose their typical bilayer feature. Enzymatic activities (glycosidase, steroilsulfatase, phospholipase, acid lipase) in the three epidermal layers thus induce chemical conversion of lipids to change the intercellular materials and cell structure in all sites.

1. In the malpighian layer, the total is made up of phospholipids together with phosphatidylcholine and, to a lesser extent, phosphatidylethanolamine and sphingomyelin. However ceramides and glycosylceramides (useful to the epidermal barrier) can be found as well as sphingomyelin decay products.

2. Although they are substantially reduced, there still are polar lipids in the granular layer with a clear increase of ceramides, cholesterol, and fatty acids.

3. In the deep stratum corneum there are only traces of phospholipids and glycosphingolipids; these are completely absent from surface horny

layer, where, conversely, the content of ceramide, sterol, and free fatty acid is substantial [26–31].

Although natural moisturizing factors (NMFs) undoubtedly play a major role in skin hydration, a tight link exists between lipids and the water needed for keeping the skin supple. The surface lipidic film is vital to protect the NMFs from attack by external agents, while intercellular lipidic lamellas have recently been shown to contribute considerably to water retention in the stratum corneum. In addition, polyunsaturated fatty acids seem to have a specific role in the skin's nutrition: arachidonic acid is thought to regulate the regenerative biorhythm, proliferation, and maturation of the stratum corneum and stratum lucidum. Linoleic, linolenic (especially gamma-linolenic) and columbinic acids possibly regulate cutaneous permeability and affect the moisture level in the stratum corneum as well.

Finally, the huge rate of ceramides among lipids (approximately 50% of lipids located in stratum corneum cells) is crucial for regulating the horny barrier. Ceramide 1, the major one, is bound to a long-chain hydroxy acid (30 to 32 carbon atoms) and is esterified with linoleic acid. Ceramide 1 is thought to modulate the moisture level in the skin by regulating the water flow and acting as a reserve of essential fatty acids [32].

III. ORIGIN AND ROLE OF NMFs

Qualitative and quantitative changes in the composition of both the surface lipidic film and NMF components were shown in many skin diseases. These changes bring about a high cutaneous dehydration, often associated with desquamation and chapping. Undoubtedly, these water-soluble substances in the stratum corneum—the natural moisturizing factors or NMFs—play a major role in regulating the water level and protective function of the skin. The NMFs in the skin's surface ensure an optimum diffusion of water from inside the body outward, in compliance with the usual action of the dermis, hypodermis, and underlying tissues. The changes in the hygrometric degree of an external medium, with the resulting effects on the body, are also regulated by NMFs, which play a major role in keeping keratinized cells intact. Under normal conditions, the skin produces enough NMFs to preserve the right moisture level in the stratum corneum, whereas hardly any NMF production occurs in pathological conditions, such as psoriasis. Although the mechanism through which the skin produces NMFs has not yet been unraveled, a loss of NMFs was shown to completely prevent the stratum corneum from binding water and noticeably reduced the uptake of moisture by keratins. However, recent studies [33–36] reportedly showed that filaggrin*

---

* Filaggrin is a cationic protein, which favors the aggregation in macrofibrils of keratin filaments.

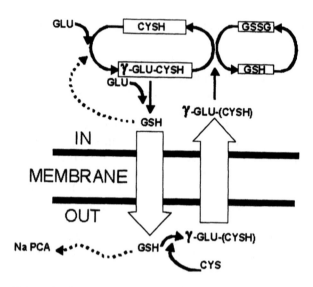

Figure 3   Probable origin of 2-pyrrolidine-5-carboxylic acid. (From Ref. 37.)

degradation results in the formation of NMFs, whose major component, 2-pyrroli-dine-5-carboxylic acid (PCA), could be produced through Meisner's glutathione cycle [37], as summarized in Fig. 3. In point of fact, profilaggrin, which is synthe-sized as a highly phosphorylated protein, was reportedly shown subsequently to accumulate in keratohyalin granules and turn into filaggrin by dephosphorylation and degradation [38,39].

IV. COMPOSITION AND CHEMICAL AND PHYSICAL FEATURES OF NMF

Although the first studies on NMFs date back to 1949, only a few authors have analyzed and studied their chemical composition. The first studies were carried out by Szakall [40,41]. More thorough analyses, as shown in Table 1, were later carried out by Spier and Pascher as well as Laden and Spitzer [10,11].

More elaborate analytical methods made it possible also to show the exis-tence of 16 different amino acids, with a higher rate of serine and citrulline (about 20%) (Table 2). According to Padberg's research [42], the ability of NMFs to retain moisture also depends on a neutral sugar fraction, which has been isolated in the stratum corneum of the human skin. This sugar complex seems to be based mainly on fructose, lactose, glucose, mannose, galactose, and small amounts of

Table 1 Natural Moisturizing Factors

Factors	%
Free amino acids	40
Pyrrolidine-5-carboxylic acid	12
Urea	7
Azo compounds (uric acid, glucosamine, creatinine)	1.5
Sodium	5
Calcium	1.5
Potassium	4
Magnesium	1.5
Phosphates	0.5
Chlorides	6
Lactates	12
Citrates and formates	0.5
Unidentified portion	8.5

Table 2 Amino Acid Content of Natural Moisturizing Factors

Serine	20–33%
Citrulline	9–16%
Alanine	6–12%
Threonine	4–9%
Ornithine	3–5%
Asparagine	3–5%
Glycine	3–5%
Leucine	3–5%
Valine	3–5%
Histidine	3–5%
Arginine	3–5%
Lysin	3–5%
Phenylanine	3–5%
Tyrosine	3–5%
Glutamic acid	0.5–2%

psicose. Thus, the roughness, fragility, and moisture level of the stratum corneum are also due to a lack of this sugar fraction.

According to Laden and Spitzer [10,11], amino acids are not very important as moisturizing agents, unlike 2-pyrrolidine-5-carboxylic acid. In fact, a direct link exists between this acid, which is found in the stratum corneum in the form of sodium salt (highly hygroscopic), and the ability of the keratinized tissue to bind water. Sodium PCA was also assumed to form through the enzymatic cyclization of glutamic acid in the skin.

According to Szakall, under normal conditions, water-soluble substances represent 30% of the stratum corneum. Upon exposure to the sun's rays, they decrease to 23%, and they may drop to 7% in psoriasis [40,41]. Furthermore, it was shown that the diffusion of sebum in the skin is up to ten times lower than normal in individuals lacking in NMFs and that seborrhoea, acne, and hyperhidrosis are related to the moisture-regulating factor.

A. Characteristics of NMFs

NMFs dissolve in water and alcohol at 50°C. The pH value of a 2% NMF solution is 4.7. They are stable at pH 4 to 8 and at a temperature of 100°C for 1 h. In addition, they are hygroscopic and show a mild surface-active property.

Because of their characteristics, NMFs applied to the skin have three effects:

1. They absorb environmental moisture thanks to their hygroscopic power.
2. They reduce the skin's surface tension and control the normal water-repellent power of keratin.
3. They absorb the water present in the skin due to insensible perspiration or from external sources.

V. CUTANEOUS DEHYDRATION

The possible causes for cutaneous dehydration are as follows:

1. Water diffusion from the dermis and epidermis outward (increase in TEWL)
2. Achievement of a certain balance in the stratum corneum's moisture level, which can be shifted toward dehydration
3. Decrease in environmental relative humidity
4. Increase in outdoor temperature and intensity of the sun's radiation
5. Currents of air

6. Chemical agents of various origins
7. Specific pathological conditions
8. Lack of ceramides, especially ceramide-1, and a resulting decrease in linoleic acid and water

Upon being degreased with strong solvents as well, the skin remains water-repellent and can be wetted only to some extent or in a longer term. In the event of water acting on the skin over a long time—for example, during a bath—the superficial skin lipids turn from a W/O emulsion into an O/W emulsion, which can be easily removed from water together with part of lipid lamellae and NMFs. This causes the skin to become highly dehydrated, a condition known as xerosis. Dry, xerotic skin is rough, damaged, wrinkled, dull, and flabby. This usually happens also in winter, because of low environment humidity, and in summer, when some skin areas become thinner and some thicker as a result of prolonged exposure to the sun's rays or to water while bathing in the sea. Since sebum production also decreases in winter, the decrease in these water-repellent compounds may result in a heavy NMF loss in the poorly protected stratum corneum. Also the skin's drying up due to aging can be ascribed to a lack of NMFs.

The first researches, carried out by Blank on strips of stratum corneum, showed that the use of soaps and detergents reduces the water level in the skin. These chemical compounds partly or wholly remove sebum as well as some of the hydrophilic substances in the stratum corneum. The skin becomes rough and dry. The lamellas in the stratum corneum are no longer tightly interwoven and fissures occur, which Jager called "V-shaped" fissures (Fig. 4). Here microorganisms penetrate that not even water can easily remove. Therefore, these fissures may become actual foci of infection [43–45].

Even an excessive use of toilet soaps and shaving creams—associated with the mechanical action of razor blades—can cause the skin to become more dry. Furthermore, contact dermatitis often develops, which may be caused in men's faces by shaving soaps and in women's hands by home detergents. Also, disinfecting with alcohol and subsequently washing one's hands with soaps for professional reasons creates conditions favorable to NMF removal from the stratum corneum. Likewise, a chronic use of strong detergents, such as sodium lauryl sulfate, in shampooing, especially if associated with the application of alcoholic lotions, removes NMFs and causes dryness of the hair and scalp. In fact, NMFs are highly soluble in alcohol at 40–50°C. For this reason, everyday use of eau de cologne containing alcohol should be avoided, especially in newborns.

However, a light, cutaneous dehydration is physiological: the skin is made up of a proteic gel, which is hydrated in its inner layers and highly dehydrated in the outer ones. A certain cutaneous dehydration minimizes the superficial proliferation of bacteria and helps the skin to remain intact. The continuous desqua-

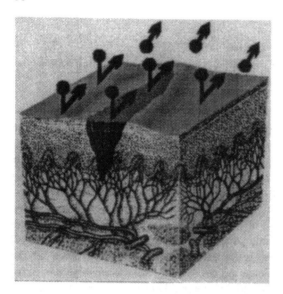

Figure 4 V-shaped fissures.

mation resulting from dehydration serves physiologically to remove all the catabolites and foreign substances that have settled on the skin's surfaces and fissures.

VI. SKIN MOISTURIZERS

In order to retain and bind water at the skin surface, hygroscopic substances are used. Examples are the active principles of NMF and particularly the sodium salt of 2-pyrrolidone-5-carboxylic acid (sodium PCA)—a physiological moisturizer found in various organs, organic fluids, and particularly the epidermis, including the stratum corneum [46–48].

Mixtures of sodium PCA, amino acids, urea, lactic acid, sodium lactate, and trace elements, known as "reconstituted NMFs," are also frequently used. Sodium PCA as well as glycine, lysine, and arginine are recognized regulators of skin hydration [49–52].

Urea (around 2–5%) is also capable of increasing hydration of the corneum by 100%, both by an osmotic effect due to its low molecular weight and for its ability to solubilize insoluble proteins. However, when used for too long, urea compounds cause harmful disaggregation of the horny layer's lamellae and damage to the skin barrier. Besides, it must not be used in atopic children under 5 years of age, where it can cause an intense burning sensation [53–55].

VII. GLYCINE

Glycine is the only amino acid with antioxidative and preserving properties that has been experimentally proved to have an interesting rehydrating effect on the skin in both topical and systemic use.

Recently, it has also been proved that the rehydrating activity of the stratum corneum due to glycine is increased by the bonds this amino acid forms with gelatin subjected to special chemicophysical treatments [37,50,51,56–58].

Sodium chloride can have a hygroscopic effect, as can several polyhydric alcohols such as sorbitol, propylene glycol, and glycerol.

The lack of water can also be improved by local application of products containing native collagen (a water-soluble protein molecule with an unaltered trihelical structure), which is capable of retaining between 200 and 300% of its own weight in water. The polypeptide chains of this collagen adhere to the surface of the corneum and create a hydrating film.

Hyaluronic acid also plays an important role owing to its great hygroscopic effect [59–61].

VIII. ALPHA HYDROXY ACIDS

Alpha hydroxy acids (AHAs) (Table 3) are used in cosmetic formulations for xerotic, hyperkeratotic, and aged skin in general. They act as modulators of keratin synthesis, improving water absorption among cells in the outer layers of the

Table 3 Alpha Hydroxy Acids

GLYCOLIC ACID	OH CH_2-COOH	mw 76,05
LACTIC ACID	CH_2-CH-COOH OH	mw 90,08
MALIC ACID	HOOC-CH-CH_2-COOH OH	mw 134,09
TARTARIC ACID	HOOC-CH-CH-COOH OH OH	mw 150,09
MANDELIC ACID	CH-COOH ◁▷—OH	mw 152,14
CITRIC ACID	COOH HOOC-CH_2-C-CH_2-COOH OH	mw 192,12

stratum corneum. The mechanism of activity of the alpha hydroxyacids is still unknown, although their capacity for disaggregating corneocytes is established.

These substances are thus useful in treating hyperkeratosis and seem to have an interesting "antiaging" activity as well [62–72].

According to recent studies, gelatin-glycine and gelatin-arginine seem to synergize the remoisturizing activity of AHAs, also reducing their stinging capacity [69–70].

IX. LINOLEIC AND GAMMA-LINOLENIC ACIDS

Atrophic, aged, xerotic, and/or hyperkeratotic skin requires emulsions containing vegetable oils rich in linoleic acid (soy, corn, and grapeseed oils, etc.) and in gamma-linolenic acid (borage and evening primrose oils).

Such unsaturated oils appear to promote mitosis and regulate cell differentiation as precursors of eicosanoids.

Furthermore, dry skin shows an increase (doubling) in palmitic acid (C 16: 1) together with a great reduction (of 50%) in linoleic acid (C 18:2). Both acids return to normal values with the local application of creams rich in gamma-linolenic acid [20,73,74].

A buildup over time of gamma-linolenic acid has been observed on the stratum corneum; this is believed to promote epidermal cell differentiation, thus improving local hydroxylation of unsaturated acids. Finally, linoleic and gamma-linolenic acids are believed to improve the exchange of water-soluble compounds, which affects cell permeability, with rehydrating effects on the stratum corneum [75–77].

X. VITAMINS

The inclusion of some vitamins is proposed in moisturizing and hydrating emulsions on the grounds that they are necessary for normal keratinization [78].

Vitamin B_5, in the form of panthenol (an active and biologically stable form of pantothenic acid), is deemed to be effective in treating dry, wrinkled skin.

Vitamin A (retinol, vitamin A alcohol) acts mainly as a regulator of the growth of epithelial tissue. It produces protein synthesis in ribosomes, labilizes lysosomal membranes, regulates epithelial mitosis, and affects the metabolism of acid mucopolysaccharides. Therefore its deficiency causes dry and wrinkled skin. From a cosmetic point of view, the local application of vitamin A seems to be very useful, especially in winter.

However, its derivate, retinoic acid (vitamin A acid), has proved to be much more active, since it is capable of penetrating the dermis (even though in small quantities). Specific proteins bind retinol and retinoic acid and occur in dermis and epidermis in equal quantities. In the dermis, retinoic acid is believed to stimulate mast cells for angiogenesis and fibroblasts for the synthesis of new elastic fibers and collagen.

Vitamin B_6 intervenes in fatty acid metabolism and as a coenzyme (pyridoxal phosphate) for the oxidative deamination of the free amino group of lysine and oxylysine.

Beta carotene and other carotenoids are believed to protect against ultraviolet rays, slowing down the degradation of dermal collagen and inhibiting the oxidation of proline to oxyproline. After being taken per os, they are, in fact, deposited in the dermis to act as free radical scavengers.

Vitamin E acts as an antioxidant to such compounds as are essential to cellular metabolism and therefore has a protective action for vitamin A, carotenoids, and polyunsaturated fatty acids. Furthermore, synergizing with vitamin A, vitamin E would prevent the formation of peroxides and thus of toxic free radicals. In the form of linoleate, vitamin E is also indicated to help restore the surface lipid barrier. Both as the ester and in the nonesterified form, this vitamin has proved to be a precious therapeutic factor in the treatment of dermatoses or of highly pruritic cutaneous afflictions. From a cosmetic point of view, it is worth remembering that vitamin E helps in the reabsorption of fatty acids, promotes metabolism, peripherally protects the natural fat content of the epidermis, and activates the cutaneous microcirculatory system.

Vitamin B_{12} improves atrophic and pigmented skin. Vitamin B_{12} deficiency causes folic acid deficiency (a coenzyme in the catabolism of amino acids such as methionine, histidine, serine, and glycine), thus stopping intracellular penetration.

Vitamin C intervenes in collagen formation by proline hydroxylation and corrects follicular keratosis, which is a typical sign of its deficiency.

Vitamin B_1 or thiamine acts as a coenzyme in many cellular processes and is indispensable for lipid synthesis.

Vitamin H or biotin is a constituent of many enzymes (carboxylase type) and effects the metabolism of aminoacids and pyruvic acid. Its deficiency is indicated by a grey, dry and wrinkled skin.

Vitamin D is indispensable for the maintenance of calcium and phosphates in the plasma. Together with the vitamin from the food supply it is also synthesized in the skin by proper exposure to ultraviolet rays. Its precursor, 7-dehydrocholesterol, is converted into vitamin D_3 (in the Odland bodies). Vitamins D_2 and D_3 speed up metabolic processes and increase the turgidity and tonicity of epidermal tissues. Their influence on the granulation process of an healthy skin and on the reepithelization of "burnt" skin is particularly known.

XI. HOW MOISTURIZING CREAMS ACT

Applying a cream to the skin has the following effects: (1) the aqueous phase of an emulsion evaporates almost completely into the outer environment and (2) the oily phase adheres to the skin. For these reasons, the oily phase of an emulsion is believed to act as a lubricant to the skin. In order to act as actual lubricants, cosmetic oils should spread among the lamellas of the stratum corneum and be allowed to slide one against the other. Certainly, this is not the main mechanism by which moisturing creams act. It has been shown that the topical application alone of oils or fats is not sufficient to make the stratum corneum flexible and soft [7,8,12,83]. Water is the only real cutaneous plasticizing agent. Applying an emulsion to the skin's surface is thought to reduce skin dryness through the following five stages:

1. Replenishing the surface lipidic film and some of the intercellular lipids
2. Slowing down the evaporation of water through insensible perspiration by the formation of a semipermeable film
3. Regulating the water exchange between the dermis and the stratum corneum
4. Enhancing filaggrin synthesis
5. Rebalancing cell turnover

XII. INDICES OF HYDRATION AND SOOTHING

Recent experimental work [79,80] shows that the amounts of surface lipid (casual level) and water (skin hydration) can be quantified. Numerical values can be used to express how much any skin is dehydrated (deprived of water) or alipic (deprived of fats). This new experimental approach makes it possible to calculate the amounts of water (hydration) and lipid (surface hydrolipic film) necessary to balance the skin of the investigated subject. Thus cosmetics can be suggested or specifically prescribed, just as sun-protection products have been for many years. For example, the hydrating index (HI) expresses the moisturizing capacity of the cosmetic in the same way as the sun-protection factor expresses ultraviolet (UV) screening capacity. The capacity to balance the surface lipidic film is known as the soothing index (SI).

XIII. THE SKIN HYDRATION INDEX

There are many well-known problems in determining the amounts of sebum (casual level) and water (skin hydration) in the stratum corneum. The methods so far used have been complicated and research-oriented rather than suitable for use with a large number of subjects. The use of hydrating and soothing indices through Dermotest has made possible a first Italian nationwide screening program by age and climatic area. Some 6,000 subjects in various geographic areas and at different seasons have been tested under conditions of controlled humidity and temperature. The first results showed high skin dehydration levels for people aged 26 to 45. High skin dehydration levels were always related to a constant lack of surface lipids, regardless of age, in northern, central, and southern Italy. These somewhat unexpected data seem to relate to climatic changes. As the relative humidity rate increases (southern Italy), in fact, skin dehydration decreases as well in people aged 26 through 45, while the surface lipid film tends to remain at a normal value [81,82].

XIV. METHODS OF TESTING MOISTURIZING ACTIVITY

As stated above, the effects of different agents (such as chemical solvents and detergents) on the water-binding power of the stratum corneum have been dealt with in a number of studies [83–86]. Hygroscopic, water-binding materials in cells seem to be protected by semipermeable membranes and the stratum corneum seems to release water by osmosis [85]. Lipid solvents, such as ether or anionic tensides, are thought only to remove hydrophobic lipids (ceramides), which act as semipermeable membranes. As a result, the natural sponges binding these membranes to the stratum corneum (NMFs and sugars) are removable by water; thus environmental pollutants easily penetrate the skin and xerosis develops. The study of the stratum corneum's capacity to bind and retain water is a useful means of testing the activity of both active moisturizing components and cosmetic products containing them.

In studying the skin's water-binding power and testing the moisturizing activity of cosmetics, environmental conditions and the skin biotype of patients should be considered first. It is necessary to carry out tests at a temperature of approximately 20–22°C with a maximum relative humidity of 50–55% in acclimatized subjects at complete rest, all having the same skin type.

In order to test the effectiveness of active components and moisturizing finished products, a number of methods were developed; these are both subjective (control groups often coordinated by experienced dermatologists) and objective (noninvasive instrumental techniques). There follows a review of some of the most important and commonly used techniques.

A. Electrical Methods

Electrical methods test the conductivity of the stratum corneum on the basis of the phenomenological resemblance between Fick's and Ohm's laws. The flow of electric current through the skin is correlated with the water level in the stratum corneum and with the following three factors:

1. Dipolar orientation of the various components of the stratum corneum, such as keratin
2. Movement of ions in the stratum corneum
3. Mobility of water and proton exchange with the stratum corneum

Clearly, water directly affects point 3 and indirectly promotes the flow of current by enhancing the dipole moment and the ion mobility linked with the decreased viscosity of moisturized stratum corneum.

Since other substances may also reduce flow resistance or impedance and can affect findings, dry electrodes were developed that are highly sensitive if used with high electrical frequencies (3.5 Mhz) [87–92].

A number of in vitro and in vivo tests widely showed that impedance decreases with an increase in the moisture level. In addition, cosmetic creams were shown to remoisturize the stratum corneum over more or less long periods, thus proving clinically effective. For these reasons, electrical measurements are used to test both the soothing index and the moisturizing power of cosmetics.

The *soothing index* can be measured by associating it with protein denaturation and the increase in skin conductivity.

The *hydration index* can be measured by associating it with the increase in skin conductivity due to the hydration of the stratum corneum.

Many authors use the methods based on the electrical properties of the stratum corneum and measure either the moisture level or the soothing index by means of special electrodes.

According to other authors, the water level in the upper layers of the skin can also be measured by the dielectric constant or capacitance. This method, which seems to be poorly related to the dynamics of stratum corneum hydration, proved to be sensitive mostly to the changes in moisture level occurring in extra-dry, dehydrated skin (as in psoriasis). Measurement of capacitance is thought to detect even small amounts of water bound to the stratum corneum.

Thus, both bound and free water in the stratum corneum can be detected by alternatively using capacitance and conductance (impedance) electrical methods, which are extremely useful especially when investigating, for example, hyperkeratosis [92–97].

Another computerized method, also based on capacitance measurements, is affected by neither ion conductivity, polarization, nor environmental humidity and temperature. This newly adopted method, known as the 3C System, includes a sensor detecting environmental data (temperature and humidity). This sensor automatically correlates the detected skin parameters with constant values (22°C temperature and 50% relative humidity) (Fig. 5) [98].

Figure 5 3C System Dermotech.

B. Photoacoustic Methods

Photoacoustic methods are based on the absorption of optical radiation by the stratum corneum, which results in heat generation. Periodic heat waves spread through the stratum corneum and, once on the skin's surface, turn into pressure waves, which are detected by an acoustic sensor. The higher the hydration index, the lower the value of the transmitted resonance.

The advantage of this complicated method is that no direct contact is required between the device and the skin. Thus, occlusion of the skin is avoided [99–101].

C. Infrared Spectroscopy

Infrared spectroscopy is based on the capacity of water to absorb infrared (IR) radiation very easily and rapidly. In order to measure IRs, attenuated total reflectance (ATR), also known as frustated multiple internal reflectance (FMIR), is used [102–106].

D. Measurement of Transepidermal Water Loss (TEWL)

This method, which measures both physiological and pathological degrees of insensible perspiration, is very useful in evaluating how water is distributed between the stratum corneum and the environment. It can be used with two different modes: ventilated and nonventilated [107–110].

In the first mode, a closed cylinder applied to the skin measures the humidity changes of a gas that is caused to pass through the cylinder. In the second, simpler mode, an open cylinder is directly placed on the skin. The changes in relative humidity that occur inside the cylinder are then measured. Measuring TEWL makes it possible to monitor the soothing and moisturizing power of cosmetic emulsions. For example, in irritated skin, the stratum corneum has a lower protective action, which results in increased TEWL. On the other hand, the tendency of a cosmetic cream to occlude the skin usually increases the moisture level in the stratum corneum while reducing TEWL. These measurements are affected by environmental humidity and temperature and, above all, also by sudden drafts of air [111–121]. For all of the above reasons, TEWL measurement is today considered a useful parameter for testing the efficiency of the skin barrier, and also because the link between percutaneous absorption and variability of the skin site has been shown under the most differing conditions [122–126].

E. Chemical and Mechanical Methods

There are many chemical but especially mechanical methods that can indirectly detect the moisture level in the skin and the cohesion of keratin-filled cells, which, in turn, can be correlated with skin ''roughness.'' The cosmetic literature deals extensively with such methods. Thus, readers are referred to more specific reviews and treatises [127–129]. This chapters deals with only some examples of methods used in the past or still in use in applied research laboratories.

Padberg [42] talks about a method modified by methylene blue to detect the degree of roughness of the stratum corneum, which decreases with an increase in the moisture level. The fading away of methylene blue, which is retained by the stratum corneum and then eluted, is measured in patients treated with a moisturizing cosmetic emulsion and in a control group, since moisturized skin retains a greater amount of dye.

Rieger and Deem [130] compare the elasticity of the stratum corneum in patients treated with cosmetic moisturizers and in a control group, since moisturized skin is more elastic than aging skin.

According to some authors, the free water in the stratum corneum, which can be transferred into environmental humidity through drying, can be measured by Cahn's electrobalance. A strip of stratum corneum from the hind leg pad of a guinea pig is placed between two special wheels having a central hole. This assembly is sealed to be airtight and placed on a vessel containing water, which is hung on the beam of the electrobalance in atmosphere at controlled temperature and humidity and undergoes continuous weighing. A 10-mg strip of skin loses exactly half of its weight in 10 min, and this weight loss is measured in both the treated stratum corneum and in a control group.

However, according to W.M. Cheng [131], the epidermis from a guinea pig's back should be used, since, as a substrate, it proved more sensitive than the stratum corneum from the hindleg pad.

Many authors [132–138] use *profilometry*, a method that measures the skin's topography and detects any changes that occur before and after the application of cosmetic or pharmaceutical products. This method can be used with or without contact. In the contact mode, a duplicate of the skin's surface is directly made; its bi- or tridimensional profile is then detected by means of mechanical and optical devices.

When without contact, *densitometric optical scanning* of a skin photograph taken under proper lighting conditions is used instead of mechanical scanning of the cutaneous duplicate [139]. Profilometry, which is widely used in cosmetic dermatology, requires very expensive equipment, including sophisticated hardware and software [140–144].

Naturally, in order to measure the skin's *softness*, peculiar to well-moisturized skin, static methods, such as profilometry, are not sufficient. Increasingly, new parameters and methods are being developed that can register the many properties of the stratum corneum. Above all, the concurrent use of different methods to detect the continuous skin changes is deemed useful. Among them, mention should be made of *gas-pad electrodynamometry* [147,148] as well as *torsion* [183–185], *friction* [149], and *viscosity* [150,151] *measurements*, all of which measure the *degree of softness, sensation of smoothness*, or the *degree of roughness* of the skin [152].

As stated above, the available methods of testing [review in Refs. 153 and 154] the moisturizing action of cosmetics are many. All of them ensure objective, repeatable results, which are vital in the development of effective products.

REFERENCES

1. Flesh P. Chemical basis of emollient function in horny layers. Proc Sci Sec TGA 1963; 40:12–16.
2. Middleton JD. The mechanism of water binding in stratum corneum. Br J Dermatol 1968; 180:437.
3. Flesh P, Esoda EC. Deficient water-binding in pathologic horny layer. J Invest Dermatol 1957; 28:5,93.
4. Sweeney TM, Downing DT. The role of lipids in the epidermal barrier to water diffusion. J Invest Dermatol 1970; 55:135.
5. Warner RR, Lilly NA. Water and the stratum corneum. In: Elsner, Berardesca, Maibech, eds. Bioengineering of the Skin. Boca Raton, FL: CRC Press, 1994:3.

6. Potts RO, Francoeur ML. The influence of stratum corneum morphology on water permeability. J Invest Dermatol 1991; 96:495–499.
7. Blank IH. Factors which influence the water content of the stratum corneum. J Invest Dermatol 1952; 18:443–440.
8. Blank IH. Further observation on the factors which influence the water content of the stratum corneum. 1953; 21:259–269.
9. Singer EJ, Vinson J. The water binding properties of skin. Proc Sci Sect Toilet Goods Assoc 1966; 46:29–33.
10. Speir HW, Pascher PG. Zur analitischen und funktionelen Physiologie der Hautberflache. Asutarzt 1966; 7:55.
11. Laden K, Spitzer R. Identification of a natural moisturizing agent in skin. J Soc Cosmet Chem 1967; 18:351.
12. Middleton JD. The mechanism of water binding in stratum corneum. Br J Dermatol 1968; 80:437–450.
13. Anderson RL, Cassidy JM, Hansen JR, Yellin W. Hydration of stratum corneum. Biopolymers 1973; 12:2789–2802.
14. Lampe MA, Burlingame LAL, Whitney J. Human stratum corneum lipids: characterization and regional variations. J Lipid Res 1983; 24:120.
15. Imokawa G, Hettori A. A possible function of structural lipid in the water-holding of the stratum corneum. J Invest Dermatol 1985; 84:282–284.
16. Landmann L. Epidermal permeability barrier: transformation of lamellar granule-disks into intercellular sheets by a membrane fusion process—a freeze-fracture study. J Invest Dermatol 1986; 87:202–209.
17. Green SC, Stewart ME, Downing DT. Variation in sebum fatty acid composition among adult humans. J Invest Dermatol 1984; 83:114.
18. Prottey VC. Essential fatty acids and the skin. Br J Dermatol 1976; 94:579.
19. Lampe MA, Williams ML, Elias PM. Human epidermal lipids: characterization and modulations during differentiation. J Lipid Res 1983; 24(2):131.
20. Morganti P, Randazzo SD. Essential fatty acids and skin ageing. J Appl Dermatol 1985; 3:211–222.
21. Long SA, Wertz PW, Strauss JS, Downing DT. Human stratum corneum polar lipids and desquamation. Arch Dermatol Res 1985; 277:284.
22. Elias PM. Epidermal lipids, barrier function, and desquamation. J Invest Dermatol 1983; 80:44.
23. Odland GP, Holbrook K. The lamellar granules of the epidermis. Curr Probl Dermatol 1987; 9:29.
24. Grubauer G, Feingold KR, Elias PM. Lipid content and lipid type as determinants of the epidermal permeability barrier. J Lipid Res 1988; 30:89.
25. Wertz PW, Downing DT. Glycolipids in mammalian epidermis: localization and relationship to cornification. Science 1982; 217:1261.
26. Downing DT, Stewart ME, Wertz PW, Strauss JS. Lipids of the epidermis and the sebaceous glands. In: Fitzpatrick TB, Eisen AZ, Wolff K, et al, eds. Dermatology in General Medicine, 4th ed. New York: McGraw-Hill, 1993:210–220.
27. Matolsky AG, Downes AM, Sweeney TM. Studies of the epidermal water barrier: Part II. Investigation of the clinical nature of the water barrier. J Invest Dermatol 1968; 50:19–26.

28. Wertz PW, Downing DT. Glycolipids in mammalian epidermis structure and function in the water barrier. Science 1982; 217:1261–1262.
29. Wertz PW, Downing DT, Frenkel RK, Traczyk TN. Sphingolipids of the stratum corneum and lamellar granules of fetal rat epidermis. J Invest Dermatol 1984; 83: 193–195.
30. Critchley P. Skin lipids. In: Priestley GC, ed. Molecular Aspects of Dermatology. New York: Wiley, 1993:147.
31. Elias PM. Advances in Lipid Research, vol 24: Skin Lipids. New York: Academic Press, 1991.
32. Rawlings AV, Scott IR, Harding CR, Bowser PA. Stratum corneum moisturization at the molecular level. J Invest Dermatol 1994; 103:731–740.
33. Scott IR. Factors controlling the expressed activity of histidine ammonia-lyase in the epidermis and the resulting accumulation of urocanic acid. Biochem J 1981; 194:829–839.
34. Scott IR, Harding CR, Barrett JG. Histidine rich protein of the keratohyalin granules: Source of the free amino acids, urocanic acid and pyrolidone carboxylic acid in the stratum corneum. Biochim Biophys Acta 1982; 719:110–117.
35. Barrett JC, Scott IR. Pyrrolidone carboxylic acid synthesis in guinea pig epidermis. J Invest Dermatol 1983; 81:122–124.
36. Scott IR. Alterations in the metabolism of filaggrin in the skin after chemical and ultraviolet induced erythema. J Invest Dermatol 1986; 87:460–465.
37. Morganti P, Randazzo SD. Enriched gelatin as skin hydration enhancer. J Appl Cosmetol 1987; 5:105–120.
38. Fleckman P, Dole BA, Holbrook KA. Profilaggrin, a high-molecular-weight precursor of filaggrin in human epidermis and cultured keratinocytes. J Invest Dermatol 1985; 85:507–512.
39. Kubilus J, Scott IR, Harding CR, et al. The occurrence of profilaggrin and its processing in cultured keratinocytes. J Invest Dermatol 1985; 85:513–517.
40. Szakall A. Seife in Lichte der Physiologie der obersten Hautschichten. Fette Seifen 1950; 52:171.
41. Szakall A. Sauremantel oder Pufferhülle der Hautoflache. Fette Seifen 1951; 53: 285.
42. Padberg G. Modifizierte methylenblau-Methode zur Prüfung des Rauhig-Keitsgrades der Hornschicht. J Soc Cosmet Chem 1969; 20:719.
43. Blank IH. Penetration of anionic surfactants into skin: II. Study of mechanisms which impede the penetration of synthetic anionic surfactants into skin. J Invest Dermatol 1962; 37:311.
44. Blank IH. Penetration of anionic surfactants into skin: III. Penetration from buffered sodium laurate solutions. J Invest Dermatol 1962; 37:485.
45. Jaeger AO, Die Haut. In: Czetsch-Lindenwald, Schmidt-La Baume, eds. Salben, Puder, Externa, 3rd ed. Berlin: Springer Verlag, 1950:457.
46. Jacobi O. About the mechanism of moisture regulation in the horny layer of the skin. Proc Sci Sec TGA 1959; 31:22.
47. Clar EJ, Fourtanier A. L'acide pyrrolidone carboxylique (PCA) et la peau. Symposium National Les substances naturelles. Orleans, France, April 21–24, 1980.

48. Tatsumi S. Pyrollidone carboxylate acid. Am Cosmet Perfum 1972; 22:119–126.

49. Muscardin L, Morganti P, Bruno C. Integrazione carboidrati scleroproteine e potere idrolegante dello strato corneo: determinazioni in vitro ed in vivo. Cosm News 1978; 1(3):114–123.

50. Morganti P. EUR PAT 77430, 1986. USA PAT 4863950, 1989.

51. Morganti P, James B. Gelatin-glycine: improved cutaneous water retention capacity. J Appl Cosmetol 1989; 7:103–109.

52. Savermann G. L-Arginin in dry skin. Symposium on dry skin. Munich: December 10, 1994.

53. Wohlrab W. Use and efficacy of urea in dermatological preparations. J Appl Cosmetol 1991; 9:1–7.

54. Raab W. Biochemistry, pharmacology and therapeutic use of urea. J Appl Cosmetol 1991; 9:15–17.

55. Swanbeck GPE. Urea in atopic skin—International Symposium on dry skin. Munich: December 10, 1994.

56. Morganti P. The future of cosmetic dermatology. Oral cosmesis: a new frontier. In: Morganti P, Ebling FJ, eds. Everyday Problems in Dermatology: The Cosmetic Connection. Rome: International Ediemme Ed, 1991:291.

57. Morganti P, James B, Randazzo SD. The effect of gelatin-glycine on skin hydration. J Appl Cosmetol 1990; 8:23.

58. Morganti P. The protective and hydrating capacity of emollient agents in the anti-aging treatment of the skin. J Appl Cosmetol 1994; 12:25–30.

59. Morganti P, Muscardin L, Fabrizi G. Protective effect of collagen upon irritation response of detergents. Int J Cosmet Sci 1983; 5:7.

60. Chvapil M, Eckmayer Z. Role of proteins in cosmetics. Int J Cosmet Sci 1985; 7:41–49.

61. Morganti P, Randazzo SD, Cardillo A. Role of insoluble and soluble collagen as skin moisturizer, J Appl Cosmetol 1986; 4:141–152.

62. Van Scott EJ, Yu RJ. Substances that modify the stratum corneum by modulating its formation. In: P Frost, SN Horwitz, eds. Principles of Cosmetics for the Dermatologists. CV Mosby Co., St. Louis, 1984:70–74.

63. Van Scott EJ. The unfolding therapeutic uses of alpha hydroxy acids. Med Guide Dermatol 1988; 3:1–5.

64. Van Scott EJ, Yu RJ. Alpha hydroxy acids: therapeutic potentials. Can J Dermatol 1989; 1:108–112.

65. Dahl MV, Dahl AC. 12% lactate lotion for the treatment of xerosis. Arch Dermatol 1983; 119:27–30.

66. Van Scott EJ, Yu RJ. Alpha hydroxy acids: treatment of the aging skin with glycolic acid procedures for use in clinical practice. Cutis 1989; 43:122–128.

67. Morganti P, Persechino S, Bruno C. Effects of topical AHAs on skin xerosis. J Appl Cosmetol 1994; 12:85–90.

68. Van Scott EJ. Biological function of alpha hydroxyacids. J Appl Cosmetol 1995; 13:87.

69. Morganti P, Fabrizi G, Randazzo SD, Bruno C. Decreasing the stinging capacity

and improving the antiaging activity of AHAs. J Appl Cosmetol 1996; 14:79–91.

70. Morganti P. Alpha hydroxyacids in cosmetic dermatology. J Appl Cosmetol 1996; 14:55–41.

71. Morganti P, Randazzo SD, Bruno C. Alpha hydroxy acids in the cosmetic treatment of photo-induced skin aging. J Appl Cosmetol 1996; 14:1.

72. Ditre CM, Griffin TD, et al. Effect of alpha-hydroxy acids on photoaged skin: a pilot clinical, histologic and ultrastructural study. J Am Acad Dermatol 1996; 34: 187–195.

73. Coupland K. Essential fatty acids: applications in cosmetics and toiletries. Parf Cosmet Arom 1985; 66(12):65.

74. Horrobin DF. Gamma-linolenic acid, Rev Con Pharmacother 1990;1:1–41.

75. Sprecher H. Biochemistry of essential fatty acids, Prog Lipid Res 1981; 20:13–22.

76. Stubbs CD, Smith AD. Essential fatty acids in membrane: physical properties and function, Bioch Soc Trans 1990; 18:779–781.

77. Cook HW. Fatty acid desaturation and chain elongation in eukaryotes. In: Vance DE, Vance JE, eds. Biochemistry of lipids, lipoproteins and membranes. Amsterdam: Elsevier, 1996:129.

78. Randazzo SD, Morganti P. Skin and water: an update. J Appl Cosmetol 1990; 8: 93–102.

79. Morganti P, Randazzo SD. Skin Moisturizing factors: method of determination, J Appl Cosmetol 1990; 8:23.

80. Morganti P, Randazzo SD. Skin hydration control and treatment: recent updates. J Appl Cosmetol 1990; 8:103–112.

81. Fabrizi G, Morganti P. Skin hydration and lipid casual level: a quantitative monitoring in Italy. J Appl Cosmetol 1994; 12:41.

82. Morganti P. Clinical relevance of cutaneous check-up. Cosmet Toil Manuf Worldwide 1996; 6:227–230.

83. Middleton JD. The mechanism of water binding in stratum corneum. Br J Dermatol 1966; 80:437.

84. Blank IH, Shappirio EB. The water content of the stratum corneum: III. Effect of previous contact with aqueous solutions of soaps and detergents. J Invest Derm 1955; 25:391.

85. Middleton JD. Symposium on Skin. Society of Cosmetic Chemistry, Enstbourne, Sussex, Nov. 19, 1968.

86. Polano MK. The interaction of detergents and the human skin. J Soc Cosmet Chem 1968; 19:3.

87. Clar EJ, Her CP, Sturelle GC. Skin impendance and moisturization, J Soc Cosmet Chem 1975; 26:337.

88. Tagami H, Ohi M, Iwatsuki K, et al. Evaluation of the skin surface hydration in vivo by electrical measurement. J Invest Dermatol 1980; 75:500.

89. Serban GP, Henry SM, Cotty VF, et al. Electronic technique for the in vivo assessment of skin dryness and the effect of chronic treatment with a lotion on the water barrier function of dry skin. J Soc Cosmet Chem 1983; 34:383.

90. Tagami H, Kanamuru Y, Inoue K, et al. Water absorption: desorption test of the skin for in vivo functional assessment of the stratum corneum. J Invest Dermatol 1982; 78:425.

91. Tagami H. Electrical measurement of the water content of the skin surface. Cosmet Toilet 1982; 97:39.

92. Leveque JL, De Rigal J. Impedance methods for studying skin moisturization. J Soc Cosmet Chem 1983; 34:419–429.

93. Kohli R. Impedance measurements for the noninvasive monitoring of skin hydration: a reassessment. Int J Pharm 1985; 26:275.

94. Serban GP, Henry SM, Cotty VF, Marcus ADC. In vivo evaluation of skin lotions by electrical capacitance and conductance. J Soc Cosm Chem 1981; 32:421.

95. Werner L. The water content of the stratum corneum in patients with atopic dermatitis: measurements with the corneometer CM 420. Acta Derm Venereol (Stockh) 1986; 66:281.

96. Loden M., Lindberg M. The influence of a single application of different moisturizers on the skin capacitance. Acta Derm Venereol (Stockh), 1991; 71:79.

97. Hashimoto-Kumasaka K, Takahashi K, Tagami H. Electrical measurement of the water content of the stratum corneum in vivo and in vitro under various conditions. Acta Derm Venereol (Stockh) 1993; 73:335.

98. Cardillo A, Morganti P. A fast, non-invasive method for skin hydration control. J Appl Cosmetol 1994; 12:11–16.

99. Cahen D, Bults G, Garty H, Malkin S. Photoacoustics in life science. J Biochem Biophys Methods 1980; 3:293.

100. Rosencwaig A. Potential clinical applications of photoacoustics. Clin Chem 1982; 28:1878–1881.

101. Pines E, et al. Dermatological photoacoustic spectroscopy. In: Marks R, Payne PA, eds. Bioengineering and the Skin. Lancaster, England: MTP Press, 1981:283.

102. Puttnam NA, Baxter BH. Spectroscopic studies of skin in situ by attenuated total reflectance. J Soc Cosmet Chem 1967; 18:469–472.

103. Comaisch S. Infrared studies of human in vivo by multiple internal reflection. Bri J Dermatol 1968; 80:522–528.

104. Puttnam NA. Attenuated total reflectance studies of the skin. J Soc Cosmet Chem 1978; 23:209–226.

105. Baier RE. Non invasive, rapid characterization of human skin chemistry in situ. J Soc Cosmet Chem 1978; 29:289–306.

106. Triebskorn A, Gloorm, Greiner F. Comparative investigation of the water content of the stratum corneum, using different methods of measurement. Dermatologia, 1983; 167:64.

107. Potts RO. In vivo measurement of water content of the stratum corneum using infrared spectroscopy: a review. Cosmet Toilet 1985; 100:27–31.

108. Miller DL, Brown AM, Artz EJ. Indirect measures of transepidermal water loss. In: Marks R, Payne P, eds. Bioengineering and the Skin. MTP Press, Lancaster, 1981:161–171.

109. Rutter N, Hull D. Reduction of skin water loss in the newborn: I. Effect of applying topical agents. Arch Dis Childh 1981; 56:669–672.

110. Scott RC. A comparison of techniques for the measurements of transepidermal water loss. Arch Dermatol Res 1982; 274:57–64.

111. Grice K, et al. The effect of skin temperature and vascular change on the rate of transepidermal water loss. Br J Dermatol 1967; 79:582–588.

112. Frost P, Weinstein GD, Bothwell JW, Wildnauer R. Ichthyosiform dermatoses: III. Studies of transepidermal water loss. Arch Dermatol 1968; 98:230–233.

113. Wildnauer RH, Kennedy R. Transepidermal water loss of human newborns. J Invest Dermatol 1970; 54:483–486.

114. Cooper ER., Van Duzee BF. Diffusion theory analysis of transepidermal water loss through occlusive films. J Soc Cosmet Chem 1979; 27:555–558.

115. Weil I, Princen HM. Diffusion theory analysis of transepidermal water loss through occlusive films. J Soc Cosmet Chem 1977; 28:481–484.

116. Maibach MI. Effect of prolonged occlusion on the microbial flora, pH carbon dioxide and transepidermal water loss on human skin. J Invest Dermatol 1978; 71:378–381.

117. Idson B. In vivo measurement of transepidermal water loss. J Soc Cosmetic Chem 1978; 29:573–580.

118. Nilsson GE, Oberg PA. Measurement of evaporative water loss: methods and clinical applications. In: Rolfe P, ed. Non-invasive Physiological Measurements, vol 1. London and New York: Academic Press, 1979:279.

119. Grice KA. Transepidermal water loss in pathological skin. In: Jarrett A, ed. The Physiology and Pathophysiology of the Skin, vol 6. London and New York: Academic Press. 1980:2115–2155.

120. Finlay AY. The ''dry'' non-eczematous skin associated with atopic eczema. Br J Dermatol 1980; 102:249–256.

121. Seits JC, Spencer TS. The use of capacitative evaporimetry to measure effects of topical ingredients on transepidermal water loss (TEWL). J Invest Dermatol 1982; 78:351.

122. Spencer TS. Methods for measuring transepidermal water loss. Bioeng Skin 1985; 1:352–353.

123. Lotte C, Rougier A, Wilson DR, Maibach HI. In vivo relationship between TEWL and percutaneous penetration of some organic compounds in man: effect of anatomic site. Arch Dermatol Res 1987; 279:351.

124. Wilson DR, Maibach HI. TEWL: a review. In: Leveque JL, ed. Cutaneous investigation in Health and Disease: Noninvasive Methods and Instrumentation. New York: Marcel Dekker, 1989:113.

125. Pinnagoda J, Tupker RA, Agner T. Serup J. Guidelines for TEWL measurement. Contact Dermatitis 1990; 22:164.

126. Pinnagoda J, Tupker RA. Measurement of the TEWL. In: Serup J, Jemec GBE, eds. Non-invasive Methods and the Skin. Boca Raton, FL: CRC Press, 1995:173.

127. Barbenel JC, Payne PA. In vivo mechanical testing of dermal properties. Bioeng Skin 1981; 3:8–38.

128. Millington PF, Wilkinson R. Skin. Cambridge, England, and New York: Cambridge University Press, 1983.

129. De Rigal J, Lévêque JL. Influence of ageing on the mechanical properties of the

skin. In: Lévêque JL Agache PG, eds. Aging Skin. New York: Marcel Dekker, 1993:15.

130. Riger MM, Deem DE. Skin moisturizers: I. Methods for measuring water regain, mechanical properties, and transepidermal water loss of stratum corneum. J Soc Cosmet Chem 1974; 25:239.

131. Cheng WM. Skin: in vitro studies. VI International Seminar of Dermopharmacology. Lyon, France, 1973.

132. Cussler EL. Understanding skin texture. Cosmet Toilet 1978; 93(2):17–28.

133. Nicholls S, King CS, Marks R. Short term effects of emollients and a bath oil on the stratum corneum. J Soc Cosmet Chem 1978; 29:617–624.

134. Cook TH. Profilometry of skin: a useful tool for the substantiation of cosmetic efficacy. J Soc Cosmet Chem 1980; 31:339.

135. Cook TH, Craft TJ, Brunelle RL, et al. Quantification of the skin's topography by skin profilometry. Int J Cosmet Sci 1982; 4:195.

136. Mignot J, Chuard M, Agache P. A wide range topographical measuring technique. Bioeng Skin 1985; 1:251.

137. Makki S, Agache P, Mignot J, Zahovani H. Statistical analysis and three dimensional representation of the human skin surface. J Soc Cosmet Chem 1984; 35:311.

138. Epsen J, Hansen HN, Cristiansen S. Laser profilometry. In: Serup J, Jemec GBE, eds. Non-invasive Methods and the Skin. Boca Raton, FL: CRC Press, 1995:97.

139. Marks R. Changes in skin surface contour and intracorneal cohesion after application of emollients. Bioeng Skin 1985; 1:353–354.

140. Marshall RJ, Marks R. A photographic method for the measurement of skin surface textures. Bioeng Skin 1982; 4:7–15.

141. Corcuff P, De Rigal J, Leveque JL. Image analysis of the cutaneous microrelief. Bioeng Skin 1982; 4:16–31.

142. Corcuff P, Chatenay F, Leveque JL. A fully automated system to study skin patterns. Int J Cosmet Sci 1982; 6:167–176.

143. Spencer TS, Anderson PJ, Seitz JC. User of a phase angle meter to measure product effects on the skin surface. Bioeng Skin 1986; 2:153.

144. Grove GL, Grove MJ. Objective measures of skin surface dryness by digital image processing. Bioeng Skin 1985; 1:355.

145. Cook TH, Craft TJ. Improved method for determination of efficacy of moisturizers. Bioeng Skin 1985; 1:354.

146. De Rigal J, Lévêque JL. In vivo measurement of the stratum corneum elasticity. Bioeng Skin 1985; 1:13–23.

147. Aubert L. An in vivo assessment of the biomechanical properties of human skin modifications under the influence of cosmetic products. Int J Cosmet Sci 1985; 7: 51–59.

148. Aubert L. In vivo relationship between stratum corneum extensibility and electric conductance under influence of cosmetics, Bioeng Skin 1985; 1:359.

149. Agache P. Biomechanical properties of human skin in vivo and aging. Bioeng Skin 1980; 2:20–30.

150. Highley DR, Coomey M, Den Beste M, Wolfram LJ. Frictional properties of skin. J Invest Dermatol 1977; 69:303–305.

151. Gerrard WA, Stimpson IM. A versatile friction meter based on a viscometer. Lab Pract 1984; 33:82–83.
152. Gerrard WA, Stimpson IM. The effect of treatment on skin friction coefficient in vivo. Bioeng Skin 1985; 1:229.
153. Elsner P, Berardesca E, Maibach HI. Bioengineering of the skin: water and the stratum corneum. Boca Raton, FL: CRC Press, 1994.
154. Serup J, Jemec GBE. Non-invasive Methods and the Skin. Boca Raton, FL: CRC Press, 1994.

5

Assessing the Bioactivity of Cosmetic Products and Ingredients

Edward M. Jackson
Jackson Research Associates, Inc., Sumner, Washington

I. INTRODUCTION

This chapter focuses on assessing the bioactivity of both cosmetic products and cosmetic ingredients on the skin. Bioactivity is defined here as the biological effect of a cosmetic on the skin. This effect is called a *claim* by companies; therefore they are required to rely on what is legally termed a *reasonable basis* to support the claimed effect. A reasonable basis for bioactivity can take the form of published scientific or clinical studies, consumer evaluation of the product for its bioactivity, professional observations by trained scientists or physicians, and instrumental analyses. This chapter explores the last three of these four foundations of reasonable basis—namely, consumer evaluation, professional observations, and instrumental analyses. Recently, results from in vitro tests have been used to support the bioactivity of some cosmetics. In vitro tests are almost exclusively in the domain of cell physiology and biochemistry and comprise cell cultures exposed to cosmetics to determine the bioactive effects on these cells. Owing to the newness of this approach, the shifting world regulation of in vitro testing, and the relative rarity of its use at the present time, this chapter does not address in vitro testing to assess cosmetic product and ingredient bioactivity. In addition, any demonstrated bioactivity from an in vitro test will still need to be demonstrated on human skin by consumer evaluation, professional observation, and/or instrumental analysis.

It is not within the purview of this chapter to determine what is or what is not a cosmetic. Such a determination is a question of regulation, definition, and classification. These vary from the United States to the European Union to the Pacific Rim as well as Canada, Mexico, and South America. However, what is helpful is a listing of cosmetic product types, as found in Table 1. This listing includes cosmetics and, by way of contrast, products called over-the-counter (OTC) drug products in the United States. As one example of a contrast, some of these U.S. OTC drugs are considered cosmetics in the European Union. Cosmetics that alter the appearance or coloration of the skin, which are sometimes referred to as decorative cosmetics, do not primarily exert a biological effect on the skin and are also excluded from this chapter.

Finally, this chapter reviews how the bioactivity of cosmetic products and cosmetic ingredients is assessed without reference to delivery systems, which alter the time of delivery as well as the concentration of the cosmetic product or ingredient delivered.

II. THE SKIN

A. Anatomy

Structurally, the skin absorbs through a dual membrane system perforated with shunts [1]. The first rate-limiting membrane is the stratum corneum of the epidermis. The second rate-limiting membrane is the dermal-epidermal junction or basement membrane. The various types of shunts are the pilosebaceous units and the ducts of the apocrine and eccrine sweat glands. Assessment of bioactivity in this discussion is limited to the stratum corneum and the other four keratinocyte layers of the epidermis.

The skin is not uniform over the surface of the body. For example, the periocular, scrotal, and labial skin is notably lacking in dermis. Glabrous skin—such as the palms of the hands and soles of the feet—lacks pilosebaceous units and therefore hair follicles. In addition, percutaneous absorption rates can differ from one anatomical site, such as the back or forearm, to another.

B. Physiology

Human skin is the site of extrahepatic metabolism for certain types of chemicals [2]. Once metabolized, the metabolites of these chemicals can be absorbed by the vascular and lymphatic systems.

Chemicals can also bind to protein in the skin. When this occurs, the chemical is termed a *hapten* and the potential for an unwanted biological effect called *allergic contact dermatitis* (ACD) exists. There are other toxicological implications in assessing the bioactivity of a chemical; these are reviewed elsewhere [3].

Table 1 Cosmetic Products[a]

Decorative cosmetic products
Facial cosmetics
    Liquid makeup (foundation)
    Blush (rouge)
    Pressed powder
    Loose powder
    Makeup remover
Eye-area cosmetics
    Mascara (waterproof, water-
      based,hybrid)
    Eye shadow
    Eye liner
    Eyebrow pencil
    Eye makeup remover

Lip products
    Lipstick
    Lip liners
Nail-care products
    Nail polish
    Basecoats
    Topcoats
    Nail polish remover
    Cuticle preparations (creams, softeners)

Skin-care products
Moisturizers
    Hand and body lotions[b]
Facial products

    Facial moisturizers
    Cleansers
    Toners
    Astringents
    Cold cream

Soap
    Bar soap
    Liquid soap (regular liquid soap, anti-
      bacterial liquid soap[c]
    Body shampoo (liquid shower soap)
Bath additives
    Bath beads
    Bath foams, bath bubbles
    Bath oils
    Bath tablets

Fragrances[d]
    Perfumes
    Colognes, Deo Colognes
    Skin fresheners

Deodorants and antiperspirants[e]

Shaving products
    Foams and gels
    Brushless shave
    After-shave treatments (alcohol-based,
      lotions)

Talcum
Styptic treatments
Depilatories

Oral hygiene products
    Toothpaste (including powders and liq-
      uids)[f]
    Mouthwash and mouthrinse
    Breath fresheners

Table 1 Continued

Hair products	
Shampoo (adult, baby)	Sheen products (pomades, brilliantines)
Mousses	Styling gels
Conditioners	Sprays (fixatives)
Hair colorants (permanent, temporary)	Bleaches and lighteners
Curl activators and revitalizers	Hair thickeners
Hair treatments (hot oils)	

Sun protection products	
Sunscreen products[g]	Tan accelerators
Suntan lotions	Self-tanning products

Feminine hygiene products
    Douches
    Feminine deodorants
    Personal lubricants

Foot-and leg-care products
    Deodorants (powder, sprays)
    Depilatories
    Athlete's foot treatment[h]
    Corn removers[i]

[a] The legal definition of a cosmetic is a product which enhances appearance, aids in personal hygiene, and does not affect the structure or function of the skin. Cosmetic products, therefore, contain no active drug ingredient. Over-the-counter drug products which perform a recognized effect and are intended for daily or seasonal use are included in this listing of cosmetic products but are identified in appropriate footnotes. Over-the-counter drugs do list the active drug ingredient on the label.

[b] Almost all moisturizers are cosmetics. There are a few exceptions, such as OTC drugs under the skin protectant and analgesic monographs. There is one prescription drug product which has an alpha-hydroxy acid as the active drug ingredient.

[c] Antibacterial liquid soaps are OTC drug products containing an antimicrobial drug ingredient.

[d] Fragrances are themselves cosmetics, and fragrances can be part of cosmetic products. *Fragrance-free* means the cosmetic product contains no fragrance, contains no odor masking ingredient, or is perceived to be fragrance-free by the consumer. *Unscented* means the cosmetic product contains no fragrance but may contain an ingredient for another formulation purpose, which coincidentally gives off an odor. An example of the latter is phenylethyl alcohol (PEA), which gives off a rose odor.

[e] Deodorants mask malodor with a fragrance. Antiperspirants contain active drug ingredients such as aluminum salts to decrease sweating in the axillae.

[f] Toothpastes that contain fluoride as an active anticaries drug ingredient are OTC drug products.

[g] Sunscreen products are OTC drug products containing ingredients to either chemically or physically screen out UVB and/or UVA radiation.

[h] Athlete's foot treatments are OTC drug product powders or sprays containing antifungal ingredients.

[i] Corn remover products are OTC drug products containing keratolytic agents to effect the removal of cornified skin.

Source: From Ref. 49.

C. The Condition of Skin

Assessing the bioactivity of a chemical involves a thorough understanding of what condition the skin is in. Table 2 lists these various skin conditions. Cosmetics are applied to skin ranging from normal to environmentally exposed. Some cosmetics are designed to alleviate specific skin conditions. Moisturizers are a good example.

Skin that has been exposed to irritants, sensitizers, phototoxins, photoallergens, or even demyelinating neurotoxins is modified. Exposure to certain cosmetic ingredients such as vitamin A in retinol and retinyl palmitate or alpha hydroxy acids (AHAs) can produce thinner skin over time. Sunscreens are recommended for those using cosmetics containing these ingredients on a regular basis.

Hydration of the skin also increases percutaneous absorption potential. Externally absorbed cutaneous water increases the partitioning of certain chemicals in the skin, and there are differences in water uptake by the skin versus the hair. Water uptake by the keratin in corneocytes is a permeating phenomenon spreading out in all directions. Hair, on the other hand, swells longitudinally, lengthening the hair strand itself. Finally, the epidermal membrane or stratum corneum is more sensitive to water uptake than is hair [4].

Occlusion of the skin can result from wearing clothes, wearing patches in patch testing, bandaging or treatment modes such as plasters or masks, or application of oleagenous materials such as petrolatum. The application of occlusive, oleagenous materials or moisturizers that leave a thin film on the surface of the skin, markedly decreases normal water loss from the skin. The use of patches in

Table 2 Skin Conditions

Normal
Injured
    Abrasions
    Lacerations
    Hematomas
    Punctures
Pathological
    Disease (psoriasis,
      ichthyosis)
    Medicaments
    Occluded skin
    Hydrated skin
    History of sensitization
    Sensitive skin
Occupationally exposed skin
Environmentally exposed skin

patch testing results in enhancing the penetration of the patch-test material through the skin by both pressure and increased hydration under occlusion. Cosmetic facial masks are films cast upon the skin to enhance absorption of cosmetic ingredients in the mask.

Finally, individuals with sensitive skin have become an important group in whom to assess the bioactivity of cosmetic products. Table 3 lists the various types of sensitive-skin populations currently used to assess the bioactivity of cosmetics.

D. Modifiers of Bioactivity

Before concluding this section on the skin, the subject of modifiers of bioactivity must be addressed. The product pharmaceutics is perhaps the most important. The forms listed under pharmaceutics in Table 4 are in decreasing order of their impact on the bioactivity of a cosmetic [5]. The pharmaceutics of the cosmetic ingredient are characterized by their stability and concentration. And the bioactiv-

Table 3 Types of Sensitive Skin Populations

Defined sensitive skin panels
Fragrances
Preservatives
Nickel
Pathological condition panels
Atopics
Acne
Psoriatics
Diabetics
Ethnic panels
Chronological panels
Newborns
Elderly
Treatment/medicated panels
Vitamin A
Alpha hydroxy acids
Environmentally stressed panels
Low humidity/low temperature
Sun-exposed
Occupational panels
Solvent-exposed
Frequent hand washing

Source: From Ref. 6.

Table 4   Modifiers of
Bioactivity

Product pharmaceutics
  Emulsion
    Oil in water
     Water in oil
  Solution
  Suspension
  Gel
  Physical mixtures
Ingredient pharmaceutics
  Stability
  Concentration
Permeation enhancers
  Surfactants
  Solvents
Reservoir effect

---

ity of cosmetic ingredients can clearly be enhanced by permeation enhancers such as surfactants and solvents. Finally, certain chemicals can pool in the skin and therefore can exert a reservoir effect. In this situation, the release of these cosmetic ingredients in the epidermis over time is key to assessing their bioactivity.

III. BIOACTIVE EFFECTS

The term *bioactive effect* is used in this chapter—instead of *toxic effect, pharmacological effect, dermatological effect*, or *therapeutic effect*. The latter are more narrow terms. The wider term *bioactive* permits discussion of effects from cosmetics that are not found in the toxicological, pharmacological, or clinical literature.

A. Negative Effects

At times there are unwanted bioactive effects, or toxic effects. These can be called *negative bioactive effects*. Table 5 provides a complete listing of the various types of subjective and objective negative bioactive effects that chemicals can exert on both the skin and the eye. The eye is included in this section because of its ability to react to vapors from cosmetics or from oleagenous ingredients in cosmetics applied to the periocular skin. The dermatological and toxicological literature provides ample explanation for each. Some comment is needed on four areas covered by Table 5 [6,7].

First, the subjective cutaneous reactions are not well understood at this time. Of the four listed here, only stinging is assessed by testing [8,9].

Second, the difference between the subjective ocular reaction of watering and the objective ocular reaction of tearing is whether or not patient or panelist experiences this negative reaction or whether it is objectively observable.

Next, the negative bioactive effects listed as comedones and papules, pustules, and nodules in Table 5 are acne lesions. However, folliculitis can be an unwanted negative effect from cosmetics as well.

Finally, the localized eye area sensitive syndrome (LEASS) is a newly described negative ocular effect [10,11]. It is objectively observable as both inflammation and tearing, but often a subjective response of foreign-body sensation is reported.

Table 5   Negative Bioactive Effects

Subjective cutaneous reactions
   Burning
   Itching
   Stinging
   Tingling
Objective cutaneous reactions
   Allergic contact dermatitis
   Irritant contact dermatitis
   Photocontact dermatitis
   Folliculitis
   Pigmentation alterations
   Comedones
   Papules, pustules, nodules
   Urticaria
Subjective ocular reactions
   Foreign-body sensation
   Watering
Objective ocular reactions
   Conjunctivitis
   Keratitis
   Iritis
   Tearing
   Localized eye-area sensitivity syndrome

B. Physical Effects

There are cosmetic effects that are physical effects, not biological. Table 6 lists some of these physical cosmetic effects. Many have to do with suface coloration, such as camouflaging, coloring, and whitening. Hair or eyelashes can be physically thickened or lengthened. Cleansing and ease of removal are also physical cosmetic effects. Overall, decorative cosmetics alter the appearance of the skin, and skin-care products can nourish, recondition, rejuvenate, repair, replenish, restore, and revitalize the skin's surface. These physical cosmetic effects are excluded from consideration in this chapter.

Likewise, palliation of the skin, protection of the skin with the use of sunscreens, or alteration of any pathological condition of the skin are also excluded.

C. Positive Effects

Examples of positive bioactive effects are listed in Table 7. Antisepsis is the reduction of surface flora or fauna with a microbiological agent. Antioxidation can be preventive; in cosmetics, however, it is claimed to be the scavenging of free radicals from the surface of the skin. Astringency is a firming and tightening of the skin through alcohol or witch hazel. The relief of itching is the alteration of a neurosensory response in the skin. While sebum reduction can be physical removal of sebum through the use of alcohol or surfactants, sebum control can also result from the use of certain cosmetic ingredients.

Table 6   Physical Cosmetic
Effects

Alter appearance
Camouflage
Cleanse
Color
Ease of removal
Lengthen
Nourish
Recondition
Rejuvenate
Repair
Replenish
Restore
Revitalize
Thicken
Whiten

Table 7   Positive Bioactive Effects

Antisepsis
Antioxidation
Astringency
Cellulite reduction
Diminishment of wrinkles
Girth reduction
Itch control
Keratolysis
Moisturization
Reduction of flaking and fissuring
Retexturization
Sebum reduction
Skin clarification

Detailed descriptions of how the other eight examples of positive bioactive effects in Table 7 can be assessed are given in the following section.

IV.   ASSESSING THE POSITIVE BIOACTIVE EFFECTS OF COSMETICS

Positive bioactive effects are those that are intended to be produced. They are the desired effects that a cosmetic can have on the skin; therefore they underlie the claims made by the manufacturer of a cosmetic product.

Three areas of research are used to assess the positive bioactive effects of cosmetics. These are *consumer evaluation, trained professional observations,* and *instrumental analyses*. When results from consumer evaluations, professional observations, and/or instrumental analyses are used to demonstrate that a particular positive bioactive effect resulted from the use of a cosmetic product, this is called *advertising claim substantiation*.

A.   Consumer Evaluation

Of all the tests that can be performed, successful consumer evaluations provide the most powerful support for a positive bioactive effect from the use of a cosmetic. This field has developed rapidly over the past 20 years and is now characterized by objective testing with statistically significant analytical potential [12,13].

Consumer understanding of the effect is a critical component to how the effect is communicated. One interesting example of how consumer understanding shapes this communication is the term *hypoallergenic* [14]. Hypoallergenicity is a claim of relative safety. From a dermatological viewpoint, *hypoallergenic* means a lower than normal incidence of allergic contact sensitization (ACD). However, the consumer has broadened the scope of this claim [15]. For the consumer, the claim of hypoallergenicity means the absence of any negative bioactive effect (Table 5). This is an example of the absence of negative bioactive effects resulting in a claim for a cosmetic product.

An example of consumer evelution that demonstrates a postive bioactive effect is aromatherapy [16,17]. Here the enhancement of positive mood or the diminishment of a negative mood can currently be evaluated only by consumers. This does not mean that trained observations and perhaps some instrumental applications—possibly in the form of blood analyses for certain biochemical markers—will not be developed [18].

B. Professional Observation

Trained professional observations can document positive bioactive effects from cosmetics. The professionals making these observations are toxicologists, estheticians, cosmetologists, scientists, and physicians. In addition to having the required background, it is important that they practice their craft on a regular basis. Physicians in practice who participate in cosmetic product testing only occasionally are best utilized to make a determination of contact dermatitis when this occurs in cosmetic product testing. Toxicologists, estheticians, and cosmetologists constitute a group of skin-care professionals who are trained and have experience in making professional observations. Technicians trained in the art and science of professional observation by toxicologists, estheticians, or cosmetologists must also be mentioned here.

Several methods of cosmetic testing lend themselves to providing proper exposure to the cosmetic and enable the trained professional observer to capture all potential data that the test can generate. These are the controlled-use test [19], the treatment-regression method [20], and what might be called the treatment-effect method [21]. In the controlled-use test, a group of test subjects are instructed to use the cosmetic as directed and examined by trained professionals periodically during the test period. The treatment-regression method uses periodic observations over a treatment period, followed by observations over a subsequent nontreatment period. This permits a determination of both how quickly the skin returns to its initial state as well as how long the treatment lasts. The treatment-evaluation test method varies slightly by having the trained observations occur periodically while the treatment continues. This method permits documentation of the real-time positive bioactive effect of the cosmetic.

Scoring scales are essential to aid the professional observation. Such scales permit quantification of the observations and some provide a direct reference guide for the observer [21].

Moisturizers are cosmetic skin-care products that literally heal dry skin by a four-step physiological process. This process consists of (1) repair of the barrier function of the skin [22–24], (2) alteration of the surface cutaneous moisture partition coefficient [25–27], (3) an increase in dermal-epidermal moisture diffusion [28–30], and (4) restoration of the intercellular lipid function [31–33]. These biochemical and biophysical events in the epidermis of the skin have now been elucidated as the physiological process of restoring the natural moisturization process in the skin regardless of its history [34]. Further, moisturizers have been used and continue to be promoted as adjunctive dermatological therapy. And yet the biochemistry, biophysics, and overall physiology of their effects are not subject to trained observation. What is observable is the improved skin, which is characterized by a moisturized look and feel; a renewal of its texture, with a consequent reduction in flaking; and—in extreme situations, reduction in fissuring from environmental stress. These effects can be enhanced by utilizing special sensitive-skin panelists, with the observations of the trained professionals documented by actual photographs [35].

C. Instrumental Analysis

Consumer-perceivable differences in the effects of positive bioactive cosmetics constitute the strongest evidence of the cosmetic's effect. As a kind of paradox however, cosmetics manufacturers spend more research and development monies on instrumental analyses than they do on consumer evaluations. And it is also true that instrumental analyses often show differences that are not readily perceivable to the consumer or trained observer.

Instrumental analyses fall into four main categories: biophysical measurements, simple mechanical measurements, optical analyses, and visual documentation. Often, combinations of these are used, as in profilometry, which is a mechanical measurement that is electronically analyzed and transformed into a visual image.

Biophysical measurements can be as simple and economical as squametry [36] or surface sebum evaluation [37] and as complex as measuring and interpreting transepidermal water loss (TEWL) [38,39].

Mechanical measurements are straightforward physical assessments of girth and the skin's plasticity or elasticity.

Optical analyses are best exemplified by profilometry [40–42] and are continually improving as computerized analyses and visualization technologies are applied [43].

Visual documentation of positive bioactive effects are powerful but require standardization [44] to maximize their value. Advances in visual documentation are exemplified by in vivo polarized videomicroscopy [45].

Instrumental analyses have been used to document the three types of positive bioactive effects on the skin of various alpha hydroxy acid moisturizers [46]. These positive bioactive effects are diminution of fine lines and wrinkles; skin clarity, which often resolves melasma and chloasma; and retexturization, which results in a smooth, supple skin feel. These positive bioactive effects can be assessed by profilometry, TEWL, squametry, and in vivo videomicroscopy.

Instrumental analyses are also important to document the effects of products on the improvement and control of cellulite [47]. Here clinical photography, profilometry, and mechanical measurements for girth, elasticity, and plasticity have been used.

V. FUTURE ASSESSMENT OF POSITIVE BIOACTIVE EFFECTS

As technologies continue to advance and synergize, refinement of existing instrumental analyses will continue. These advances will produce ever more detailed and statistically significant differences, which will be proof of the positive bioactive effects of cosmetics on the skin. However, the touchstone will be whether or not the consumer or the trained observer can also experience this bioactivity.

In vitro tests have the potential to probe cellular biochemical and physiological changes from exposure to cosmetics [48]. But once again, the question is: "Will the trained observer also detect this bioactivity?"

VI. CONCLUSION

Cosmetics produce negative bioactive effects, physical effects, and positive bioactive effects. These effects can now be identified, observed, and interpreted. The ultimate result will be an improvement in current cosmetics, especially in the category of skin-care products which will lead to targeted bioactive effects from new cosmetic products.

REFERENCES

1. Malkinson FD, Gehlman L. Factors affecting percutaneous absorption. In: Drill VA, Lazar P, eds. Cutaneous Toxicity. New York: Academic Press 1977.
2. Bickers DR. The skin as a site of drug and chemical metabolism. In: Drill VA, Lazar P, eds. Cutaneous Toxicity. New York: Academic Press, 1980.

3. Jackson EM. Toxicological aspects of percutaneous absorption, In: Zatz JL, ed. Skin Permetation. Wheaton, IL: Allured, 1993.

4. Robbins CR, Fernee KM. Some observations on the swelling of the human epidermal membrane, J Soc Cosmet Chem 1984; 34:21.

5. Larsen WG. Jackson E.M, et al. A primer on cosmetics. J Am Acad Dermatol 1992; 27:469–484.

6. Jackson EM. The use of sensitive skin populations to test hypoallergenic products. In: Proceedings of the First International Symposium on Cosmetic Efficacy (a supplement to Cosmet Dermatol) November 1996:22–23.

7. Jackson EM. Understanding mascaras. Cosmet Dermatol 1996; 9(10):43–45.

8. Frosch P, Kligman AM. A method for appraising the stinging capacity of topically applied substances, J Soc Cosmet Chem 1977; 28:197–209.

9. Green BG. Regional and individual differences in cutaneous sensitivity to chemical irritants: capsaicin and menthol. J Cutan Ocular Toxicol 1996; 15:277–295.

10. Stephens TJ, McCulley TJ, et al. Localized eye area sensitivity syndrome (LEASS). J Cutan Ocular Toxicol 1990; 8:569–570.

11. Jackson EM, Stephens TJ, Establishment of the international registry of localized eye area syndrome (LEASS). Cosmet Dermatol 1996; 9(6):64–65.

12. Moskowitz HR. Cosmetic Product Testing: A Modern Psychological Approach. New York: Marcel Dekker, 1985.

13. Moskowitz HR. Consumer Testing and Evaluation of Personal Care Products. New York: Marcel Dekker, 1996.

14. Jackson EM. Hypoallergenic claims. Am J Contact Derm 1993; 4(1):47–49.

15. Jackson EM. "Hypoallergenic"—the clinical and consumer definition, Cosmet Dermatol 1994; 7(2):34–35.

16. Jackson EM. Aromatherapy. Am J Contact Derm 1993; 4:240–242.

17. Jackson EM. A review of bath products and their potential adverse reactions. Cosmet Dermatol 1995; 8(5):38–42.

18. Cousins N. The Healing Heart. Studio City, CA: Dove Books on Tape, 1987.

19. Jackson EM, Robillard NF. The controlled use test in a cosmetic safety substantiation program, J Cutan Ocular Toxicology 1982; 1(2):109–132.

20. Kligman AM. Regression method for assesing the efficacy of moisturizers, Cosmet Toilet 1978; 93:27–35.

21. Stanfield JW, Levy J, et al. A new technique for evaluating bath oil in the treatment of dry skin. Cutis 1981; 28:458–460.

22. Pitz EL. Skin barrier function and use of cosmetics. Cosmet Toilet 1984; 99:30–35.

23. Wu M, Yee DJ, et al. A moisturizer's effect on water distribution in the stratum corneum. Drug Cosmet Ind October 1984.

24. Madison KC, Swarzendruber DC, et al. Presence of intact intercellular lipid lamellae in the upper layers of the stratum corneum. J Invest Dermatol 1987; 88:714–718.

25. Flynn GL. Mechanisms of percutaneous absorption from physicochemical evidence. in Percutaneous Absorption In: Bronaugh RL, Maibach Hl, eds. New York: Marcel Dekker, 1985.

26. Batt MD, Davis WB, et al. Changes in the physical properties of the stratum corneum following treatment with glycerol. J Soc Cosmet Chem 1988; 39:367–381.

27. Rieger MM. Skin, water and moisturization. Cosmet Toilet 1989; 104:41–45.
28. Idson B. In vivo measurement of transepidermal water loss. J Soc Cosmet Chem 1978; 29:573–580.
29. Leveque JL, Garson JC, et al. Transepidermal water loss for dry and normal skin. J Soc Cosmet Chem 1979; 30:333–343.
30. Pillsbury DM, Heaton CL. A Manual of Dermatology. Philadelphia: Saunders, 1980.
31. Meczel E. Skin delipidization and percutaneous absorption. In: Bronaugh RL, Maibach Hl, eds. Percutaneous Absorption. New York: Marcel Dekker, 1985.
32. Downing DT, Stewart ME, et al. Skin lipids: an update. J Invest Dermatol 1987; 88:2s–6s.
33. Pugliese PT. Physiology of skin: behavior of normal skin. Skin Inc 1988; 1:28–37.
34. Jackson EM. Moisturizers: what's in them? How do they work? A J Contact Derm 1992; 3(4):162–168.
35. Jackson EM, Stephens TJ, et al. The use of diabetic panels to test the healing properties of a new skin protectant cream and lotion. Cosmet Dermatol 1994; 7(10):44–48.
36. Miller DL. Application of squametry in the assessment of dry skin. In: Proceedings of the First International Symposium on Cosmetic Efficacy (a supplement to Cosmet Dermatol). November 1996: 14–16.
37. Sebutape. Hermal Pharmaceutical Laboratories, Oak Hill, NY.
38. ServoMed Evaporimeters. ServoMed AB, Vallingby, Sweden.
39. Skin Surface Hydrometer, Skicon-200, I.B.S. Co., Ltd., 33-19, Motochama-cho, Hamamatsu-shi, Shizoia-ken, 430, Japan.
40. Cook TH. Profilometry of the skin: a useful tool for the substantiation of cosmetic efficacy. J Soc Cosmet Chemi 1980; 31:339–356.
41. Cook TH, Craft TJ, et al. Quantification of the skin's topography by skin profilometry. Int J Soc Cosmet Chem 4:195–205, 1982.
42. Makki S, Agache P, et al. Statistical analysis and three dimensional representation of the human skin surface. J Soc Cosmet Chem 1984; 34:311–325.
43. Hart J. 2-dimensional and 3-dimensional topographical analysis of the skin. In: Proceedings of the First International Symposium on Cosmetic Efficacy (a supplement to Cosmet Dermatol) November 1996:8, 13.
44. Seitz JC. Rizer RL, et al. Photographic standardization of dry skin. J Soc Cosmet Chem 1984; 34:423–437.
45. Dorogi PL, Jackson EM. In vivo video microscopy of human skin using polarized light. J Cutan Ocul Toxicol 1994; 13(1):97–101.
46. Jackson EM. Supporting advertising claims for AHA products. Cosmet Dermatol 1996; 9(5):40–47.
47. Jackson EM. Substantiating the efficacy of thigh creams. Cosmet Dermatol 1995; 8(2):31–41.
48. Jackson EM. Stephens TJ, et al. Supporting advertising claims: reviewing a 3-dimensional in vitro human cell test. Cosmet Toilet 1993; 108(12):41–42.
49. Jackson EM. Consumer products: cosmetics and topical over-the-counter drug products. In: Chengelis CP, Holson JF, Gad SC, eds. Regulatory Toxicology. New York, Raven Press, 1995: 107–108.

6

Stability Testing of Cosmetic Products

Perry Romanowski and Randy Schueller
Alberto Culver, Melrose Park, Illinois

I. INTRODUCTION

Products formulated by cosmetic chemists are intended to perform a variety of "miracle" functions, such as reshaping hair, delivering fragrance, smoothing and softening skin, imparting color to the face, and cleansing the entire body. Chemists can deliver many of these miracles by using the variety of technologies described elsewhere in this book. In using these technologies to develop products, chemists must be aware of formulation issues that might prevent the product from performing optimally. Assessing product stability is a critical part of this formulation process. This chapter discusses the basic principles of stability testing of cosmetic delivery systems. We will begin with a general definition of stability testing and move on to problems encountered by specific formula types. We will conclude this section with a discussion of stability issues that are not necessarily directly related to the formulation, such as processing and packaging.

II. A PRACTICAL DEFINITION OF STABILITY TESTING

Stability testing may be defined as the process of evaluating a product to ensure that key attributes stay within acceptable guidelines. In order to make this testing meaningful, it is important to accurately establish the nature of these critical product attributes, to measure how they change over time, and to define what degree of change is considered acceptable. Defining which parameters are crucial requires a combination of chemical knowledge about the formula and common sense about product usage. The chemist should be aware that cosmetic products must not only continue to function over time but must also look, feel, and smell the same each time the consumer uses them. Therefore, testing must evaluate esthetic characteristics in addition to functional properties. This is an important consideration because cosmetic products can change in a number of different ways, which may affect consumer perception. For example, fragrances become distorted, colors may fade or darken, and consistency may change, resulting in a thicker or thinner product. Chemists must determine which of these product characteristics will change over time and design appropriate testing to measure the extent of the changes. Nacht cites several technical issues to be considered, including compatibility between the delivery system and the active ingredient, compatibility with the overall formula, appropriate mechanism of release for the particular application, the rate of release of the active ingredient, and overall safety for the user [1]. This chapter discusses some of the key tests that the chemist can use to measure the changes in these characteristics. An important fact to remember is that no product remains unchanged forever. Depending on the intended use of the product and its anticipated shelf life, a small change over time may be inconsequential or devastatingly detrimental. In general, if a change is consumer-perceptible, the product may not be considered stable.

III. USEFUL INFORMATION PROVIDED BY STABILITY TESTING

Stability data are useful as an "early warning system" that can alert the chemist to potential formulation/package-related problems. Such advance information can be helpful in many ways.

A. Guiding the Chemist During Product Development

While you are formulating a product, preliminary testing of its stability can guide you in making modifications to ensure that it is stable. If you determine, for example, that an emulsion shows separation after exposure to freeze/thaw conditions, you may elect to modify the surfactant system to correct the problem and then repeat the test on the modified formula to determine whether it performs better or worse. Preliminary stability test data are an important parts of the trial-and-error development process.

B. Ensuring That the Product Will Continue to Be Esthetically Acceptable to the Consumer

More than other products, cosmetics are intended to be esthetically pleasing to the consumer. For this reason consumers are likely to notice subtle changes in the odor or appearance of their favorite products. Since no product remains 100% unchanged as it ages, it is critical that the chemist anticipate the changes that may occur and makes sure that they stay within limits that are not consumer-perceptible. Stability testing allows you to see how the product will behave over time.

C. Determining That the Product Will Perform as Intended and Remain Safe to Use

Studying the performance of samples that are exposed to accelerated aging lets you assess how the product will function over time. This is particularly important for cosmetic products that use the technologies described in this book to deliver "active" ingredients. If the formula is not stable, the delivery of the active ingredient may be impaired. Take, for example, the case of an antiperspirant stick with an encapsulated fragrance that is released upon exposure to moisture and heat. If the delivery system is poorly designed, the fragrance may be released too soon or not at all. Properly designed stability testing can reveal such problems so that corrective action can be taken.

D. Forewarning the Company About Problems That Might Occur After Consumer Purchase

For example, testing can show that the product may thicken somewhat over time and may be difficult to dispense from the package. Realizing this beforehand is important to the company because it will allow the company to anticipate consumer reaction.

Even though stability testing provides much useful information, it is not an exact science and will not guarantee a trouble-free product, but it can give an idea of the risks involved and help provide a solid scientific foundation for evaluation of future problems.

IV. STABILITY TEST DESIGN

When faced with a situation where testing might be appropriate, ask some basic questions about the task ahead.

A. Why Is Testing Being Done?

Why is testing necessary? Are you concerned with product appearance or do you want to determine if specific performance characteristics change over time? The reasons for doing the tests will determine what kind of tests are required. Therefore it is critically important to approach this testing with a scientific mind set and to have a clearly defined hypothesis to be tested. Take, for example, the case of a skin lotion formula that develops an unpleasant odor. The reason for the test is to determine what is causing the odor. Your hypothesis may be that the fragrance you have selected is reacting with the formula ingredients to cause this problem. To test this hypothesis appropriately, you will need to assess the odor of the unfragranced base to determine how the fragrance affects the overall smell of the product. In this example, the unfragranced samples are the controls because the fragrance, which is the scientific variable, has been removed. Evaluation of appropriate control samples can prove or disprove the hypothesis—i.e., that the fragrance is causing the problem.

Another example illustrating the importance of conducting a properly controlled study is the case of an emulsion that separates after prolonged storage in its plastic bottle. In this case the reason for the test is to determine what is causing the separation. One hypothesis may be that the package is allowing water vapor to escape, thus leading to emulsion instability. To test this hypothesis, you will need to screen out the variable of concern: the packaging. Therefore, control samples could be packaged in glass to eliminate the possibility of moisture loss. If the control samples do not show the same instability that the packaged samples show, you have demonstrated that the packaging material is indeed having a negative effect on the product.

Finally, consider a case where the variable of interest is the viscosity of the product. If you are concerned that the product may become too thick over time and will not dispense properly, you could design a study to track product batches with varying initial viscosity. Suppose the target viscosity is 20,000 cps. You could monitor the viscosity of a series of batches with viscosities ranging from low to high. You may make batches which are initially at 5000, 10,000, 15,000, and 20,000 cps, respectively. You would then monitor the viscosity of these batches as a function of time and temperature. You may learn that viscosity does not change significantly from the initial value, which means that a very narrow specification will be required. In other words, the product must be very close to its final viscosity when it is produced. On the other hand, you may discover that as long as the initial viscosity is between 5000 and 15,000 cps, the product will build to 20,000 cps within 2 weeks and stay at that level for 2–3 years. In this case your specification can be rather broad, since—regardless of the initial value—the consumer will only be exposed to product that is 20,000 cps. In all these cases, understanding why the test needs to be done helps you establish appropriate controls, which are essential if meaningful test results are to be obtained.

B. What Is Being Tested?

Another important factor to understand is the status of the formula being tested. Is it a developmental prototype or the final production material? Consider a situation, as in the example provided above, where you are primarily concerned with the change in product viscosity. Furthermore, consider that the final color and fragrance of the product have not yet been firmly established, although there are several candidates under evaluation. You could prepare samples with every possible color/fragrance combination and measure their viscosity over time. This could involve thousands of samples and tens of thousands of measurements, which are both costly and time-consuming. So, bearing in mind that you are testing a prototype and not a finished product, you may instead opt to test the uncolored, unfragranced base formulation first. In this way you can expeditiously get data on the parameter of interest—in this case viscosity. By evaluating proto-types early on, you have given yourself more time to react to problems. Of course, the testing may have to be repeated once the final formula is established because the fragrance may affect viscosity. Similarly, if the final production package is not yet available, you may choose to evaluate formula stability in a packaging material that approximates the characteristics of the final container. Here too, the final formula and package combination must eventually be tested together, be-cause the formula may interact unfavorably with the package. Asking the "what" question will help make your testing meaningful without forcing you to go to excessive lengths.

C. Where Will Test Samples Be Stored and How Many Are Necessary?

Ideally, you could gain information on formula stability by performing exhaustive tests on every variable involved in every formulation you work with, but this is not always feasible, because proper testing requires a significant commitment of time and resources. Therefore, most companies have standardized test procedures for the storage of stability samples which depend on the objective of the study. Such procedures involve evaluations of samples stored at a variety of conditions and include enough samples to be statistically significant. Usually sample storage is done at elevated temperatures, under freeze and/or freeze thaw conditions, and with exposure to various types of light. Elevated temperature storage is critical, since the rate of chemical reactions roughly doubles for every 10°C increase in temperature. Storage at higher temperatures allows you to accelerate the aging process and to see certain problems much sooner than they would appear at room temperature. Of course, the potential drawback is that, at high temperatures, you may be forcing reactions to occur that would not happen at all at lower tempera-tures. Cold storage evaluates conditions that may negatively affect the solubility of ingredients or stability of emulsions. Sunlight and ultraviolet (UV) light expo-sure can reveal problems with ingredients that are reactive to the respective wave-lengths; fragrances and colors are particularly sensitive in this regard. The most common storage conditions used in this industry are 54°C or 50°C, 45°C, 37°C or 35°C, room temperature (25°C), 4°C, freeze/thaw, and exposure to fluorescent and natural light.

Since many of the tests that must be conducted to evaluate product performance will affect the sample physically (e.g., spraying an aerosol can), multiple samples are required at each storage condition to ensure there will be enough samples left for evaluation at the end of the test period. Depending on the protocol set by your organization, as many as one hundred or more samples may be required for a complete study. Again, you should follow your corporate guidelines to make sure that sample quantities will be enough for a thorough evaluation at all necessary conditions.

D. How Samples Are Evaluated and What to Look for— Identification of Instability

How samples are evaluated depends entirely on the type of product and the nature of the problems that might occur. Instability is typically identified by evaluating various product characteristics either by subjective observation of properties— such as color, odor and appearance—or by objective instrumental evaluation of pH, viscosity, particle size, and electrical conductivity. For instance, simply looking at a sample that has been stored at accelerated temperatures can often reveal significant changes such as color changes, emulsion separation, or rheological changes. Similarly, a quick olfactory evaluation can uncover major flaws in fragrance stability. More rigorous characterization of product attributes can be obtained instrumentally—for example, with a viscometer or pH meter. These instruments are highly sensitive and can distinguish small changes in products. Such changes are important to note since, as in the case of a change in pH, they may represent chemical reactions that are occurring in the formula.

Other specialized testing can be performed to quantify specific changes in formulated systems. For example, microscopic evaluation and light scattering are used to appraise changes in particle size and distribution of emulsions. A Coulter counter is also used for these determinations [2], as are conductivity measurements [3]. Nuclear magnetic resonance (NMR) and x-ray crystallography can also be used to reveal additional information regarding emulsion structure. In certain systems, specific assays are performed to measure the activity of functional ingredients. These types of tests are tailored for the compound in question. For instance, the bactericidal efficacy of preservatives or other antimicrobial compounds may be measured over the course of a stability test. In addition, chromatographic tests, spectroscopic measurements, titrametric evaluations, and other wet chemical methods can be used to detect signs of instability. Other indications of instability include incompatibility of product and package, which can lead to weight loss and package degradation (such as softening or cracking of container walls, clogging of orifices, corrosion of metal parts, etc. [4]). But perhaps the most important question to ask in assessing instability is to determine how much change is acceptable. Knowlton and Pearce have stated that a useful rule of thumb is to consider product rejection if the attributes being measured deviate by more than 20% of their original value [4]. This value is an interesting reference point; however, for some formulations, much smaller deviations may be critical. The impact of such changes must be assessed on a case-by-case basis.

V. SITUATIONS THAT REQUIRE STABILITY TESTING

A good chemist should have an understanding of factors that are critical to product stability, so that appropriate testing can be conducted when necessary. Situations in which stability testing is generally necessary include but are not limited to the following situations: consideration of a new formulation, qualification of new raw materials, evaluation of new manufacturing processes, and testing of different packaging components. As you will see, stability testing is not a finite, one-time task; instead, it is an ongoing, dynamic process that begins when the product is being developed and continues to evolve as the formula, packaging, or manufacturing processes change.

VI. FORMULA-RELATED REASONS TO STABILITY TEST

A. Specific Considerations Related to Development of Particular Formula Types

The process of stability testing a product is closely tied to the process of creating the formulation. As you develop formulations, you should always screen stability samples early in the process to make sure that your efforts are headed in the direction that will lead to a stable product. Every formula will have slightly different stability testing requirements, but for the sake of this discussion, we will give primary consideration to the types of cosmetic delivery systems detailed in this text.

B. Emulsions

Emulsions are among the most common types of delivery systems used for cosmetic products. They enable a wide variety of ingredients to be quickly and conveniently delivered to hair and skin. While many definitions of emulsions have been proposed, we will define them as heterogeneous systems in which at least one immiscible or barely miscible liquid is dispersed in another liquid in the form of tiny droplets of various sizes [5]. Consequently, these systems are inherently unstable and eventually, given enough time or energy, will separate into separate phases.

Emulsions used for cosmetic products are typically semisolid materials composed of an oil (hydrophobic) phase and a water (hydrophilic) phase. These phases are characterized as either the internal phase or external phase, depending on the overall composition of the emulsion. The internal phase is that which is contained inside separate discrete particles surrounded by surfactants; these particles are known as *micelles*. The external phase is the "solvent" or diluent, which surrounds the micelles. Usually, the external phase is the more abundant one. Depending on the composition of each phase, simple emulsions can be either oil in water or water in oil, the type of which depends specifically on what emulsifier is used.

Although the internal-phase particles of an emulsion are polydisperse (meaning they have various sizes), their average size is often used for emulsion classification [6]. When the average diameter of internal particles is less than 100 Å, the system is called a *micellar emulsion*. A particle diameter of 2000 to 100 Å is called a *microemulsion*. Larger particles produce macroemulsions, which are the most common types found in cosmetic formulations. More complex emulsions can have multiple internal phases. These emulsions, called *multiple emulsions*, can be oil in water in oil or some combination. For cosmetic applications, they are formed by first making a water-in-oil emulsion and then mixing that emulsion with a water phase. These types are particularly useful for encapsulating materials giving prolonged release when applied to a surface such as skin [7].

C. Stability Considerations

Since emulsions represent a mixture of two or more materials that are not miscible in each other, they are, according to the second law of thermodynamics, inherently unstable. This means that eventually the two phases will separate. The degree and speed of instability are quite variable. For example, a mixture of mineral oil and water when shaken will form a macroemulsion, which immediately separates upon standing. Other emulsions can remain stable for years, but eventually all emulsions will separate. While the second law of thermodynamics suggests that emulsions will separate over time, it does not provide a mechanism of this destabilization. Investigation into how emulsions destabilize has revealed three primary processes leading to instability: flocculation, creaming, and coalescence [8].

1. Flocculation

This process is characterized by a weak, reversible association between droplets of the emulsion's internal phase. Each individual droplet maintains its own identity; thus there is no change in the basic droplet size [8]. Flocculation represents a less serious sign of instability, which can be reversed by shaking the system [9].

2. Creaming

When particles of an emulsion aggregate, there is a tendency for upward sedimentation. This causes a partial separation of the emulsion into two emulsions, one of which is richer in the internal phase and the other richer in the external phase [9]. As in the case of flocculation, this stability problem can be reversed by agitation.

3. Coalescence

An aggregation between two particles can, if the two particles combine, lead to the formation of one larger particle. This process, known as *coalescence*, represents a more serious stability problem. A related phenomenon is that of Ostwald ripening, in which the particles all tend to become the same size. Both of these processes are irreversible and can eventually lead to complete separation of the internal and external phases of the emulsion [10]. An alternative consequence of these forms of instability is phase inversion, in which the internal phase becomes the external phase and vice versa [9]. For stability considerations, this change is typically undesirable, since it will change the physical properties of the product.

All emulsions are potentially subject to all of these destabilizing processes simultaneously, and the resulting effects on any given emulsion will vary. For example, microemulsions and micellar emulsions are initially transparent. Over time, the size of their internal-phase particles may increase, and they will develop translucent appearance. Since macroemulsions are opaque, a similar change in appearance will not be notable; however, there may be changes in viscosity and measurable separation. Multiple emulsions are typically less stable than monoemulsions. Over a short period of time, the number of multiple emulsion particles tend to be reduced. This results in the "leaking out" of some of the encapsulated material and reduces the duration of prolonged release.

In addition to the inherent processes that destabilize emulsions, other factors may be involved. Storage temperature has been shown to affect emulsion product stability. Generally, elevated temperatures result in destabilization, while reduced temperatures improve emulsion stability. Aqueous-phase evaporation may also contribute to instability over the life of a product. Microbial contamination can also cause a breakdown of emulsion stability. Finally, chemical reactions within the emulsion can lead to a change in the stability of the emulsion. While these types of reactions can be initiated by temperature increases, they can also be prompted by UV light or other types of electromagnetic radiation.

VII. VESICULAR SYSTEMS—LIPOSOMES AND NIOSOMES

A. Definition/Description

Vesicular systems encompass a number of delivery technologies, including liposomes and niosomes. Both of these systems employ a "vessel" to contain active ingredients within a formula and to provide controlled delivery of these ingredients. Nacht defines controlled delivery as a "system that would result in a predictable rate of delivery of its active ingredients to the skin" [1]. Liposomes are a classic example of this technology, in which phospholipids are used to create lipid "capsules" that can be loaded with various ingredients. Although liposomes are enjoying tremendous popularity in cosmetics today, they have their roots back in the early 1960s. At that time Professor Bangham, at the Institute for Animal Physiology in Cambridge, U.K., was one of the first to speculate that lipids such as phosphatidyl choline could be used to create sealed vesicles with bilayer membranes similar to cell membranes [1]. Niosomes are another delivery technology related to liposomes; the difference is that, unlike liposomes, niosomes are based on nonionic surfactants. L'Oréal pioneered the development of nonionic liposomes using nonionic surfactants such as polyoxyethylene alkyl ethers combined with fatty alcohols or fatty acids [1].

B. Stability Considerations

Liposome and niosome stability may be referred to in terms of leakage of contents, presence of oxidation products, or changing particle size due to aggregation formation and fusion. They are rather fragile capsules, and certain precautions must be taken to make sure that they remain intact and are able to deliver their contents. Leakage can be caused by mechanical forces like high-shear processing, which should be avoided. Similarly, excessive heat, which may destabilize the lipid bilayers, should be avoided. Perhaps most notably, liposomes may be solubilized by surfactants that may be present, and therefore they are not suitable for use in detergent systems. This is particularly true of systems such as shampoos and body washes, which contain strong anionic surfactants that can dissolve the lipid walls. In fact, even though liposomes are often used in creams and lotions, the emulsifiers used in these formulas may also be enough to disrupt the fragile walls. For these reasons, many formulators believe that gels are the ideal vehicle for liposomes because they lack the high HLB (hydrophilic lipophilic balance) surfactants present in many conventional emulsions, which might disrupt the lipid bilayers [10]. There is hope for using liposomes in emulsion. K. Uji et al. report that stable liposome suspensions can be prepared by using a cross-linked acrylic acid/alkyl acrylate copolymer at very low concentrations, because it can effectively stabilize lecithin liposomes in o/w emulsions [11]. Furthermore, there is some evidence in the patent literature that the addition of collagen, albumin, or gamma globulin to the liposomes can decrease the harmful effects of detergents [10].

In addition to leakage, vesicle systems may fuse together and no longer be available as discrete units for the delivery of active agents. According to Weiner, such fusion can occur for several reasons, including preparation below their transition temperature, the presence of contaminants such as fatty acids and divalent cations, changes in pH, or the addition of nonelectrolyte hydrophobic molecules [12]. Furthermore, phase separation of bilayer components can occur upon extended storage. In an excellent review on the subject, Fox refers to an article by Crommelin et al., that reports on preserving the long-term stability of liposomes. Crommelin discusses the chemical pathways by which phospholipids can degrade: by hydrolysis of the ester groups or oxidation of the unsaturated acyl chains. This research points to an optimal pH for liposome stability. For phosphotidylcholine liposomes, the pH for the lowest hydrolysis rate was found to be 6.5. The stability of liposomes was further enhanced by using phospholipids with fully saturated acyl chains (like those made from hydrogenated soybeans, so the opportunity for oxidation is reduced) [10]. Similarly, liposomes may be stabilized by sugar esters, for example, maltopentose monopalmitate have been used to improve stability of cosmetic systems [13].

For a more detailed discussion of the morphology of liposomal bilayers, we refer the reader to Liposomes: From Biophysics to Therapeutics [12]. The author provides an excellent discussion of the elastic properties and tensile strength of liposomes as well as the effect of solvents and osmotic effects on liposomal structures.

VIII. MOLECULAR CARRIERS

A. Definition/Description

Molecular carriers represent a delivery system in which one compound is used to bind another compound to a substrate, thereby changing the former's characteristics. This allows the bound material to be delivered to a surface and released when conditions are appropriate. One example of this type of technology is cyclodextrin chemistry. Cyclodextrins are created from starch-derived glucopyranose units and are classified as cyclic oligosaccharides. When formed, they contain a hydrophobic cavity capable of entrapping molecules of different sizes, shapes, and polarities. Molecules entrapped as such are found to be more resistant to environmental stresses and therefore more stable [14]. They can be used to entrap various types of compounds such as fragrances, vitamins, pigments, and dyes. Cyclodextrins have been used in cosmetic products for a variety of reasons, such as to reduce odor in mercaptan-containing systems [15], improve the stability of hair dyes [16], and as an active ingredient to treat acne [17].

B. Stability Issues

The complex of the cyclodextrin with a guest molecule is typically quite stable under ambient temperatures and dry conditions. However, in the presence of certain materials the guest molecule can be prematurely displaced thereby reducing the effectiveness of the delivery system [18]. This factor is of major concern when developing and particularly when assessing the stability of a formula.

IX. PARTICULATE SYSTEMS—MICROCAPSULES, BEADS, AND MICROSPHERES

A. Definition/Description

Microcapsules are one of the oldest controlled release technologies. They were developed to produce carbonless carbon paper and are composed of a core with the active ingredient surrounded by a shell, analogous to an egg. Microcapsules may have a multilayer construction with multiple cores containing the active. The active ingredients are released either by rupture of the capsule walls or by diffusion/permeation of the contents [1]. Fairhurst and Mitchnick list a range of materials that are typically used in this regard including adhesives, drugs, colors, fragrances, flavors, agricultural chemicals, solvents and oils. Classic shell materials include gelatin or gum arabic, cellulosic polymers, or synthetic polymers [19]. Starch based capsules are often used to deliver fragrance and cosmetic ingredients.

Beads and microspheres are small solid particles onto which other ingredients can be adsorbed for later delivery. Nylon particles, for example, are useful for delivery of certain active ingredients. Antiperspirant salts are said to be more efficacious when delivered via nylon spheres, and the esthetics of the product are said to be improved. Coloring agents may be delivered in this manner as well; Schlossman discloses a patented method (U.S. patent 5,314,683) of coupling cosmetic pigments to microspheres to provide uniform reflectivity, improved dispersion, and superior viscosity characteristics [10]. Tokubo et al. describe a process for preparing spherical hectorite particles, with a diameter of about 100 Å, which can be used to deliver glycerin and solid pigments such as titanium dioxide, zinc oxide, and ferric oxide.

B. Stability Considerations

Microcapsules are somewhat fragile physically and care must be taken to avoid premature rupture and release of the contents. Excessive temperature should be avoided by adding microencapsulated ingredients late in the manufacturing process. Likewise, refrain from formulating with materials that may act as solvents on the capsules walls. Finally, avoid high-shear processing, such as milling and homogenizing, which can physically disrupt the capsules. Additional techniques for enhancing the stability of microcapsules can be found in the technical literature. Fox refers to an interesting Shiseido patent for improving the stability of gelatin microcapsules by coating the surface of the capsule with a basic amino acid or its polymer [10]. In general, microcapsules are a stable, efficacious method of delivering chemicals in cosmetics. In fact, when properly formulated, microcapsules can actually enhance stability of systems by protecting the ingredients they carry from external forces. For instance, in an example provided by the Mono-Cosmetic Company, ascorbic acid particles are coated with silicone or a polymer—e.g., ethyl cellulose, to protect the ascorbic acid against oxidation [10]. Similarly, in delivering cosmetic materials via beads and microspheres, care must be taken not to disturb the matrices physically. As with microcapsules, excessive shear can be a problem, for if the capsules are broken, their ability to retain the ingredient to be delivered will be impaired.

X. GENERAL CONSIDERATIONS RELATED TO FORMULA MODIFICATION

Regardless of which delivery technology you choose to utilize in a formulation, there are certain fundamental stability considerations that you must deal with. For each of the technologies discussed above, factors such as raw material sources, manufacturing process, and packaging composition all play a role in product stability.

A. Raw Material Substitution

Often it becomes necessary to substitute one raw material for another similar material. This frequently occurs because a supplier discontinues one of the raw materials used in your formula. In exchange, a different, yet supposedly "identical," material may be offered. Depending on the chemistry of the materials involved, there is no way to anticipate if such a change will affect formula stability. Therefore, in such situations you must conduct testing to ensure your formula will remain stable. Similarly, you may wish to substitute another material that is cheaper but is not anticipated to function differently. For example, in a shampoo formula, you may substitute sodium lauryl sulfate for ammonium lauryl sulfate. Given the functional similarities between the two, you would not anticipate significant problems; nonetheless, some degree of stability testing would be prudent.

B. Alternate Vendor Qualification

You may also elect to qualify alternate raw material suppliers for ingredients in the formula. It is desirable to have secondary sources for most raw materials to ensure a steady supply and competitive pricing. Unfortunately, even though raw materials from different suppliers may have the same CTFA (Cosmetics, Toiletries, and Fragrance Association) designation, they may not be chemically identical, because chemical feedstocks and processing conditions vary between suppliers. Therefore, a raw material from one supplier cannot always be automatically inserted into a formula developed with a different supplier's raw material. The impact of even seemingly inconsequential changes in raw materials must be established by stability testing.

XI. NON-FORMULA-RELATED REASONS

A. Processing Issues

In addition to the formulation and raw material issues described above, there are processing issues that can affect product stability. For example, stability testing is typically required the first time a new formulation is made on a large scale. This is because the way in which the product is made on a large scale can have a dramatic effect on its stability. This is particularly true of emulsions, because the energy used in processing determines particle size and distribution, which helps determine product stability. The only way to fully assess the impact of the chosen manufacturing method on product stability is to evaluate samples made under actual production conditions. This may require that a trial production batch be made prior to commercialization of the formula. At the very least, stability testing should be done on the first production batch of any new product, so that the impact of actual production processing conditions may be evaluated.

Once a manufacturing process has been shown to be successful, any changes to that process may require additional testing. Alterations in the order of raw material addition may be necessary to reduce processing time; changes in heating and cooling rates may occur due to differences in heat transfer in large batches; and different mixing conditions will all affect the amount of shear the product experiences. Any one of these changes can cause stability problems.

B. Packaging Issues

Even with the formulation and manufacturing processes held constant, variations in packaging material can cause problems that require stability testing. Not all packages are created equal: glass and plastic behave differently, and different kinds of plastic vary in properties such as oxygen permeability, color fastness, and thermal resistance. Certainly a new combination of formula and package should be tested, and even a change in an existing packaging material or the supplier of that material merits evaluation. The stability of aerosol systems for example, is extremely package-dependent, since the package composition will help to determine how resistant the final product is to corrosion. The overall objective is to be alert for changes to the formulation/manufacturing/packaging system that may necessitate additional testing, so that you can be confident that your product will remain stable. Of course, your observations should not be limited to the formula itself. Changes that result from formulation and packaging interaction may be critical to total product integrity. To this end, weight loss, changes in plastic color and odor, and other package-related observations are important. The objective is to gain as much knowledge as possible regarding the behavior of the product over time.

XII. CONCLUSION

This chapter is intended to provide insight into the issues associated with the stability testing of cosmetic products. For the beginning chemist, we stress the importance of careful, methodical observation to ensure that as many stability problems as possible are identified. For the veteran formulator, we urge periodic review of the latest technical literature so that it will be possible to keep pace with new developments in stabilizing the specific delivery systems discussed in this book. Hopefully the references we have provided will be helpful in this regard.

REFERENCES

1. Nacht S. Encapsulation and other topical delivery systems. Cosmet Toilet 1995; 110(9):25–30.
2. Rieger M. Stability testing of macroemulsions. Cosmet Toilet 1991; 106(5):59–66.
3. Jayakrishnan A. Microemulsions: evolving technology for cosmetic applications. J Soc Cosmet Chem. 1983; 34:343.
4. Knowlton J, Pearce S. The Handbook of Cosmetic Science and Technology. Oxford, England: Elsevier Advanced Technology, 1993:436–439.
5. Becher P. Emulsions: Theory and Practice. New York: Reinhold, 1965:2.
6. Prince L. Microemulsions: Theory and Practice. New York: Academic Press, 1977: 1–2.
7. Fox C. An introduction to multiple emulsions. Cosmet Toilet 1986; 101(11):101–102.
8. Becher P. Encyclopedia of Emulsion Technology. New York: Marcel Dekker, 1983: 133–134.

9. Eccleston GM. Application of emulsion stability theories to mobile and semisolid O/W emulsions. Cosmet Toiletr 1986; 101(11): 73–135.
10. Fox C. Advances in cosmetic science and technology: IV. Cosmetic vehicles. Cosmet Toilet 1995; 110(9):59–68.
11. K Uji K et al. J Soc Cosmet Chem Jpn 1993; 27:206–215.
12. Ostro MJ, ed. Liposomes: From Biophysics to Therapeutics. New York: Marcel Dekker, 1987: 343.
13. Fox C. Cosmetic raw materials literature and patent review Cosmet Toilet 1991; 106(8):78.
14. Dalbe B. Use of cyclodextrins in cosmetics. 16th IFSCC Meeting, New York, 1991. pp. 635–639.
15. Kubo S, Fumiaki N. US patent 4,548,811. Shiseido Company Ltd.
16. Oishi T et al. US patent 4,808,189, Hoyu Co.
17. Koch J. US patent 4,352,749.
18. Duchene D. New Trends in Cyclodextrins and Derivatives. Dermal Uses of Cyclodextrins and Derivatives. Paris: 1991:473–474.
19. Fairhurst D, Mitchnik M. Submicron encapsulation of organic sunscreens. Cosmet Toilet 1995; 110(9):47.

7

Quantitation of Penetrant Molecules Within the Skin

Victor M. Meidan and Elka Touitou
The Hebrew University of Jerusalem, Jerusalem, Israel

I. INTRODUCTION

Many skin-care products—such as deodorants, moisturizing creams, after-sun lotions, and antiaging creams—lie in the borderline area between cosmetics and pharmaceuticals. Such products are frequently termed *cosmeceutics*. Skin-care cosmeceutics have been formulated with penetrants such as vitamin E, vitamin E lineolate, and retinoic acid [1]. These molecules have been claimed to exhibit antiaging, moisturizing, and healing properties, respectively. In order to understand and validate the exact mode of action of such cosmeceutic penetrants, it is necessary to study their distribution and localization behavior within the skin. Important parameters include the amount of penetrant accumulated within the different layers of the skin as well as the flux through the skin into the vasculature. Moreover, it is important to determine whether penetration is mediated *via* the bulk strata or through the appendages. A number of different in vivo and in vitro techniques are available for quantitating permeants within the skin. These are reviewed in this chapter.

II. IN VIVO QUANTITATION TECHNIQUES

A. Disappearance Measurements

This technique involves measuring the amount of penetrant remaining on the skin surface at the end of the application period. The quantity absorbed corresponds to the initial penetrant quantity minus the final penetrant quantity. This method has been used to quantify hydrocortisone absorption into rat skin in vivo [2]. Since in any clinically relevant model nearly all of the formulation remains on the skin surface, small inaccuracies in the quantification of permeant at the treatment site can lead to large differences in the estimated absorption rates. Consequently, these disappearance measurements require the use of an extremely sensitive analytical technique, such as scintillation counting of radiolabeled molecules. The method cannot be used if the penetrant molecule is removed from the application site by physical sloughing, spreading, evaporation, precipitation, or adsorption to the stratum corneum. Because of these disadvantages, some authors have argued that disappearance measurements should not be employed to evaluate percutaneous absorption [3]. In any case, this approach exhibits the obvious disadvantage that it does not yield any penetrant localization data.

B. Tape Stripping

By far the most convenient in vivo technique for evaluating penetrant localization in human stratum corneum is to physically detach this layer with repeated applications of adhesive tape. The protocol involves the topical application of the penetrant formulation to the application site for a set exposure period, after which the excess formulation is washed off. It is important that the application site be clean, dry, and free of formulation before stripping is initiated. Skin stripping involves pressing the adhesive tape to the skin surface and removing it with a single fluid motion. The amount of penetrant in each tape strip is then measured by employing an appropriate analytical method. When radiolabeled molecules are used, as they are in virtually all the studies of this type, the analytical method is generally digestion in solubilizer followed by liquid scintillation counting. From the data obtained, it is possible to build a profile of concentration versus stratum corneum depth for the investigated molecule—a procedure pioneered by Schaefer [4].

Several practical points need to be considered during skin stripping. The first one or two tape strips are usually discarded, as they may contain superficial residues of formulation. Another important consideration is that any given adhesive tape will detach different quantities of horny layer at different depths and sites of the skin. Furthermore, the amounts removed can be affected by factors such as the contact time and formulation type as well as by interindividual variability. It is therefore essential to quantify the mass of stratum corneum removed with each strip by weighing the tape before and after each stripping procedure.

In human volunteer studies, tape stripping has been used to measure the stratum corneum distribution profile of various molecules including benzoic acid [5,6], caffeine [5,6], acetylsalicylic acid [5,6], sodium benzoate [5], 4-cyanophenol [7], and betamethasone dipropionate [6]. In an important series of human volunteer trials [5], Rougier and coworkers applied formulations containing 1 mmol of ^{14}C-labeled penetrant—either benzoic acid, sodium benzoate, caffeine, or acetylsalicylic acid—to a 1-cm^2 skin area. Two identical applications were performed on each volunteer. The first was designed for measurement of total absorption as determined by analysis of urine samples collected over 24 h. The second application was performed 48 h later, at the contralateral body site, for the quantitation of penetrants in the stratum corneum. Tape stripping (15 strips) was initiated 30 mins after application. By plotting the total amounts penetrating against the amounts measured in the horny layer, the researchers were able to demonstrate a linear correlation. Significantly, the correlation is unaffected by parameters such as the physicochemical properties of the permeant, the dose applied, the vehicle used, and the application time. Thus, tape-stripping measurements can be used to predict the total penetrating amount.

The skin-surface biopsy method can be used as an alternative to tape stripping. The technique involves pressing a glass slide coated with cyanoacrylate resin to the skin surface. After the adhesive has set, the slide is removed, thus detaching the horny layer. It has been claimed that this method is faster, more versatile, and easier to standardize than tape stripping, as the stratum corneum is removed after only three slide applications [8].

C. Mechanical Sectioning of Layers

Following tape stripping, molecular localization data for the residual skin can be obtained by repeatedly sectioning this tissue parallel to the skin surface and measuring the amount of permeant in each derived skin layer. Again, since radiolabeled molecules are invariably used, the analytical technique is digestion in tissue solubilizer or combustion, followed by scintillation counting. From the results, a concentration–depth profile is obtained for the penetrant in the investigated skin. For human in vivo studies, it is necessary to use a healthy skin area that has been designated for surgery. The permeant is applied to the skin for an application period that terminates with the onset of the operation. Just before surgery, the excess substance is removed and the stratum corneum detached. After surgery, the tissue is frozen and punch biopsies are taken, which are subsequently sectioned in a cryotome. During slicing, it is important to section through the deepest region of tissue first and subsequently work upward, harvesting the uppermost skin layer at the end of the procedure. This reduces the likelihood of contamination, as penetrant concentrations generally decrease with tissue depth. The blade should also be cleaned between cuts. Penetrant concentrations in the plasma or urine can also be measured periodically.

For over two decades, the combined use of tape stripping and mechanical sectioning has been employed in order to study the in vivo penetration kinetics of several different agents through human skin. These have included hydrocortisone [9], dithranol [10], triacyl dithranol [10], econazole [11], and 8-methoxypso-ralen [4]. One limitation of parallel sectioning is that excision will obviously not follow the undulations of the dermoepidermal interface. Fortunately, it is possible to induce separation at this junction by immersing the skin for 1 min in water maintained at 60°C and then prying the layers apart with a sharp spatula [12].

One typical sectioning and stripping study was conducted in order to evaluate the absorption and distribution of the antifungal agent econazole in human skin [11]. An ointment containing 1% tritated econazole was applied in vivo to a 28-cm² area of skin for 30, 100, 300, or 1000 min. At the end of the exposure period, the ointment was washed off. The stratum corneum was detached by repeated stripping with adhesive tape and the residual tissue was sectioned in a freezing microtome. The epidermis was sliced into 16 sections 10 μm wide and the dermis was sliced into 40-μm sections. Following scintillation counting, a gradient profile for econazole was obtained. It was found that about 90% of the penetrant remained on the skin surface, 7–9% formed a reservoir in the cornified layer (constant, regardless of application period), and no more than 2% penetrated into the epidermis and dermis. Effective inhibitory econazole concentrations, as determined from literature values, were detected down to a depth of approximately 700 μm. This corresponds to the middle region of the dermis.

In another in vivo sectioning study, the penetration kinetics of the antipsoriatic agent dithranol and its triacetate derivative were measured in human skin [11]. Each radiolabeled compound, formulated in one of four vehicles, was rubbed for 50 s to a 28-cm² skin area. Following an application period of 10, 30, 100, 300, or 1000 min, the treated stratum corneum was detached by 15 strips of adhesive tape. Punch biopsies were then taken, and these were sliced by cryotome to produce tissue sections 15 μm thick. Following scintillation counting, the researchers were able to show that for both compounds, maximal concentrations were gained after about 300 min and that penetration was optimized with lipophilic vehicles such as petroleum jelly. However, dithranol was found to accumulate mainly in the epidermis, while triacetyl dithranol accumulated in the dermis. Since the two penetrants exhibited fundamentally different and distinct localization behavior, the authors were able to deduce that triacetyl dithranol was not transformed in the skin to dithranol in any significant quantity.

If mechanical separation and stripping are conducted in conjunction with qualitative autoradiography, information about penetrant concentrations in the pilosebaceous structures can be obtained. One group [13] studied the in vivo localization of ¹⁴C-labeled flutrimazole in minipig skin by conducting a parallel sectioning study. The gradient profile for flutrimazole showed that flutrimazole was retained mainly in the viable epidermis. These workers also sectioned the skin vertically and performed autoradiographic experiments. Qualitative observations of the images indicated that the autoradiography signals were much more intense at the exterior of the corneocytes than in the interior. This suggests that flutrimazole penetrates intercellularily rather than transcellularily. Qualitative autoradiography has also been used to investigate ibuprofen localization in guinea-pig skin [14]. Visual analysis of the autoradiographs suggested that ibuprofen was concentrated in the hair follicles and that these shunts were the main penetration route. Qualitative autoradiography enables visualization but not quantification of the penetrant in the tissue. Consequently, this methodology provides only a general indication of penetrant deposition and transport in the integument.

A more precise way of evaluating follicular penetrant concentrations is to remove the follicles from the skin mechanically. This approach has recently been employed in Syrian golden hamster ear skin [15]. The hair follicles and sweat glands of this skin are anatomically and physiologically similar to those found in human skin. Following the in vivo topical application of a formulation composed of tritiated cimetidine, the isolated skin specimens were sectioned by gentle peeling or scraping into three layers—epidermis, ventral dermis, and dorsal dermis. The workers found that the sebaceous glands in the ventral dermis could easily be removed by gentle scraping with a blunt scalpel across the bottom surface of this layer. The procedure was evaluated by light microscopy. The process was assessed as complete when the skin areas previously containing the glands appeared void. The scraped pilosebaceous units and the skin layers were assayed separately for cimetidine by scintillitation counting. It may be possible to duplicate this technique with human skin.

D. Quantitative Skin Autoradiography by Computerized Image Analysis

Until several years ago, the use of autoradiography in transdermal research was limited to qualitative analysis as a supplement to mechanical sectioning, as discussed above. Recently, Touitou and coworkers developed the use of microcomputer-based image analysis of autoradiograms to quantify penetrant localization within the skin [16–19]. With this technique, it is possible to visualize and measure penetrant concentrations in the various skin strata and appendages without resorting to stripping or horizontal slicing. Quantitative autoradiography initially requires calibration [17]. This is achieved by mixing serially increasing amounts of tritiated penetrant with skin homogenate. These mixtures are frozen at $-130°C$ and vertically sectioned by freezing microtome to produce slices 6 µm thick. These sections are fixed on glass slides and then placed in film cassettes and covered by tritium-sensitive autoradiography film. Following 4–7 weeks exposure at room temperature, the autoradiograms are developed. The calibration curve can then be built by plotting the optical densities against the penetrant concentration per gram of tissue. The technique exhibits good reproducibility between the optical density readings of standard sections containing the same penetrant concentration.

Although no human in vivo studies have yet been carried out, quantitative autoradiography has been used to quantify the in vivo uptake and localization of tetrahydrocannabinol and oleic acid in hairless rat skin [17]. To this end, a formulation containing one of the tritiated penetrants was applied to the dorsal region of each animal and covered with a Hill Top Chamber. The application period lasted between 2 to 24 h. The rat was then sacrificed and excess formulation was wiped off the skin surface. The relevant skin area was then excised with scissors and scalpel. Each skin sample was frozen, vertically sectioned, and processed for autoradiography in the same manner as the standards. In addition, the skin sections were stained with hematoxylin and eosin dye. The autoradiograms were analyzed using an image analysis system comprising a video camera, high-resolution light microscope, and IBM-compatible computer. A software system measured the gray level of each autoradiogram and converted it into a penetrant concentration value according to the calibration curve. The derived digitized images were projected on to a TV screen. The software displayed the images in pseudocolor mode. This produced a more visually meaningful gradation of penetrant localization. The equipment was then used to superimpose the autoradiography image with the histological image. The software was then used to quantify penetrant concentrations in selected regions of skin.

The researchers used the new technique to investigate the effect of application time and vehicle on the localization of both oleic acid and tetrahydrocannabinol in rat skin. It was found that after a 2-h application period in a Transcutol-containing vehicle, the localization patterns of both tritiated penetrants were similar, with only low concentrations accumulated in the skin. For both molecules, the epidermis and the appendages contained twice the concentrations detected in the dermis. After 24 h, each penetrant exhibited a distinct localization pattern. For tetrahydrocannabinol, a concentration gradient was observed from the epidermis through the appendages to the dermis. In contrast, the concentration of oleic acid localized in the epidermis was not significantly different from that in the appendages. This indicates that at longer application times, the molecule's preference for a specific absorption pathway becomes enhanced. When tetrahydrocannabinol was incorporated into formulations containing polythylene glycol 400 or propylene glycol and ethanol, entirely different distribution patterns were seen. Polythylene glycol 400 and propylene glycol were less effective than Transcutol in delivering the penetrant to the epidermis, appendages, and papillar dermis.

Computerized image analysis has also been used to study in vivo caffeine delivery to rats from liposomal formulations [16]. Formulations containing [^{3}H]caffeine-entrapped liposomes were applied to the dorsal skin of hairless rats. After a 24-h exposure period, the animals were killed and the treated skin area excised. Quantitative autoradiography demonstrated that the greatest levels of caffeine were in the epidermis (280 µg/g) and the lowest levels were in the retinal dermis (50 µg/g). A substantial concentration of penetrant was also detected in the appendages (110 µg/g). The results of these experiments do not give any indication as to the state of the caffeine measured, whether it was free or entrapped within liposomes. This study did show, however, that liposomes can be employed in order to obtain a large penetrant reservoir in the skin.

E. Microdialysis

Originally devised as a research tool for experimental neurophysiology, microdialysis is a bioanalytical sampling technique that can now be used to assess percutaneous absorption in the skin of human volunteers [20–22]. The technique is used to quantify the interstitial concentration of penetrants in the dermis. Single dialysis fibers are glued to nylon tubing using cyanoacrylate adhesive and inserted into human skin in vivo with a fine needle. The fibers are connected to a peristaltic pump and perfused with isotonic saline. Each dialysis fiber requires initial in vivo calibration, and this is achieved by perfusing it with a range of different penetrant concentrations. Regression analysis of the resultant calibration curve yields the in vivo relative recovery of penetrant. Microdialysis has been used to assess dermal ethanol [20], histamine [21], and glucose [22] concentrations after topical administration. C-mode ultrasound scanning has been utilized to validate the intradermal positioning of the probe tip. Although probe implantation does induce some pain, no discomfort was reported when local anesthetic was preappied to the treatment site [21]. Furthermore, laser Doppler velocimetry studies have shown that traumatic hyperemia becomes negligible 1 h after implantation. The main drawback of microdialysis seems to be that the sensitivity of detection may limit the procedure to a rather limited number of molecules.

F. Spectroscopic Approaches

Over recent years, fourier transform infrared (FTIR) skin analyzers equipped with attenuated total reflectance (ATR) accessories have become commercially available for the detection of cosmetics and other penetrants in the skin [23]. With ATR-FTIR spectroscopy, the measuring distance is 0.8–2 µm which represents some 10–20% of the depth of the horny layer. Alternatively, analysis can be performed on tape-stripped skin. This results in a successive series of ATR spectra, each representing absorption at incrementally increasing depths within the cornified layer [24]. The technique requires initial control experiments to be performed in order to establish that the detected intensity signals are indeed proportional to penetrant concentration. ATR-FTIR spectroscopy has been employed to quantify the in vivo uptake of oleic acid [24], 4-cyanophenol [7], and various other penetrants [25] into human stratum corneum. Importantly, ATR analysis of 4-cyanophenol permeation through human skin has been validated by direct comparison with tape stripping [7].

Remittance spectroscopy involves measuring backscattered UV light from the interior of the horny layer. Since the presence of UV-absorbing molecules in the tissue decreases backscattering, it is thus possible to quantify penetrants within the skin. The measuring distance in the skin is determined by the wavelength-dependent optical penetration depth. Consequently, the technique can be used to derive a penetrant concentration–depth profile at the treated skin site. This approach has been applied in order to follow the penetration of the sunscreening agent, Uvinul T150, through human forearm skin in vivo [26]. Remittance spectroscopy is clearly a highly sensitive and reproducible technique which can be employed to detect small quantities of UV-absorbing penetrants through the depth of the entire stratum corneum.

Like remittance spectroscopy, fluorescence spectroscopy is a technique that involves the detection of remitted light, with the measuring depth dependent upon the beam wavelength. However, in fluorescence spectroscopy, only one emission wavelength is detected. Importantly, this approach is not limited to the quantitation of self-fluorescent molecules. It is possible to label the skin with a fluorescent compound such as dansyl-chloride and then measure the reduction in fluorescence produced by a nonfluorescent, UV-absorbing penetrant. Such a method was employed to quantify the uptake of the sunscreen agent, Eusolex 8020, through the skin of human volunteers [27]. The researchers applied the sunscreen agent, in a 2% isopropyl myristate vehicle, to the volunteers' skin. Control formulations were composed of pure vehicle. A fluorescence spectrometer was employed to direct a beam of monochromatic UV light to the skin surface and analyze the remitted fluorescence (wavelength 280 nm). The procedure was then repeated with the incident UV beam set at different wavelengths (300 to 400 nm). From the resulting graph of fluorescence intensity versus incident beam wavelength, the amount of penetrant at any depth could be determined from the intensity difference between the Eusolex 8020 and placebo curves.

In photothermal spectroscopy, the investigated skin site is irradiated with a modulated monochromatic UV beam. The energy interacts with any UV-absorbing molecule in the skin to raise the oscillating surface temperature. This produces pressure fluctuations at the skin surface which can be analyzed with an acoustic sensor. This approach has been used to quantify the in vivo penetration of sunscreening agents through human skin [27,28]. The measuring depth, which is inversely proportional to the modulation frequency, varies between 3 and 7 μm. One way of increasing it is to measure variations in the temperature-dependent air refraction index caused by the photothermal effect. One group developed a "mirage detector," which analyzed fluctuations in the air refractive index from the changed position of a helium-neon laser beam [27]. This refinement extended the measuring distance to a depth of 30 μm, which is well into the viable epidermis.

III. IN VITRO QUANTITATION TECHNIQUES

A. Skin Extraction Techniques

The simplest way of measuring penetrant concentration in whole skin in vitro involves removal of the excess formulation, followed by extraction in an appropriate solvent and the use of a suitable analytical method, such as HPLC or scintillation counting, to quantify the penetrant [16,29,30]. Touitou and coworkers investigated the retention of caffeine in full-thickness hairless mouse skin [16]. The abdominal skin samples were mounted in either Valia-Chien or Franz diffusion cells in which the donor compartment was filled with a specific caffeine-containing solution. The system was maintained at 37°C and diffusion was allowed to proceed for 24 h. At the end of the diffusion period, the skin was removed from the chamber, washed, and homogenized. Caffeine retained in the skin was extracted in distilled water and quantified by HPLC.

Although these whole-skin measurements can be easily and rapidly carried out, their limitation is that they offer no molecular localization data. It is possible to improve this aspect by conducting separate extraction experiments on the isolated epidermis and dermis. Such a methodology has been described in two reports documenting retinoid uptake into human skin in vitro [31,32]. Human cadaver skin samples derived from the leg or abdominal sites were inserted into diffusion cells. A retinoid-containing formulation was applied on to the skin surface. Following a 24-h application period, the skin was removed from the diffusion cell and washed with isopropyl alcohol. The epidermis was detached from the dermis by gentle teasing with needle-tip tweezers. The isolated dermis was thoroughly minced with scissors. Penetrant extraction was performed by immersing each tissue in an acidified solution of methylene chloride and isopropyl alcohol. Following mixing and centrifugation, the organic phase was recovered and evaporated under dry nitrogen. Samples were reconstituted in isopropyl alcohol for quantitative analysis by HPLC.

B. Tape Stripping and Mechanical Sectioning

Although stripping and sectioning should ideally be conducted following in vivo application, such studies are expensive to perform and require ethical approval, which can be difficult and time-consuming to obtain. In vitro experimentation is therefore an alternative option that is particularily suitable for preliminary investigations into penetrant skin distribution. In vitro tape stripping and slicing studies involve the administration of the permeant to samples of mastectomized or cadaver skin, usually mounted in Franz-type diffusion cells maintained at a physiological temperature. At the end of the application period, the excess formulation is removed, the horny layer is stripped off, and the remaining skin is excised, frozen, and sliced as described for the in vivo studies. Penetration through the skin barrier can also be quantified by repeated sampling of the receiver phase. This general approach has been employed to quantitate the influx of hydrocortisone [9], econazole [11], fatty acids [33], and piroxicam [34] into human skin. These studies have generally demonstrated that there is a good correlation between in vivo and in vitro data for diffusion periods not exceeding 4 h. For longer diffusion periods, the in vitro models show elevated dermal penetrant concentrations; this effect is due to the absence of vasculature in the in vitro system.

One in vitro technique for determining the role of the follicles is to slice the skin horizontally at the level of the hair follicles and then reposition the upper section on top of the lower section. It is extremely unlikely that the follicles will be realigned in such split skin. This approach has been documented in an investigation into parathion penetration through porcine skin [35]. Following diffusion, penetrant concentrations in each of the the two sections were determined from scintillation counting. Parathion levels in the lower section were significantly diminished in the presectioned specimen compared to control. This indicates that parathion penetration was largely mediated via the hair follicles.

C. Quantitative Skin Autoradiography by Computerized Image Analysis

Quantitative autoradiography has been used to analyze [³H] tetrahydrocannabinol localization in human skin [18]. The skin samples, originating from human breast tissue, were mounted in diffusion cells for either 2 or 6 h. The test formulation was composed of oleic acid, propylene glycol, ethanol, and water. The technique demonstrated that the highest penetrant concentration was in the epidermis, followed by the papillary dermis, followed by the retinal dermis. After 6 h of exposure, much more cannabinoid had penetrated the skin, but the regional localization pattern was relatively unchanged. Tetrahydrocannabinol concentration in the epidermis was approximately 20 pmol/g at 2 h and 44 pmol/g at 6 h. In further experiments, formulation factors were investigated. It was found that the use of vehicles that contained water resulted in enhanced cannabinoid retention in the skin.

In another in vitro study, quantitative autoradiography was used to quantify the localization of tritiated papaverine hydrochloride following its iontophoresis into rabbit pinnae [19]. A current density of 0.5 mA cm^{-2} was applied at 20-kHz frequency and an on:off ratio of 1:2. Quantitative autoradiography was performed after both 10 and 30 min of iontophoresis. The duration of iontophoresis was not found to significantly affect penetrant concentrations in the epidermis and dermis. However, penetration was greatly facilitated by the use of an aqueous vehicle as opposed to one containing ethanol.

D. The Use of Reconstituted Skin

As a readily available alternative to human skin, Testskin LSE (living skin equivalent) has recently been developed as a model membrane for in vitro permeation studies [36,37]. The membrane is specifically designed to mimic the properties of human skin. It consists of an organotypic coculture of human dermal fibroblasts embedded in a collagen-containing matrix, a stratified epidermis composed of human epidermal keratinocytes, and a well-differentiated stratum corneum. Penetrant localization in Testskin LSE can be evaluated by using some of the in vitro quantitation approaches discussed above—i.e. mechanical sectioning, quantitative autoradiography, and the spectroscopic techniques. It should be mentioned that this membrane cannot be harvested by tape stripping because of the low adherence of its cornified layer. In the near future, improvements in the quality of artificial skin are expected, so that the morphology of these membranes will be much closer to that of human skin.

IV. CONCLUSIONS

There are currently a number of methods for quantifying cosmeceutics localized within the skin or various skin strata. The available methods can be conveniently divided into those relatively noninvasive approaches that can be performed on healthy human volunteers and those more invasive techniques that can be conducted only on skin samples derived from tumors, cadavers, animals, or skin cultures.

Of the methods which are employable on healthy volunteers, disappearance measurements from the skin are easily and rapidly facilitated, although they do not offer any penetrant distribution data. Tape stripping can be used to obtain molecular localization data for the stratum corneum. The procedure is relatively easy, accurate, rapid, and relatively noninvasive in humans. Tape stripping also exhibits the key advantage that short-term exposure measurements can be used to predict the total systemic absorption of molecules. Although skin stripping has been used in the safety evaluation of cosmeceutics, it has not yet obtained widespread acceptance [38]. It has not yet been determined whether the tape-stripping protocols vary between different laboratories and different cosmeceutic formulations and whether this significantly affects safety assessment. Furthermore, since stripping does not remove the follicles, the technique is unsuitable for quantitating the systemic absorption of molecules that undergo transfollicular penetration. The spectroscopic methodologies represent very powerful noninvasive approaches for tracking penetrant transport in the stratum corneum. They offer the advantage that they are relatively sensitive and provide real time penetration data. However, these techniques are limited to permeants exhibiting suitable light-absorption properties as well as by the restricted penetration of radiation into skin tissue. In addition to quantitation, ATR-FTIR spectroscopy can be used to study the structural and conformational status of stratum corneum constituents. Microdialysis can be used to determine dermal penetrant concentrations, although the system is currently limited to a small number of molecules.

Of the more invasive techniques, mechanical sectioning can be used to determine the molecular concentration versus depth plot from the viable epidermis to the subcutis or beyond. Since this approach will not provide information about retention and transport in the appendages, it is helpful to apply it together with either a follicle-removal method or qualitative autoradiography. Quantitative autoradiography is a new, highly refined technique that permits visualization and quantification of the permeant throughout a transverse cross section of the tissue. This technique is advantageous in that quantification can be carried out down to the depth of the lower dermis as well as within the pilosebaceous units. The system requires adequate instrumentation and it is limited to the use of radiolabeled molecules. Although the autoradiography exposure period is 1–2 months long when [^{3}H]molecules are used, this process can be significantly shortened with the use of [^{14}C]molecules.

Further research in this field requires a greater effort to be placed on validating the different quantitation techniques available.

REFERENCES

1. Djerassi D. The role of vitamins in aged skin. J Appl Cosmetol 1993; 11:29.
2. Panchagnula R, Ritschell WA. Development and evaluation of an intracutaneous depot formulation of corticosteroids using transcutol as a cosolvent: in-vitro, ex-vivo and in-vivo rat studies. J Pharm Pharmacol 1991; 43:609–614.
3. Schaefer H, Redelmeier TE. In: Schaefer H, Redelmeier TE, eds. Skin Barrier: Principles of Percutaneous Absorption Basel: Karger, 1996:124.
4. Schaefer H, Stuttgen G, Zesch A, et al. Quantitative determination of percutaneous absorption of radiolabelled drugs in vitro and in vivo by human skin. Curr Probl Dermatol 1978; 7:80–94.
5. Rougier A, Lotte C, Maibach HI. In vivo percutaneous penetration of some organic compounds related to anatomic site in humans: predictive assessment by the stripping method. J Pharm Sci 1987; 76:451–454.
6. Lotte C. Wester RC. Rougier A, Maibach HI. Racial differences in the in vivo percutaneous absorption of some organic compounds: a comparison between black, Caucasian and Asian subjects. Arch Dermatol Res 1993; 284:456–459.
7. Higo N, Naik A, Bommannan DB, et al. Validation of reflectance spectroscopy as a quantitative method to measure percutaneous absorption in vivo. Pharm Res 1993; 10:1500–1506.
8. Montenegro L, Ademola JI, Bonina FP, Maibach HI. Effect of application time of

betamethasone-17-valerate 0.1% cream on skin blanching and stratum corneum drug concentration. Int J Pharm 1996; 140:51–60.

9. Zesch A, Schaefer H. Penetration of radioactive hydrocortisone in human skin from various ointment bases: II. In vivo experiments. Arch Dermatol Res 1975; 252:245–256.

10. Kammerau B, Zesch A, Schaefer H. Absolute concentrations of dithranol and triacetyl-dithranol in the skin layers after local treatment: in vivo investigations with four different types of pharmaceutical vehicles, J Invest Dermatol 1975; 64:145–149.

11. Schaefer H, Stuttgen G. Absolute concentrations of an antimycotic agent, econazole, in the human skin after local application. Drug Res 1976; 26:432–435.

12. Tsai JC, Weiner N, Flynn GL, Ferry JJ. Drug and vehicle deposition from topical applications: localization of minoxidil within skin strata of the hairless mouse. Skin Pharmacol 1994; 7:262–269.

13. Conte L, Ramis R, Mis R, et al. Percutaneous absorption and skin distribution of [^{14}C] flutrimazole in mini-pigs. Drug Res 1992; 42:847–853.

14. Giese U. Absorption and distribution of ibuprofen from a cream formulation after dermal administration to guinea pigs. Drug Res. 1990; 40:84–88.

15. Lieb LM, Flynn G, Weiner N. Follicular (pilosebaceous unit) deposition and pharmacological behaviour of cimetidine as a function of formulation. Pharm Res 1994; 11:1419–1423.

16. Touitou E, Levi-Schaffer F, Dayan N, et al. Modulation of caffeine skin delivery by carrier design: liposomes versus permeation enhancers. Int J Pharm 1994; 103:131–136.

17. Fabin B, Touitou E. Localization of lipophilic molecules penetrating rat skin in vivo by quantitative autoradiography. Int J Pharm 1991; 74:59–65.

18. Touitou E, Fabin B. A new method for determination of drugs in the skin region and appendages, image computerized quantitative autoradiography. 6th International Conference on Pharmaceutical Technology, 1992:296–302.

19. Touitou E, Alkabes M, Fabin B, et al. Quantitative skin autoradiography: an efficient tool in measuring drug localized in skin layers, hair follicles and glands. Proceedings of the International Symposium on Controlled Release Bioactive Materials. 1994:431–432.

20. Anderson C, Andersson T, Molander M. Ethanol absorption across human skin measured by in vivo microdialysis technique. Acta Derm Venereol 1991; 71:389–393.

21. Petersen LJ, Stahlskov P, Bindslev-Jensen C, Soendergaard J. Histamine release in immediate-type hypersensitivity reactions in intact human skin as measured by microdialysis. Allergy 1992; 47:635–637.

22. Petersen LJ, Kristensen JK, Bulow J. Microdialysis of the interstitial water space in human skin in vivo: quantitative measurement of cutaneous glucose concentrations. J Invest Dermatol 1992; 99:357–360.

23. Klimich HM, Chandra G. Use of Fourier transform infrared spectroscopy with alternate total reflectance for in vivo quantification of polydimethylsiloxanes on human skin, J Soc Cosmet Chem 1986; 37:73–87.

24. Naik A, Pechtold LARM, Potts RO, Guy RH. Mechanism of oleic acid-induced skin penetration enhancement in vivo in humans. 1995; J Contr Rel 37:299–306.

25. Mak VHW, Potts RO, Guy RH. Percutaneous penetration enhancement in vivo measured by attenuated total reflectance infrared spectroscopy. Pharm Res 1990; 7:835–841.

26. Kolmel KF, Sennhenn B, Giese K. Investigation of skin by ultraviolet remittance spectroscopy, Br J Dermatol 1990; 122:209–216.

27. Sennhenn B, Giese K, Plamann K, et al. In vivo evaluation of the penetration of topically applied drugs into human skin by spectroscopic methods. Skin Pharmacol 1993; 6:152–160.

28. Kolmel K, Sennhen B, Giese K. Evaluation of drug penetration into the skin by photosacoustic measurement. J Soc Cosmet 1986; 37:375–385.

29. Touitou E, Abed L. Effect of propylene glycol, azone and n-decylmethylsulphoxide on skin permeation kinetics of 5-fluorouracil. Int J Pharm 1985; 27:89–98.

30. Lafforgue C, Carret L, Falson F, et al. Percutaneous absorption of a chlorhexidine digluconate solution. Int J Pharm 1997; 147:243–246.

31. Lehman PA, Slattery JT, Franz TJ. Percutaneous absorption of retinoids: influence of vehicle, light exposure, and dose. J Invest Dermatol 1988; 91:56–61.

32. Lehman PA, Malany AM. Evidence for percutaneous absorption of isotretinoin from the photo-isomerization of topical tretinoin. J Invest Dermatol 1989; 93:595–599.

33. Schalla W. Ph.D. thesis, Berlin, 1978.

34. Pellet MA, Roberts MS, Hadgraft J. Supersaturated solutions evaluated with an in vitro tape stripping technique. Int J Pharm 1997; 151:91–98.

35. Reifenrath WG, Hawkins GS, Kurtz MS. Percutaneous penetration and skin retention of topically applied compounds: an in vitro–in vivo study. J Pharm Sci 1991; 80:526–532.

36. Ernesti AM, Swiderek M, Gay R. Absorption and metabolism of topically applied testosterone in an organotypic skin culture. Skin Pharmacol 1992; 5:146–153.

37. Hager DF, Mancuso FA, Nazareno JP, et al. Evaluation of Testskin as a model membrane. Proceedings of the International Symposium on Controlled Release Bioactive Materials. 1992:487–488.

38. Schaefer H, Redelmeier TE, eds. Skin Barrier: Principles of Percutaneous Absorption Basel: Karger, 1996:246.

8

Multiple Emulsions

Shlomo Magdassi and Nissim Garti
The Hebrew University of Jerusalem, Jerusalem, Israel

I. INTRODUCTION

Multiple emulsions are termed *emulsions of emulsions*, the droplets of the dispersed phase themselves containing even smaller dispersed droplets. As presented schematically in Fig. 1, a water-in-oil-in-water (W/O/W) emulsion consists of small water droplets separated by an oil membrane from neighboring droplets and from a continuous aqueous phase. This structure may lead to formation of a unique delivery system, in which water-soluble components are entrapped inside the inner droplets and can be released slowly or instantly by a suitable chemical trigger.

More than seventy years ago, William Seifriz reported on the formation of multiple emulsions during an inversion of an O/W emulsion into a W/O emulsion [1]. The inversion of an O/W emulsion using olive oil was induced by the addition of barium hydroxide and, as described:

> Both types of emulsion ramify among each other, with the *oil* globules of the one containing dispersed water, and the *water* droplets of the other containing dispersed oil; or the system may consist of a conglomerate of large and small oil droplets scattered in the aqueous matrix, with most of the *oil* droplets being themselves smaller water-in-oil emulsions....After 3 hours this double emulsion still exists as such;

Seifriz also reported the spontaneous formation of more complex emulsions: "Such type of double emulsions, where one emulsion is within the other, may be called bimultiple systems. The term is convenient since these multiple systems may still be more complex, becoming trimultiple, quatermultiple and even quinquemultiple."

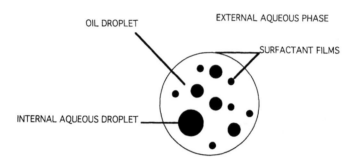

Figure 1   Schematic illustration of a multiple emulsion droplet.

At present, the common terminology refers to two types of multiple emulsions (often called *double emulsions*): W/O/W emulsions, in which the external continuous phase is aqueous, or O/W/O emulsions, in which the external continuous phase is an oil.

Since this early publication, numerous reports on the formation and utilization of multiple emulsions have been published. Basically, there are two methods for the preparation of multiple emulsions: (1) spontaneous formation, which is obtained by approaching the inversion of a simple emulsion [2], and (2) the two-step emulsification process [3,4], which is based on formation of a primary O/W or W/O emulsion followed by emulsification of the primary emulsion in a phase that is incompatible with the external phase of the primary emulsion.

As will be demonstrated in the following sections, such emulsions, as well as the reverse system O/W/O emulsion, can easily be prepared without any sophisticated instrumentation. Although applications and formulations of multiple emulsions have been reported in numerous patents and reports, including approximately fifty publications related to cosmetics, to the best of our knowledge there is no commercial cosmetic product to date based on such formulation. This is probably because of the inherent instability of the system, which, in addition to the common instability of macroemulsions, also results from their unique structure. These factors include, for example, migration of the active ingredient from the internal droplets, breakage of the emulsion due to changes in osmotic pressure, changes in viscosity due to penetration of the external water into the internal droplets, etc. A detailed description of transport phenomena and stabilization was recently published by Garti et al [52–54].

It appears that during the last few years most researchers focused on the possibilities of improving the stability of multiple emulsions, mainly by the use of polymeric surfactants or combinations of monomeric and polymeric surfactants in the presence of macromolecules in the aqueous phases [55–58]. For example, Yazan et al. [59] reported on the addition of PVP into multiple emulsions containing vitamin A, which were believed to give elasticity and moisture to the skin. In another study, various polymeric surfactants were used to stabilize multiple emulsions containing urea and glycolic acid in order to obtain a formulation leading to prolonged moisturizing activity [60]. The use of multiple emulsions in cosmetics for enhancing moisturizing effect in hair and skin treatment was suggested in several patents assigned to Helene Curtis, Inc [61,62]. A rinse-off W/O/W emulsion that can deposit water-soluble active compounds onto the hair or skin was prepared. This compound can be released during drying or rubbing of the hair or skin. Thus, such a composition may contain two separate active components dissolved in the external and internal aqueous phases. This type of formulation has the advantage of combining two incompatible components (e.g., anionic surfactant and a quaternary ammonium conditioner) and preventing wastage of one of the active components during rinsing.

The purpose of this chapter is to describe the various possibilities of preparing multiple emulsions and to give the reader the basic tools with which to form and evaluate the resulting multiple emulsions.

We focus mainly on the two-step preparation method, since it has a great potential for practical applications.

II. PREPARATION OF MULTIPLE EMULSIONS

A. Inversion Method

Early reports describe the formation of multiple emulsions as an incidental event during the preparation of simple W/O or O/W emulsions [1,39]. The multiple emulsions were usually obtained while the simple emulsions were being prepared by the inversion method [40], involving unstable O/W or W/O emulsions, in conditions favoring their inversion to W/O or O/W emulsions, respectively. For example, an O/W emulsion can be prepared by the inversion method as follows: the aqueous phase is added stepwise to an oil phase, which contains a hydrophilic emulsifier and is unsuitable for stabilization of W/O emulsions. At first, a W/O emulsion may be obtained, but with further addition of the aqueous phase, the viscosity of the system increases significantly until a sudden reduction in the viscosity at a certain phase fraction of the dispersed phase is observed. At this point, the system has inverted from a W/O emulsion into an O/W emulsion. The inversion process may lead to metastable conditions; these favor formation of multiple emulsions, which are usually unstable.

This process was explained by migration of the emulsifier between the phases [39,41]. At first, a simple W/O emulsion is formed, but this is unstable, since the emulsifier is not suitable for formation of such emulsions [e.g., high hydrophil-lipophil balance (HLB]. Further addition of water while shearing would lead to separation of large droplets from the bulk of the W/O emulsion and migration to the newly added aqueous phase, as described in Fig. 2. Later on, the hydrophilic emulsifier will migrate to the external aqueous phase, the

W/O	Inversion	O/W
Emulsion	Point	Emulsion

Figure 2 Schematic presentation of formation of multiple emulsion via inversion of W/O emulsion to O/W emulsion.

internal water droplets will no longer exist, and a simple O/W emulsion will be obtained.

It is possible to obtain multiple emulsions by inversion of either O/W to W/O emulsions or W/O to O/W emulsions by using an emulsifier with an HLB value that does not particularly favor a certain type of emulsion, such as sorbitan monopalmitate [42].

Due to changes in emulsifier solubility, inversion of an emulsion can also be achieved by increasing the temperature, mainly with ethoxylated emulsifiers. It was found that while heating an O/W emulsion, the number of multiple droplets observed increased as the phase inversion temperature (PIT) was approached [43].

The inversion can also be approached by the formation of emulsions that have a high fraction of the dispersed phase [44]. Theoretically, the maximal internal volume phase fraction is 0.74 if the droplets are rigid and monodispersed. Since this is usually not the case for simple emulsions, it might be possible to form emulsions with higher internal-phase fractions. In these cases, the emulsion droplets are closely packed and the external phase may be entrapped within the droplets, thus forming a multiple emulsion. Matsumoto [45] described a systematic study aimed at forming multiple emulsions by the inversion method. He found that it is possible to formulate procedures based on proper surfactants that would lead to the formation of multiple emulsions.

The overall conclusion is that it is indeed possible to prepare multiple emulsions by the inversion method, but the multiple emulsion should be viewed as an unstable transition state between O/W and W/O emulsions. Recently, Gohla et al. [63] proposed a new one-step process to produce stable W/O/W emulsions. This process involves a partial solubilization-inversion technology (PPSIT) based on controlled salting-out of the hydrophilic emulsifier during droplet formation. However, it should be emphasized that the practical use of the inversion methods for cosmetic products is limited owing to the fact that, by this method, it is impossible to entrap active materials in the internal phase only.

B. Two-Step Method

The two-step process for the preparation of a W/O/W emulsion is schematically presented in Fig. 3. This process is very simple to conduct and offers a great variability of multiple emulsions.

The two-step process reflects the idea that the multiple emulsion can be regarded as two separate emulsions: a W/O emulsion and an O/W emulsion.

In a W/O/W emulsion, each oil droplet contains many tiny water droplets. Microscopically, each oil droplet can be considered as a continuous oil phase; therefore, such a droplet can be viewed as a simple W/O emulsion. At the external boundaries of the droplet, the oil is in contact with the external aqueous phase of the multiple emulsion. Therefore, while ignoring the internal content of the oil droplets, we may view this emulsion as a simple O/W emulsion. Having this in mind, we can apply our knowledge of the formation and stabilization of simple

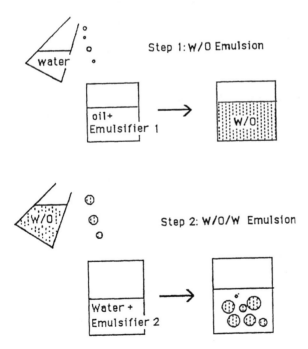

Figure 3 Schematic presentation of formation of multiple emulsions by the two-step process.

O/W and W/O emulsions to the formation of multiple emulsions provided that we divide the preparation into two steps. The first step is basically aimed at forming a simple W/O emulsion (the primary emulsion). In the second step, the primary emulsion is used as an *oil phase* in a simple O/W emulsion. Therefore, the preparation method is restricted to compositions suitable for such preparation—e.g., the internal-phase volume fraction in the primary emulsion should not exceed 74% v/v and the emulsifier for the primary emulsion must be suitable for W/O emulsion (low HLB value, high oil solubility, etc.). The emulsifier for the second emulsification must be suitable for O/W emulsions (high HLB value, high water solubility, etc.).

It should be noted that by using the two-step procedure, one can *set* different requirements for each emulsion, such as droplet size distribution and viscosity. For example, the primary emulsion may be prepared by a high-shear-rate homogenizer, thus obtaining an internal droplet size below 1 μm [46,47], while the second emulsification step may be prepared by a low-shear homogenizer, forming much larger droplets.

Obviously, such isolation of the two steps is based on oversimplification of the whole system: the emulsifiers can migrate from one interface to the other, and interaction between emulsifiers from step 1 and step 2 may occur. Therefore, the two-step method could be very useful, but one might expect that the development of a stable multiple emulsion begins only after a suitable emulsion is formed at each step; still, the final multiple emulsion can be unstable.

The primary emulsion can be prepared with various oil phases and surfactants and various internal phase-volume ratios, and it may contain water-soluble additives (or oil-soluble additives in O/W/O) such as drugs, fragrances, etc. Similarly, the second emulsification step may involve other surfactants, viscosity modifiers at the external phase, etc.

The simplicity of W/O/W emulsion preparation can be demonstrated in the following typical example: The primary W/O emulsion is prepared by dropwise addition of the aqueous phase into an oil phase, which contains the surfactant, followed by 20 min of homogenization until the droplet size is about 1–3 μm (Fig. 2, step 1).

The composition of this typical emulsion is: 10% w/w Brij 92, 30% sodium chloride solution (1% w/w), and 60% w/w light mineral oil [5]. The primary W/O emulsion is then added dropwise to a water phase, which contains the second emulsifier at a constant stirring rate (magnetic stirring) (Fig. 2, step 2). The addition is completed within 10 min. The composition of the resulting multiple emulsion is 5% w/w Span-80–Tween-80 (HLB = 11), 20% w/w primary W/O emulsion, and 75% w/w deionized water. A photomicrograph of the resulting multiple emulsion is presented in Fig. 4.

However, as shown by Florence et al. [3] in addition to the emulsion type presented in Fig. 3, two other types of multiple emulsions can be prepared by a

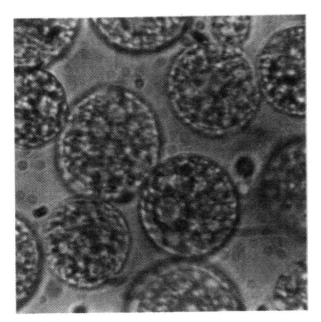

Figure 4 Microscopic observation of multiple emulsions prepared by nonionic surfactants, Span-80–Tween-80.

suitable selection of the surfactants: *type A*, in which each oil droplet contains only one water droplet, and *type B*, in which each oil droplet contains a few distinguishable water droplets. *Type C* contains a very large number of water droplets (Fig. 5) dispersed in the oil droplets.

In addition, as discussed later, unique emulsions—such as gel-in-oil-in-water [6] or water-in-solid-oil-in-water [7]—can be formed by a suitable selection of the various components.

Most of the applications of multiple emulsions are aimed at obtaining a system in which a substance is trapped in the internal droplets and should be released to the external phase during a given period of time. Such applications are required in cosmetic formulations, pharmaceutical and food applications, etc.

Therefore, from a practical point of view, three main subjects should be considered upon formulation of a multiple emulsion:

1. Yield of preparation, which is a measure for the ratio between the amount of a substance entrapped in the internal droplets and external phase, or the percentage of multiple emulsion droplets immediately after preparation.

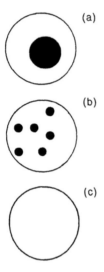

Figure 5   Presentation of the three common types of multiple emulsion droplets.

2. Stability of the multiple emulsion, a parameter that indicates the break-
    age and coalescence of the emulsion droplets, which eventually lead
    to phase separation or formation of simple O/W or W/O emulsions.
3. Characteristics of release of the entrapped substance from the internal
    phase to the external phase.

This chapter concentrates on the yield of preparation and the parameters
that should be considered while formulating a multiple emulsion.

III. EVALUATION OF YIELD OF PREPARATION

All the evaluation methods are based on quantitative measurements of the amount
of a marker released from the internal phase to the external phase. Such markers
could be electrolytes (NaCl, ephedrin·HCl [5,8]; water-soluble additives such as
glucose; or fluorescent probes, which may be determined by potentiometric titration [8]; specific electrodes [10]; microanalysis [9]; electrical conductivity
[5,10,59]; or fluorescence measurements [48].
    The yield of preparation is defined in Eq. (1):

$$\frac{M_{in}}{M_{in} + M_{out}} \times 100 \tag{1}$$

Where M_{out} and M_{in} are the amounts of marker in the external and internal phases, respectively; therefore, the yield of preparation is the percentage of a marker released a short time after preparation.

The determination of the released marker could be followed directly in the multiple emulsion [5], only after dialysis against water [9], or after filtration of the multiple droplets [47]. We believe that in using dialysis tubes, the measurement indicates not only the yield of preparation but also the amount released during the dialysis period, which could be as long as 20 h. It should be emphasized that the markers may affect the formation of the multiple emulsions via an increase in the osmotic pressure, interactions with the surfactants, etc. Therefore it is advisable to conduct microscopic observations as a qualitative guide for the yield of preparation. Some investigators have used the microscopic observation as a quantitative measure of the yield of preparation by counting the empty and the filled emulsion droplets. However, since this method is time-consuming, it is not recommended for routine use in practical applications of multiple emulsions [11,12]. Matsumoto and coworkers made use of a relation between the viscosity and the dispersed phase volume fraction to estimate the stability of multiple emulsions, but when measurements were conducted immediately after preparation, they gave a measure for yield of preparation.

A direct method was recently suggested by Eads et al. [49] using unique direct measurements of the emulsions by nuclear magnetic resonance (NMR).

In conclusion, the first stage in preparing multiple emulsions should focus on developing a suitable method to measure the yield of preparation. This method, will obviously also be used to determine the stability and release characteristics of a specific multiple emulsion.

Since the addition of various probes may also affect the properties of the multiple emulsions, it is advisable that the measurement of yield of preparation be conducted by using the same required internal additive (such as drugs, hemoglobin, humectant) as the probe itself.

IV. FORMULATION

The formation of multiple emulsions via the two-step procedure involves the use of two incompatible phases and several surfactants that function as O/W and W/O emulsion stabilizers. In addition, various materials may be added to each phase in order to overcome problems such as high osmotic pressure gradient, rapid release of markers, leakage of the internal phase, etc. (The variability of multiple emulsions preparation methods and compositions is demonstrated in Table 1.) It is obvious that each parameter may be varied according to the final application—e.g., edible oil phase for food applications or pharmacologically

Table 1 Examples of Multiple Emulsion Compositions

Oil	Emulsifier I	Emulsifier II	ϕ_1	ϕ_2	Marker	Misc.	Additives	Ref.
Kerosene	Extract of sp. pps-II	Ethanol precipitate of sp. pps-II + Tw 20	50% vol.	50% vol.	NaCl	Microbial emulsifier		13
Peanut oil	Span-65	Propylen glycolalginate	40% vol.	40% vol.	Naltrexone		NaCl or sorbitol	14
Peanut oil	Span-80 + Tween-20	Span-80	80% w/w	20% w/w	Naltrexone	O/W/O emulsion	NaCl	14
Vaseline	AGOND polymer (ICI)	Synperonic PE 127 (ICI)	72% vol.	80% vol.	NaCl or MgSO$_4$		MgSO$_4$ or NaCl	15
Isopropyl myristate	Span-80	Triton X-165, Span-80 + Tween 80, Pluronic F-87, Pluronic F-88	50% vol.	50% vol.		γ-Irradiation: gel-oil-water, or gel-oil-gel	Acrylamide	16
Soybean	Monolaurin	Sodium caseinate	25% vol.	40% vol.	Methylene blue	Micellar solution in water		17
Isopropyl myristate	Acrylolyl modified pluronic (VL-44) + Span-80	Acryloyl modified pluronic VF-68)	50% vol.	50% vol.		Interfacial polymerization	Ammonium persulfate (initiator)	18
Light paraffin oil	Span-80 + Tween-80 (low HLB)	Span 80 + Tween 80 (high HLB)	30% w/w	20% w/w	NaCl			19
P-xylene	EMU-09 (nonyl phenol diethylene glycol ether)		75% vol.	40% vol.		Stabilization by liquid crystals		20

Oil								Ref
Olive oil	Span-80	Tween-80	75% vol.	40% vol.	Glucose		Glucose, sucrose, acetic acid, citric acid, ascorbic acid, NaCl	14
Hexane	Span-80 + soy lecithin	Tween-80			Glucose or NaCl	Liposome-like system		21
Light paraffin	Arlacel 83 or Arlacel A	Tween-80	50% vol.	50% vol.	Methotrexate			22
Isopropyl-myristate	Span-80 + BSA	Tween-80	50% vol.	50% vol.	NaCl	Interfacial complex	Sorbitol in external phase	23
Isopropyl-myristate	Poloxamer 331 + BSA	Poloxamer 403	50% vol.	50% vol.		Interfacial complex		
Light paraffin oil	Brij 92	Span-20 + Tween-80	30% w/w	20% w/w	NaCl		Glucose	25
Paraffin oil + paraffin wax	Hexaglycerol dioleate + decaglycerol decaoleate	Span-80 + Tween-80	30% w/w	20% w/w		Semisolid oil phase		7
Liquid paraffin	Arlacel C	CTAB	50% vol.	50% vol.	Salicyclic acid		NaOH, HCl	26
Sesame oil	Hydrogenated castor oil (H-CO-10)	Hydrogenated castor oils (H-CO-60)	30% vol.	20% vol.	5-fluorouracil (5-FU)			27
Sesame oil	Span 80	Pluronic F-68 + Span-80	30% vol.	20% vol.	^{3}H-5-FU	Sonification		28

safe surfactants for pharmaceutical applications. The role of each parameter is discussed further on.

A. Emulsifiers

Generally, the role of emulsifiers in the preparation and stabilization of multiple emulsions is similar to their role in simple O/W and W/O emulsions. Formulation of W/O/W emulsions requires the formulation of a suitable W/O emulsion in the first step (droplet size, stability) and, in the second step, the application of an emulsifier suitable for the preparation of O/W emulsions. By using such an approach, it is obvious that one should match the emulsifier to the oil/water interface. A hydrophobic emulsifier such as Span-80 is required for preparation of W/O emulsion and a hydrophilic emulsifier such as Tween-80 for preparation of O/W emulsions [4,23].

The optimal surfactant to emulsify a given oil can be determined by using the hydrophil-lypophil balance (HLB) approach. In general, in a W/O/W emulsion the optimal HLB value of the primary emulsifier is in the range of 3–7 and 8–16 for the secondary emulsifier.

As shown in Fig. 6, the optimal HLB of the second emulsifier in a multiple emulsion composed of water and light mineral oil is 11, which is approximately the required HLB for the oil [5].

Surfactant concentration also plays a significant role in the yield of a preparation: increasing the concentration of the secondary emulsifier leads to a decrease of yield at preparation at each HLB studied (Fig. 7), including at the required HLB of the oil [25].

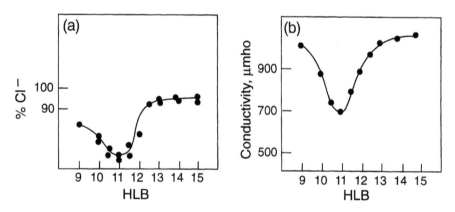

Figure 6 Determination of the optimal HLB of the second emulsifier by chloride titration (a) and electrical conductivity (b).

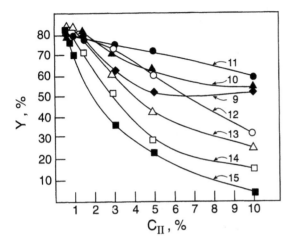

Figure 7 Dependence of the yield of preparation on the concentration of emulsifier at the external water-phase C_{II}, at various HLBs.

This effect is caused by a decrease in the size of the multiple droplets. Obviously, formation of small droplets would increase the amount of broken inner droplets of the primary W/O emulsion and, therefore, would lead to a decrease in the yield of preparation. However, at very low concentrations of the secondary emulsifier, the yield of preparation is very high. This probably results from the significant increase of the multiple droplet size until approaching the very large droplet size often found in liquid membranes, in which there is no secondary emulsifier. This phenomenon was first observed by Mastumoto and coworkers. [9]: a series of multiple emulsions were prepared while changing the concentration of the first or second emulsifier. It was found that when the primary emulsion was prepared with a high concentration of hydrophobic emulsifier—30% w/w Span-80—the yield of multiple emulsion prepared with Tween-20 decreased with the increase in its concentration in the range of 0.5–5 w/w. The opposite effect was observed when the second emulsifier concentration was kept constant (0.5% w/w), and the concentration of the primary emulsifier was changed from 10 to 50% w/w; the yield of preparation increased while increasing the concentration of the primary emulsifier.

However, these observations contradict other findings. For example, while preparing multiple emulsions with a sugar ester as the second emulsifier and soy lecithin together with Span-80 as the first emulsifier [50].

Since the correlation between yield of preparation and emulsifier concentration appears to be dependent on the type of emulsifier, we offered a general explanation describing the correlation between emulsifier type, its HLB, and its concentration during each step of emulsification.

The correlation of emulsifier concentration and the optimal HLB is a unique phenomenon in multiple emulsions. Since the primary emulsifier can migrate to the external oil surface, it may interact with the secondary emulsifier, therefore decreasing the HLB of the secondary surfactant mixture. This would lead to a shift of the optimal HLB towards higher HLB [5]. The shift toward higher HLBs should be dependent on the concentration of the free emulsifier at the primary emulsion. This behavior is demonstrated in Fig. 8, which describes the optimal HLB of the second emulsifier, at constant concentration (5% w/w), while changing the concentration of the first emulsifier from 5 to 30% w/w.

It is clearly seen that the higher the concentration of the primary emulsifier—the larger the HLB shift, that is to say—either higher concentration or higher HLB of the hydrophilic surfactant is needed to achieve optimal multiple emulsions.

The opposite but yet related phenomenon is observed while changing the concentration of the second emulsifier: increasing the concentration of the second emulsifier requires either an increase of first emulsifier concentration or an equivalently decrease of its HLB. A detailed analysis of the relation between emulsifier concentrations and HLB values revealed that the optimal HLB value is inversely proportional to the concentration of the second emulsifier [5].

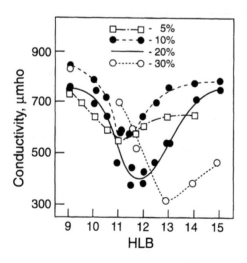

Figure 8 Optimal HLB of second emulsifier (5% w/w, Span-20–Tween-80) of multiple emulsions prepared by various concentrations of first emulsifier (Brij 92). The yield of preparation was evaluated by conductivity measurements: the lower the conductivity, the higher the yield.

In summary, while developing a multiple emulsion, one should take into account the possible interactions between the various surfactants present in the system and also the effect of various additives, such as electrolytes, on the properties of each surfactant (HLB, solubility, etc.).

Since both hydrophobic and hydrophilic emulsifiers have some solubility in the oil phase as molecules or mixed reversed micelles, they can also play an important role in transport phenomena across the oil phase [29,30]. As suggested by Kita et al. [29], water may permeate through the oil phase in W/O/W emulsions by solubilization in reverse-mixed micelles. Therefore, one should take into account the possible implications of the use of excess emulsifiers in the formulation process on the future solute release of the multiple emulsion.

The nature of emulsifiers is given only slight attention in the literature. The most widely used emulsifiers are the nonionic block copolymers (pluronic) and derivatives of sorbitan esters (Span-Tween) as shown in Table 1. In contrast, only a few ionic surfactants were studied—SDS or CTAB—or proteins, BSA and NA-caseinate, probably owing to the fact that the stabilization of the multiple droplets against coalescence was not the main formulatory problem.

Some unique preparations have also been reported: the use of microbial surfactant [13] and the use of surfactants that lead to the formation of liquid crystals around the droplets [20] or polymeric film at the interface [18].

Recently, the use of polymeric surfactants was reported by several investigators; it appears that by using suitable polymeric surfactants, very stable emulsions can be obtained [66] having very slow release profiles [53–60,64]. In this respect, the combination of proteins with monomeric surfactants, which are capable of forming a complex at the oil-water interface, was proven to be very effective in stabilizing the multiple emulsion [58].

B. Oil Nature

The nature of the oil phase can have a very significant effect on the formation and behavior of multiple emulsions. As with simple O/W and W/O emulsions, some characteristics of the oil, such as viscosity and polarity, will determine both the initial droplet size distribution and emulsion stability. Since the oil phase in the multiple emulsion is the main constituent of the membrane between internal and external phases, it may influence the release pattern in these emulsions.

Many oils were utilized for preparation of multiple emulsions, depending on the final potential application. As shown in Table 1, the oils include hydrocarbons such as kerosene [13], light paraffin oil [23], and hexane [21]; esters of fatty acids (isopropyl myristate) [24]; natural oils such as olive [4], sesame [28], and peanut oils [14]; or mixtures of various oils.

It is believed that the viscosity of the oil phase has a dominant effect on the physicochemical characteristics of the multiple emulsions. Therefore, several attempts have been made to use viscous oils (petroleum jelly) or mixtures of oils that yield varying viscosities [15]. For example, a mixture of a viscous isoparaffinic oil with a nonviscous oil leads to increased stability with the increase in viscosity and a decrease in the rate of transfer of material from the internal to the external phase [31].

Other additives that increase oil viscosity, such as beeswax [32] and palmitic acid [33], were also evaluated and found to yield more stable emulsions. Therefore, it was suggested to form emulsions in which the oil phase is very viscous or even solid during storage while becoming liquid during application [3,7]. As shown in Fig. 9, such an emulsion was prepared by using a mixture of paraffin wax and paraffin oil [7]. At room temperature, the multiple droplets are irregular, indicating the existence of a solid phase or an extremely viscous phase (Fig. 9a), but upon melting (~40°C), the *multiple* particles become liquid droplets (Fig. 9b).

The rigidity of these multiple emulsion droplets was demonstrated by dilution of the multiple emulsion in electrolyte solution. As expected, the size of multiple emulsion droplets, which contained a liquid oil phase, had decreased owing to water leakage caused by the osmotic pressure gradient (Fig. 10a) after addition of NaCl to the external phase. In comparison, no change in droplet size was observed while a semisolid oil phase was used (Fig. 10b).

The effect of oil nature on the release characteristics of the multiple emulsions was studied with the drug methotrexate [34]. It was found that the release rate decreased in the order isopropyl myristate > octadecane > hexadecane > dodecane > octane. This order is the reverse of the mean size of the internal droplet of the primary emulsion. Therefore, it can be concluded that the oil nature can affect the release pattern not only through viscosity modification but also through droplet size distribution. In view of these findings, it is clear that the preparation of the primary emulsion should be based on a compromise between optimal stability (small droplets) [34,35] and optimal release pattern (large droplets).

C. Homogenization

Since each step of preparation is based on the breakage of one phase into droplets, the importance of the emulsification method is obvious. In general, the first emulsification step is conducted by high-shear homogenizers in order to obtain fine primary droplets. In contrast, the application of a high shear rate during the second step is undesired, since it could lead to breakage of the internal droplets, hence leakage of the inner phase of the external phase and, therefore, to a decrease in the yield of preparation [25]. This effect is demonstrated in Fig. 11, in which the yield of the preparation decreases with the increase in the time of homogenization at the second step.

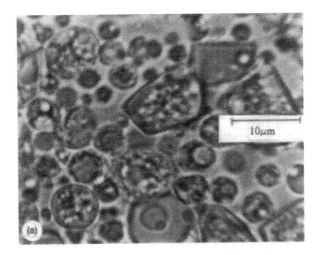

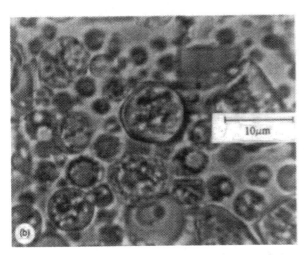

Figure 9 Water-in-solid oil-in-water multiple emulsions, at room temperature (a), and after melting the oil phase (b).

D. Phase Volume Fraction

Multiple emulsions were prepared at phase volume ratios, varying from 0.1 to 0.8 for each emulsification step, as shown in Table 1. Matsumoto [9] found that the water fraction in the primary W/O emulsion had no significant effect on the yield of preparation or on the water permeability constant in the multiple emul-

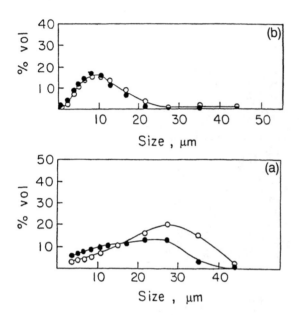

Figure 10 The effect of dilution of simple multiple emulsions (a) and solid-oil multiple emulsions (b), on droplet size distribution.

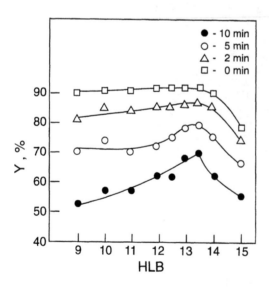

Figure 11 Effect of homogenization on the yield of multiple emulsions prepared by 1% w/w emulsifier II at various HLBs. (Each curve represents a different period of homogenization.)

sion. Contrarily, Collings [32] found that 25–50% volume was the optimal phase fraction and that the minimum release of NaCl was obtained at about 20% volume.

The fraction of primary emulsion in the final multiple emulsion has an interesting effect on the yield of preparation: below 40% volume, the yield of preparation increases with the increase of volume fraction, while above 40% volume, no significance is observed [9].

E. Water-Soluble Additives

The presence of water-soluble additives in the external and internal phases of W/O/W multiple emulsions has great importance for release rate and stability via three mechanisms:

1. An osmotic pressure gradient, which may cause leakage from the internal phase, can lead to the formation of a simple O/W emulsion (if an electrolyte is initially placed in the external phase). If the electrolyte is initially placed in the internal phase, water may penetrate into the internal droplets and a rupture of the multiple droplets occurs, leading again to formation of an O/W emulsion [32,36,37]. The rate at which water permeates the oil membrane under an osmotic pressure gradient was a subject of several studies in which the role of oil nature and possible carriers was evaluated [29,30]. Stable W/O/W emulsions were recently reported, while the osmotic pressure was adjusted by adding $MgSO_4$ to the external phase at a concentration producing an osmolarity similar to that of the internal phase [59].

2. Interaction with emulsifiers, which may lead to formation of a rigid film surrounding the initial droplets, thus altering the release kinetics in the multiple emulsion systems. (For example, addition of NaCl or sorbitol to the internal water phase could increase or decrease the release of naltrexon–HCl, depending on the electrolyte concentration [14].) This effect is probably a result of dehydration of emulsifier, which leads first to a rigid barrier towards drug release but, at higher additive concentration, to changes in the ordered structure at the interface.

3. Increase in the viscosity of the external phase may prevent the creaming phenomenon, as was observed by addition of polyvinyl alcohol [13].

V. UNIQUE FORMULATIONS OF MULTIPLE EMULSIONS

Achievement of a commercially acceptable multiple emulsion should overcome the following problems: stability regarding coalescence, leakage and creaming, prevention of substance release from the internal droplets, and appropriate control of the release kinetics.

Several interesting attempts to overcome these problems have been reported, besides the conventional formulation of emulsions:

A. Polymerization

Polymerization of the internal and external phases in such a way that the resulting system is either gel-in-oil-in-water emulsion or water-in-oil-in-gel emulsion may lead to very stable emulsions. The first attempt was made by Yoshioka et al. [28], who used gelatin in the internal phase and thus obtained a sphere-in-oil-in-water emulsion.

Other attempts include the use of poloxamer surfactants, which were cross-linked by γ-irradiation or by cross-linking of acryloyl derivatives of poloxamer surfactants or acrylamide [16,18] In using the poloxamers, it is clear that the polymerization also takes place at the oil-water interface, thus leading to a rigid barrier at the interface.

B. Micellar Solution

An interesting approach was suggested by Pilman et al [17]. Their inner W/O phase was composed of a thermodynamically stable micellar solution of the inverse type—an L_2phase. This W/O phase was emulsified in water and a new multiple emulsion, a micellar solution-in-oil-in-water, was formed.

C. Microcapsules via Multiple Emulsions

A unique method for the preparation of liquid controlled-release formulations was suggested by Bohm [51]. The method is based on the formation of a complex between a water-soluble active ingredient and a polymer, formation of a multiple emulsion with a polymer-containing solvent (water-solvent-water emulsion), and then removing the solvent. The result is an aqueous dispersion of microcapsules.

D. Stabilization by Colloidal Silica

Dupuis recently reported on the two-step preparation of a cosmetic multiple emulsion. The procedure is based on the use of colloidal silica to stabilize the oily film [65], while urea was incorporated in the primary and the final emulsion. The effect of the presence of urea was studied on pigskin in vivo. It was found that the best moisturization results were obtained when urea penetrated into the stratum corneum. Urea, encapsulated in the internal aqueous phase, reacted like urea brought on the skin by a simple W/O emulsion. Unencapsulated urea did not hydrate nor penetrate the skin.

This example also demonstrates the advantages of using multiple emulsions for cosmetics: while the urea functions as delivered by a W/O emulsion, it is obvious that the organoleptic properties of the formulation are more of an O/W emulsion—e.g., without the oily feel associated with the application of a W/O emulsion.

VI. SUMMARY

In this chapter we described the various methods for preparation of multiple emulsions. Since multiple emulsions contain several separate phases and interfaces, there are numerous combinations that can lead to the formation of multiple emulsions. For example, one may form a W/O microemulsion in aqueous gels or nonaqueous multiple emulsions simply by the proper use of the various phases and surfactants. Therefore, we expect that future work in the formation of multiple emulsions will focus on new combinations of components leading to tailor-made multiple emulsions having specific physicochemical properties. This will produce new cosmetic products owing to the possible entrapment of an active material and its slow release while the formulation adheres to the skin or hair. These products may be in liquid or semisolid forms and may be combined in formulations as specific components used to deliver and protect the active ingredients.

REFERENCES

1. Seifriz W. J Phys Chem 1925; 29:738.
2. Matsumoto S, Koh Y, Michura A. J Dispersion Sci Technol 1985; 6:507.
3. Florence AT, Whitehill D. Int J Pharm 1982; 11:277.
4. Matsumoto S, Sherman P. J Texture Studies 1981; 12:243.
5. Magdassi S, Frenkel M, Garti N, Kazan, K. J Coll Interface Sci 1984; 97:374.
6. Law TK, Whateley TL, Florence AT. Int J Pharm 1984; 21:277.
7. Magdassi S, Garti N. J Coll Interface Sci 1987; 120:537.
8. Magdassi S, Garti N. J Controlled Release 1986; 3:273.
9. Matsumoto S, Kita Y, Yonezawa D. J Coll Inter Sci. 1976; 57:353.
10. Takahashi K, Ohtsubo F, Takenchi H. J Chem Eng Jpn 1981; 14:416.
11. Davis SS, Burbage AS. J Coll Interface Sci 1977; 62:361.
12. Whitehill D., Florence AT. J Pharm Pharmacol 1979; 31(supp 3P).
13. Panchal CJ, Zazic JE, Gerson F. Coll Interface Sci 1979; 68:295.
14. Brodin AF, Kavoliunas DR, Frank SG. Acta Pharm Suecica 1978; 15:1.

15. Grossiord JL, De Luca M, Medard JM, et al. 5th Congress Intern Technol Pharm, Ed. APGI, Paris, 1989; 1:172.
16. Florence AT, Whitehill D. J Pharm Pharmacol 1982; 34:687.
17. Pilman E, Larsson K, Torhberg E. J Dispersion Sci Technol. 1980; 1:267.
18. Law TK, Florence AT., Whateley TL. Colloid Polymer Sci 1986; 264:167.
19. Magdassi S, Frenkel M, Garti N. Drug Dev Ind Pharm 1985; 11:791.
20. Kavaliunas DR, Frank SG. J Coll Interface Sci 1978; 66:586.
21. Matsumoto S. Maku 1977; 1:434.
22. Benoy CJ, Schneider R, Elson LA, Jones M. Eur J Cancer 1974; 10:27.
23. Omotosho JA, Law TK, Whately TL, Florence AT. Colloids Surfaces 1986; 20:133.
24. Florence AT, Law TK, Whately TL. Coll Interface Sci 1985; 107:584.
25. Magdassi S, Frenkel M, Garti N. J Dispers Sci Technol 1984; 5:49.
26. Morimoto Y, Sugibayashi K, Yamaguchi Y, Kato Y. Chem Pharm Bull 1979; 27: 3188.
27. Fukushima S, Kaguhiko J, Nakano M. Chem Pharm Bull 1983; 31:4088.
28. Takahashi T, Kono K, Yamaguchi T. Tohoku J Exp Med 1977; 123:235.
29. Kita Y, Matsumoto S, Yonezawa D. Nippon Kagaku Kaishi 1978; 1:11.
30. Garti N, Magdassi S, Whitehill D. J Coll Interface Sci 1985; 104:587.
31. Frankenfeld JW, Fuller GC, Rhodes CT. Drug Dev Commun 1976; 2:405.
32. Collings AM. British Patent 1235667, 1971.
33. Engel RH, Fahrenbach MJ. Nature 1968; 219:856.
34. Omotosho JA, Whateley TL, Florence AT. J Microenc 1989; 6:183.
35. Omotosho JA, Whateley TL, Law TK, Florence AT. J Pharm Pharmacol 1986; 38: 865.
36. Florence AT, Whitehill D. J Colloid Interface Sci 1981; 79:243.
37. Matsumoto S, Khoda M. J Colloid Interface Sci 1980; 73:13.
38. Yoshioka T, Ikeuchi K, Hashida M. Chem Pharm Bull 1982; 30:1408.
39. Sherman P. In: Rheology of Emulsions London: Pergamon Press, 1963: 77–90.
40. Becher P. In: Emulsions: Theory and Practice: Reinhold Publishing Corp., NY.
41. Lin TJ. J Soc Cosmet Chem 1968; 19:683.
42. Becher P. J Soc Cosmet Chem 1958; 9:141.
43. Sherman P, Parkinson C. Prog Colloid Polymer Sci. 1978; 63:10.
44. Sherman P. In: Emulsion Science London: Academic Press, 1968.
45. Matsumoto S. J Colloid Interface Sci, 1963; 94:362.
46. Prybilski C, de Luca M, Grossiord JL, et al. Cosmet Toilet 1991; 106:98.
47. Zheng S, Beissinger RL, Wasan T. J Colloid Int Sci 1991; 144:72.
48. Davis SS, Walker I. Int J Pharm 1983; 17:203.
49. Eads TM, Weiler, RK, Gaonkar AG. J Colloid Int Sci 1991; 145:466.
50. Matsumoto S, Heda Y, Kita Y, Yonezawa D. Agric Biol Chem 1978; 42:739.
51. Bohm H. U.S. Pat. No. 4857335, 1989.
52. Garti N, Aserin A. In: Benita S, ed. Microencapsulation "Pharmaceutical emulsions, double emulsions and microemulsions," New York: Marcel Dekker, 1996.
53. Garti N, Aserin A. Adv Colloid Interface Sci. 1996; 65:37.
54. Garti N, Lebensem-Wiss U. Food Sci. Technology 1997; 30:222.
55. Sela Y, Magdassi S, Garti N. Colloids Surfaces A 1994; 83:143.
56. Sela Y, Magdassi S, Garti N. Colloid Polymer Sci 1994; 272:684.

57. Sela Y, Magdassi S, Garti N. J Contr Rel 1995; 33:1.
58. Garti N, Aserin A, Cohen Y. J Contr Rel 1994; 29:41.
59. Yazan Y, Aralp U, Seiller M, Grossiord JL. Cosmet Toilet 1995; 119:53.
60. Jager-Lezer N, Denime R, Grossiord JL, et al. Cosmet Toilet 1996; 111:1153.
61. Graig H, Liang Bin C, Boyong CJ, et al. EP 717978 A2 960626.
62. Graig H, Liang Bin C, Boyong CJ, et al. EP 715842 A2 960612.
63. Gohla S, Nielsen J. SOFW J. 1995; 121:707.
64. Gruening B, Hameyer P, Weitemeyer C. EP 631774 A1 950104.
65. Dupuis, L., SOFW J. 1996; 122:658.
66. Tadros Th F, Dederen C, Taelman, MC. Cosmet Toilet 1997; 112(4):75–86.

9

Highly Concentrated Water-in-Oil Emulsions (Gel Emulsions)

Ramon Pons, Gabriela Calderó, Maria-José García-Celma,* Núria Azemar, and Conxita Solans
Consejo Superior de Investigaciones Científicas (CSIC), Barcelona, Spain

I. INTRODUCTION

A. Emulsions as Drug and Cosmetic Delivery Systems

Emulsions represent an effective formulation approach for delivery of drugs and cosmetic agents. They allow mixing immiscible ingredients into single formulations and enable the regulation of rheological properties by changing the relative proportions or the degree of dispersion of the various phases over wide ranges without significantly affecting the efficacy of the active ingredients they contain [1]. One of the most important applications of pharmaceutical emulsions is in skin therapy. A good topical formulation should be one in which the dosage form has both physical and chemical stability and cosmetic acceptability and that also provides the optimal environment for the active ingredient to reach the skin surface [2,3]. The purpose of cosmetic emulsions may be to care for the skin or to simultaneously release incorporated active ingredients onto the skin, i.e., to cause specific effects on the surface of the skin, in the epidermis, or in the stratum corneum. Usually, an emulsion is less noticeable on the skin than a nonemulsified product, and this constitutes a major factor in consumer acceptance. Emulsions are easily applied topically and can be formulated to eliminate oiliness with cleansing action. Emulsion systems also aid in carrying water, an excellent softener, to the skin. In addition, the formulator can control the viscosity, appearance, and degree of greasiness of cosmetic or dermatological emulsions [4].

The composition of the emulsion is essentially critical for the intended

---

* Current affiliation: Universidad de Barcelona, Barcelona, Spain.

application purpose. The literature deals with many types of emulsions prepared with a variety of oils and emulsifiers; for pharmaceutical or cosmetic applications, however, the choice of oil, emulsifier, and emulsion type—O/W, W/O, or multiple—is limited by its use and route of administration. Potential toxicity, cost, and chemical incompatibilities in the final formulation must be taken into account, as well as processing details that affect variables such as droplet size distributions and rheology, which control stability and therapeutic response. It is difficult to isolate these effects, as each one influences the other [5].

Emulsion type is an important consideration in the preparation of pharmaceutical and cosmetic emulsions. Frequently, the rate of release of a drug or cosmetic ingredient from a topical product is dependent of emulsion type. Emulsions of the O/W type are washable and less oily than W/O emulsions. Therefore, o/w types are more useful as water-washable bases for general cosmetic purposes. They spread easily and have the advantage that as the water evaporates they exert a cooling effect at the skin surface. The major disadvantage is that despite the fat content, they dry out. On application to the skin, much of the continuous phase evaporates and increases the concentration of a water-soluble drug in the adhering film [2,6].

The W/O emulsions exist in a wide range of consistencies depending on the components in the oil and the aqueous phase and in the emulsifier blend; the relative proportions of each phase and the properties of the various auxiliary agents also have a marked effect [7]. Water-in-oil emulsions have advantages in certain disease conditions related with dryness, where an emollient effect on the stratum corneum is required. They deposit a continuous lipid film as a protection barrier for the skin and generally provide occlusion. The occlusion effect increases the hydration of the stratum corneum and hence enhances the permeation of some drugs [2].

The importance of the topical route in the delivery of drugs for both local and systemic effects has been recognized; therefore interest has been increased in investigating the microstructure of emulsions to this effect. Although it is now well established that the structure of an emulsion can markedly influence drug bioavailability, the exact mechanisms are not clear. Most dermatological emulsions are complex, multiphase systems, and both the type of drug and the method of incorporation affect its location and release from the dosage form. Today it is realized that topical delivery provides a unique opportunity to deliver active agents directly to a desired site in higher concentration with minimum potential for the provocation of side effects [8]. The efficiency of vehicles of various types in aiding skin penetration can be reasonably predicted by the way in which the vehicle alters the activity of water in the stratum corneum and influences the stratum corneum/vehicle partition coefficient.

The interest in emulsions as topical drug delivery systems is related to their advantageous characteristics, frequently enhancing the bioavailability of the drug substance. Emulsions also offer potential in the design of systems capable of

giving controlled rates of drug release and of affording protection to drugs susceptible to oxidation or hydrolysis due to the presence of hydrophilic and lipophilic
domains [7,9]. In emulsion systems, the distribution of the drug between the
various phases and the total drug concentration will define the overall concentration gradient that exists across the skin. Also, the relative proportions of the lipid
and aqueous phases in the emulsion will affect the degree of hydration of the
stratum corneum and hence drug penetration rates. Release of active ingredients
from an emulsion system prior to its appearance at the absorption site may also
be prolonged due to either the necessity of partitioning of the species in the
emulsion system or to decreased diffusivity resulting from species interaction
with formulation components, increased inter- or intraphase microscopic viscosity, or the lengthening of the species diffusion path lenght due to obstruction by
the dispersed phase [2]. The pharmaceutical and cosmetic suitability of an emulsion for topical delivery is related to its viscosity, which will affect its consistency, spreadability, and extrudability [10]. Also, the viscosity will be of great
importance in determining the rate at which the active drug can diffuse to the
outer layers of the stratum corneum. Surfactants present in emulsion systems may
also influence the rate of release from the formulation as well as the rate of
absorption. The transfer of a drug from an emulsion involves, on a molecular
scale, the drug crossing an oil/water interface. In complex systems, the rate of
transfer will be influenced by surfactant films that are present to stabilize the
interface [2].

Recently our group has foccused its attention to the study of W/O highly
concentrated emulsions as model drug delivery systems [11,12]. Our studies are
based on a good knowledge of W/O highly concentrated emulsions concerning
their stability, formation, structure, and rheological properties (described in Sec
I. B, below).

B. Properties of W/O Highly Concentrated Emulsions Formed with Alcohol Ethoxilated Surfactants

Water-in-oil (W/O) highly concentrated emulsions (gel emulsions) constitute an
interesting class of emulsions because of their large internal phase-volume fraction (as large as 0.99), their low surfactant content (as low as 0.50%), and their
unusual rheological and optical properties. By appropriate selection of composition variables and temperature, optically transparent systems with a gel-like consistency can be obtained. These features makes them of particular interest in,
among others, the cosmetic field as formulations and novel delivery systems.

1. Formation

High-water-content (W/O) gel emulsions form in the water-rich region of ternary
water/nonionic surfactant/oil systems. It was shown [13–17] that in polyoxyethy-

lene-type nonionic surfactant systems they form at temperatures above the hydrophile-lipophile balance (HLB) temperature or phase-inversion temperature (PIT) of the corresponding system. These type of surfactants change from water-soluble to oil-soluble with the increase of temperature [18–20]. This is depicted in Fig. 1, a schematic representation of the phase behavior of a ternary water/polyoxyethylene-type surfactant/oil system. The main two-phase region extends from the water to the oil axis and at temperature T_1 consists of aqueous micellar (or O/W microemulsion) and excess oil phases, as indicated by the tie lines which converge near the oil apex. At a higher temperature, T_3, the tie lines in the two-phase region converge towards the water apex, the surfactant is mainly dissolved in the oil and the constituting phases are reverse micellar solution (or W/O microemulsion) and excess water. At an intermediate temperature, T_2, a three-phase region consisting of water, surfactant (middle phase or bicontinuous microemulsion), and oil appears. This intermediate temperature is the HLB temperature, also called PIT because inversion from oil in water (O/W) to water in oil (W/O) emulsions or vice versa is produced at this temperature [18].

Phase behavior studies [14] have shown that gel emulsions separate into two isotropic liquid phases at equilibrium: one phase is a submicellar surfactant solution in water and the other is a swollen reverse micellar solution (or W/O microemulsion). This phase equilibrium is represented in Fig. 1 by the ternary

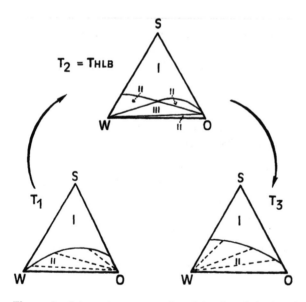

Figure 1 Schematic representation of the phase behavior of a water/nonionic surfactant/oil system.

diagram at T_3. Gel emulsions exist only in limited regions of the miscibility gap (the two-phase region). The boundaries of the gel emulsion regions depend on the system and also on the method of preparation. Gel emulsions can be prepared by the usual preparation method for W/O emulsions, dissolving a suitable emulsifier in the oil component (continuous phase) followed by addition of the aqueous component with continuous stirring. However, they can be prepared according to several other methods [14,21–24] that have been proposed specifically for highly concentrated emulsions. Among them, the so-called multiple emulsion method [14,24] and spontaneous formation method [24,25] are described in some detail.

The multiple emulsion method consists of weighing all the components at the final composition followed by shaking or stirring the sample. The time taken for gel emulsion formation ranges from a few minutes to about 1 h. The process of emulsification can be facilitated by addition of small glass beads or porous materials such as textile fabrics. The size of the container, the agitation method, and the intensity of agitation also play an important role in determining the time taken to achieve complete emulsification. It has been interpreted that the role of glass beads is of a mechanical nature, increasing the efficiency of agitation, which is specially important in the final stages of the process. In contrast, the role of a porous material could be to facilitate the aggregation of oil droplets in the earlier stages of the emulsification process. Gel emulsion formation can be achieved by shaking the sample manually or stirring with magnetic or low-shear stirrers. High-shear agitation generally produces, instead, diluted O/W emulsions. Figure 2 shows the macroscopic aspect of a mixture composed of 99 wt% water and equal weight ratios of oil and surfactant at various stages of the manual agitation process. Optical microscopy revealed the process to consist essentially of three steps. At first, after slight agitation, the mixture consists of oil droplets dispersed in water (O/W emulsion). However, the surfactant has lipophilic properties at the temperature of preparation and tends to stabilize W/O emulsions. Therefore, with further agitation, coalescence of oil droplets is observed. Moreover, the oil droplets experience growth by incorporation of water. At this step the mixture consists of a multiple W/O/W emulsion. The process of W/O emulsification inside the oil droplets can be considered analogous to that of the preparation of an emulsion by the classical method of stepwise addition of the disperse phase to the continuous phase. The final step is reached with continued agitation when all the water is emulsified, resulting in a W/O gel emulsion.

In the spontaneous formation method, emulsification is achieved by a rapid temperature change of an oil-swollen micellar solution (or O/W microemulsion). It should be noted that the system is below the HLB temperature at the starting temperature and that emulsification takes place when it is rapidly heated above the HLB temperature. The emulsions produced have finer and narrower droplet size distributions than those obtained by other methods and little or no agitation

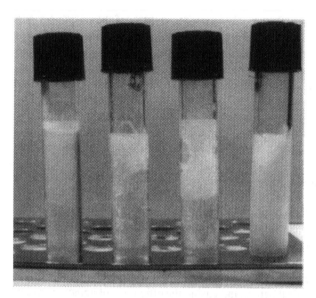

Figure 2   Picture of the various stages of gel emulsion formation by manual agitation.

is required. The emulsification process was followed conductimetrically [24] and the results are shown in the conductivity versus time plot in Fig. 3. The conductivity of a single-phase microemulsion at 8°C (time 0) experienced a monotonically drop of about four orders of magnitude in a short period of time (less than 1 min) when it was transferred to higher temperatures without stirring. The conductivity remained practically constant afterwards. The visual aspect of the sample experienced also a drastic change during the emulsification process from a transparent liquid to a milky, highly viscous emulsion with the temperature increase. It was observed that the higher the temperature (i.e., the faster the emulsification process), the lower the conductivity. These results could be interpreted when the changes in conductivity during the process of emulsification were followed while the system was under agitation [25]. Then, the decrease in conductivity was not monotonic but showed a maximum. The mechanism of spontaneous gel emulsion formation could be understood by studying the phase behavior of the system as a function of temperature [25]. It was found that the phase inversion from water-continuous microemulsion to W/O gel emulsions occurs through a lamellar liquid crystalline phase and a bicontinuous surfactant L_3 phase. The spontaneous curvature of the interfacial surfactant layer changes continuously with the temperature increase because the gel emulsion consists of a reverse micellar solution and an excess water phase. The surfactant self-organizing structure in the system is

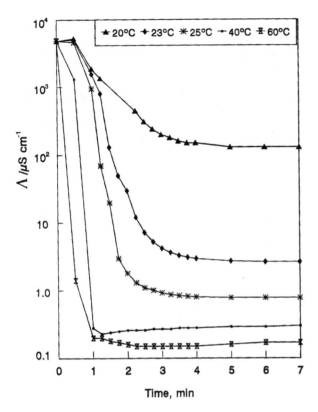

Figure 3 Conductivity changes of a microemulsion at 8°C (time = 0) when it is rapidly heated to higher temperatures.

changed from aqueous micelles to reverse micelles through vesicles, lamellar liquid crystals, and surfactant-continuous L_3 phase, as indicated in Fig. 4.

2. Structure

As the volume fraction of water (dispersed phase) in gel emulsions exceeds that of closest packing of undistorted spheres, 0.74, the water droplets are not spherical [26,27]. Indeed, optical microscopy of gel emulsions with water volume fractions near unity revealed a close-packed structure of droplets with polyhedral shape, as shown in Fig. 5. The structure of these emulsions resembles that of foams in which the water droplets are covered by a very thin layer of continuous phase. The interfacial area is very large, although the volume of the continuous phase is very small. The nature of the continuous and disperse phases, a microemulsion of the W/O type and water containing monomeric surfactant respectively, were

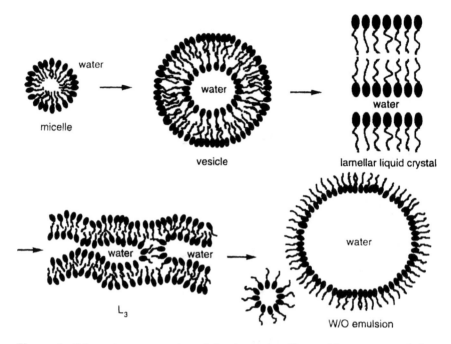

water

micelle

water

vesicle

water

lamellar liquid crystal

water   water

L₃

water

W/O emulsion

Figure 4 Schematic representation of the change in self-organizing structures during spontaneous gel emulsion formation. (From Ref. 25.)

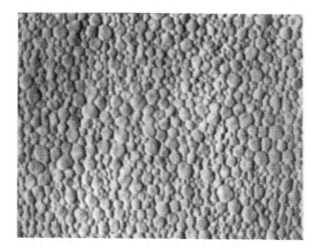

Figure 5 Photomicrograph of a gel emulsion.

assessed by means of phase behavior studies [14]. Small-angle x-ray scattering (SAXS) of gel emulsions as a function of water volume fraction, oil-to-surfactant weight ratio, and temperature confirmed that the structure of such a system is a water-in-(W/O microemulsion)emulsion [28]. However, theoretical considerations [28,29] as well as electron spin resonance (ESR) measurements [29] suggested that the number of reverse micelles or microemulsion droplets is decreased as the volume fraction of disperse phase increases and that, at values of water volume fraction close to unity, most of the surfactant molecules are adsorbed at the interface of water droplets. Estimation of mean droplets size of gel emulsions was achieved by means of optical microscopy and SAXS [28], and it was found that it is practically independent of volume fraction up to values on the order of 0.95. Above this value, it increases considerably. Typically, droplet size of gel emulsions with water volume fractions above 0.95 range from submicrometers to about 5 μm, while the upper limit of those with volume fractions of the order of 0.90 is of about 1 μm. The microstructure of the continuous phase, a reverse micellar solution or W/O microemulsion, was also confirmed by means of Fourier transform pulsed gradient spin echo (FT-PGSE) nuclear magnetic resonance (NMR) technique [30]. In addition, NMR determinations showed the decisive role of the structure of the continuous phase for both the formation and stability of gel emulsions, as explained below.

3. Stability

The stability of gel emulsions is greatly affected by the nature of the components, the volume fraction of the dispersed phase, the oil-to-surfactant weight ratio, the temperature, and the presence of additives. At equilibrium, the continuous microemulsion phase coexists with the dispersed aqueous phase. The time taken for phase separation can be significantly retarded by appropriate selection of the different composition variables and temperature; accordingly, gel emulsions with long-term stability, suitable as cosmetic formulations, can be obtained.

Like normal emulsions, in which the stability against coalescence increases with the size of the surfactant molecules, longer-chain surfactants can stabilize better gel emulsions than shorter chain surfactants. As has been described above, aliphatic hydrocarbons form gel emulsions with polyoxyethylene type nonionic surfactants provided that the temperature of formation is above the HLB temperature of the corresponding water/nonionic surfactant/oil system [14]. However, aromatic hydrocarbons do not follow this trend [17]. The reason is the high solubility of these surfactants in aromatic oils; as a result, these oils can penetrate into the array of surfactant molecules at the interface and stable gel emulsions cannot be produced. However, with surfactants, such as monoglycerides, which possess stronger intermolecular interactions than polyoxyethylene-type surfactants, gel formation is achieved [17].

Stability studies have shown that the stability of gel emulsions with aliphatic hydrocarbons is maximum at about 25–30°C higher than the HLB temperature of the corresponding system. This fact was explained on the basis of (FT-PGSE) NMR determinations [30]. Self-diffusion measurements as a function of temperature revealed that the structure of the microemulsion phase changed from bicontinuous (at temperatures close to the HLB temperature) to the reverse or W/O type (at higher temperatures), as illustrated in Fig. 6. Figure 6a shows the solubilization of water in a $C_{12}EO_4$ /decane solution (filled circles) as a function of temperature. The values of the self-diffusion coefficients for samples along

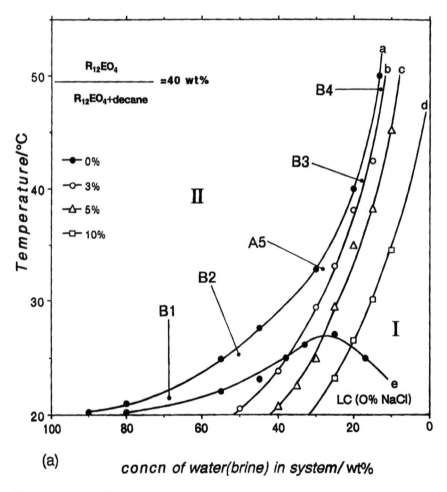

(a) **concn of water(brine) in system/wt%**

Figure 6a Solubilization of water and 3, 5, and 10% brine as a function of temperature (filled circles) in a decane solution containing 40% $R_{12}EO_4$. (From Ref. 30.)

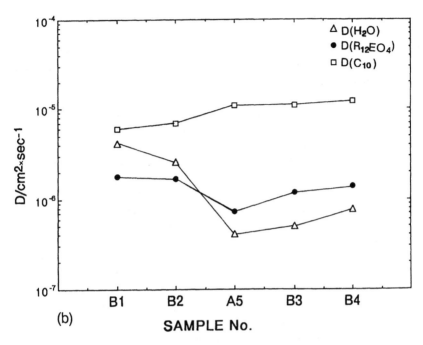

Figure 6b Self-diffusion coefficient for water, $R_{12}EO_4$, and decane in a single-phase region along the solubilization curve for water (filled circles). (From Ref. 30.)

the solubilization curve (which constitute the continuous phase of gel emulsions) are displayed in Fig. 6b. Water and oil self-diffusion coefficients are similar at low temperatures (samples B1 and B2), an indication of a bicontinuous structure. However, self-diffusion coefficient of water decreases while that of oil remains practically constant with the increase of temperature (samples A5, B3, and B4), indicating that the structure has changed to a reverse microemulsion. Similarly, a change in microemulsion structure from bicontinuous to w/o type was observed with the increase in oil-to-surfactant weight ratio [30]. The results of conductivity of gel emulsions as a function of oil-to-surfactant ratio [24] were consistent with those of NMR. These changes in structure were also related to gel emulsion stability. At low oil-to-surfactant ratios, gel emulsions either are very unstable or they are not formed, while at high oil-to-surfactant ratios, when the structure of the continuous phase is that of a W/O microemulsion, the stability is maximum. However, at very high oil-to-surfactant ratios, gel emulsions are very unstable or are not formed, owing to an insufficient amount of surfactant to stabilize the water droplets. Therefore the structure of the continuous phase has a drastic influence on the formation and stability of gel emulsions.

Additives also have an important effect on gel emulsion stability, and several types of additives have been studied in this context, including additives of cosmetic interest [31]. Among all of them, the effect of inorganic salts on gel emulsion stability has been studied most systematically. This effect was found to be dependent on the type of salt added [17]. The solubilization curve of water in a a surfactant/oil solution is shifted to lower temperatures with increasing salt concentration; in other terms, the solubilization of water in a microemulsion decreases (Fig. 6a) because the HLB temperature of the system decreases. It was found that salts with large salting-out effects are more effective in stabilizing these emulsions because they decrease more pronouncedly the cloud point in a water/nonionic system and consequently the HLB temperature of a water/nonionic surfactant/oil system [17]. One reason for the enhanced stability of gel emulsions by salt addition could be that the hydrophilic part of the surfactant dehydrates in the presence of electrolytes. The surfactant-surfactant interactions increase and therefore the interfacial films become more rigid. Figure 7 shows the effect of electrolyte concentration on the HLB temperature of the system water/$C_{12}EO_4$/hexadecane [32]. As expected from the HLB temperature plots, stability studies showed that Na_2SO_4 stabilizes gel emulsions more effectively than $CaCl_2$ and NaCl [17]. Moreover, the apparent order parameter obtained by ESR measurements in gel emulsions with Na_2SO_4 was higher as compared with those in systems with $CaCl_2$ and NaCl, as depicted in Fig. 8 [29]. These results also showed that the type and concentration of added salt change significantly the packing of the surfactant molecules at the interface. Interfacial tension measurements as well as rheological determinations agreed with these assumptions [24,33]. Therefore, the electrolytes that decrease the HLB temperature more effectively are better stabilizers because they induce a tighter packing of the surfactant at the water droplet films.

4. Rheology

One of the most striking features of the rheology of gel emulsions is their high viscosity as compared with that of the constituent phases. Knowledge of the rheological behavior of these systems is very important, not only in view of the applications but to get information concerning their structure and stability. The rheological properties of highly concentrated emulsions, analogous to those of foams, have been extensively studied [34–37]. They are non-Newtonian fluids characterized by a yield stress below which they show a solid-like behavior. The influence of temperature, oil-to-surfactant ratio, volume fraction, and salinity on a gel emulsion's rheological properties were studied by means of dynamic (oscillatory) measurements [33]. From the storage, G', and loss, G'', moduli, shear modulus, G_0, and relaxation time were obtained from fits of a Maxwell model for a viscoelastic liquid. G_0 values were fitted by means of Princen equation [37]. A maxi-

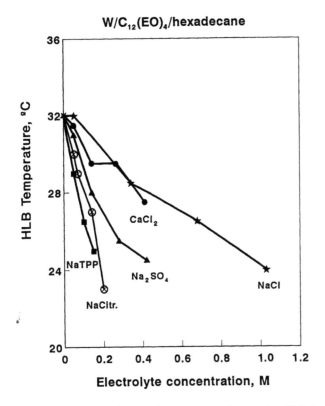

Figure 7 Effect of electrolyte concentration on the HLB temperature of the system water/$C_{12}(EO)_4$/hexadecane. (From Ref. 32.)

mum in G_0 was obtained as a function of temperature, which was interpreted as a result of a compensation of effects. The increase of both interfacial tension between aqueous and microemulsion phases (increase in G_0) and droplet size (decrease in G_0) with the increase of temperature. Similarly, a maximum in G_0 was obtained as a function of oil-to-surfactant ratio, since interfacial tension as well as droplet size increase with the oil-to-surfactant ratio. Increasing dispersed phase volume fraction produced an increase in G_0 and in some gel emulsions a maximum was obtained consistent with the strong increase in droplet size at volume fractions above 0.95. Addition of salt caused an increase of G_0 related to the increase in interfacial tension. The values of relaxation time of gel emulsions ranged from values above 500 s to 10^{-2} s and were found to be proportional to the continuous phase viscosity and inversely proportional to the continuous phase-volume fraction.

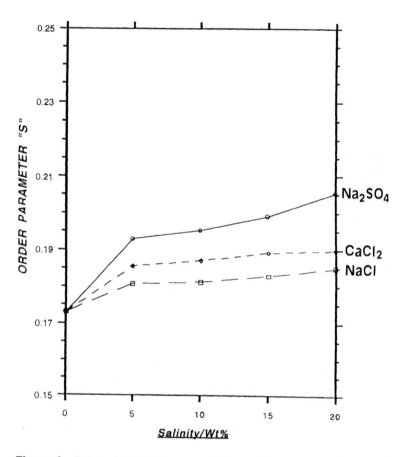

Figure 8 Effect of electrolyte concentration on the apparent order parameter of gel emulsions obtained by ESR in the system water/$C_{12}E_4$/heptane. Spin probe: 5-doxylstearic acid. (From Ref. 29.)

With this understanding of the basic properties, we have evaluated some parameters that influence the diffusion of active ingredients within the gel emulsion and its transfer to a receptor (aqueous) solution.

II. EXPERIMENTAL METHODS

Three different methods have been used in evaluating the release and diffusion coefficient of a hydrophilic molecule. These methods generate information that is complementary, and they are mathematically equivalent.

A. Determination of the Release and Diffusion Coefficients

1. Release from a Gel Emulsion Contained in a Dialysis Bag to an Aqueous Solution

A gel emulsion that contains an initial concentration of a hydrophilic molecule [α-hydroxy phenyl acetic acid (mandelic acid) in most of our studies] is placed in a dialysis bag. The geometry of the releasing system is quasicylindrical. The dialysis bags were porous cellulose membranes with a molecular mass cutoff (MWCO) of 12,000–14,000 Da—much greater than the molecular mass of the released molecule but smaller than the surfactant aggregates. Approximately 2 g of gel emulsion were used in each experiment. The dialysis bag was then inserted in a thermojacketed vessel containing 130 mL of water and well stirred by means of a magnetic stirrer. This ensures homogeneity of the receptor solution. The concentration of the releasing molecule was monitored as a function of time in the receptor solution. The analysis of mandelic acid was performed by taking a volume of the solution and measuring the absorbance at 257 nm. The data, concentration in the receptor solution as a function of diffusion time, are then fitted to master curves obtained by numerical solution of Fick's laws [11,12].

2. Release from a Gel Emulsion Directly to a Water Solution

Owing to the high viscosity of most gel emulsions, the release can be monitored without requiring the presence of a membrane. Gel emulsions can be put in contact with a water phase without dispersing in it, because the external phase is of the oil type. Provided that the viscosity is high and the agitation of the receptor solution is not too strong, the identity of gel emulsions does not change for long periods of time. The stirring of the receptor solution should be low but high enough to ensure homogeneity of the receptor solution. In this case, the geometry of the releasing gel emulsion is well defined by a surface and a depth of gel emulsion. The concentration of the diffusing molecule is monitored as a function of time in the same way as in method 1 or, alternatively, aliquots of the receptor solution are taken at several times and kept for HPLC analysis. In this case the volume that is withdrawn is replaced by the same volume of water. Analysis of the data is performed in the same way, taking into account the dilution that is introduced by replacing an aliquot of the receptor solution by an equal volume of water.

3. Diffusion Within a Gel Emulsion

The experimental setup consists of a box with a partition in the middle. A gel emulsion containing an added molecule is put in one of the compartments and a gel emulsion with the same composition except for the added molecule is put in the other compartment. The experiment starts when the partition is taken out.

At this moment both gel emulsions come into contact and diffusion starts. The box is kept at controlled temperature for a given time and then the system is sliced by introducing thin glass plates that effectively stop diffusion. The portions are taken out and their content in diffusing molecule is measured by UV analysis. The concentration of diffusing molecule is plotted as a function of the distance to the middle plane of contact between the two starting emulsions. These concentration profiles are fitted to the exact solution of Fick's laws with semi-infinite boundary conditions (if the concentration at the extremes of the box have not changed during the experimental time) or to a numerical solution of Fick's laws if the semi-infinite boundary conditions are not maintained [11].

B. Preparation and Characterization of Emulsion

1. Preparation of Emulsion

Emulsions were prepared by mixing the surfactant with the oil to give a clear solution. Water (or aqueous solutions of the releasing molecules) was added dropwise to the surfactant-oil mixture with vigorous agitation by means of a Vibromixer. The addition was pursued until the desired water weight fraction. These emulsions were kept at the aproppriate temperature to stabilize the mixture and were reagitated before starting the experiment to maintain the same conditions for all emulsions.

2. Determination of Stability

The stability of several emulsions was assessed by visual observation of phase separation. This information is complementary to the diffusion coefficients determined. A low stability of the emulsions imply a change in the structure of the system that affects the experimental determinations.

3. Determination of Partition Coefficients

In order to evaluate the possible influence of the partition of the difusing molecule between the continuous and dispersed phases of the emulsions, partition coefficients were determined between the phases separated at equilibrium. The concentration in both phases was determined by UV absorption at 257 nm.

III. DETERMINATION OF DIFFUSION COEFFICIENTS FROM EXPERIMENTAL RESULTS

Typical releasing curves as such obtained from setups 1 and 2 are shown in Fig. 9. Initially a fast release is observed that slows down with time, and finally a plateau is reached. Numerical solution of Fick's laws were determined with boundary conditions: $t = 0$, $x < 0$, $C = C_0$; $t = 0$, $x > 0$, $C = 0$, $t > 0$, $x >$ 0, $dC/dx = 0$, where t corresponds to elapsed time, x the distance from the dividing plane, C the concentration at point x, and C_0 the initial concentration in the drug charged gel emulsion. This last boundary condition implies homogeneity of the receptor solution. These numerical solutions can be scaled with the factor $x/(Dt)^{1/2}$, in which x corresponds to the depth of the gel emulsion to the diffusing plane, D the diffusion coefficient, and t the time of diffusion. The experimental results can be fitted by master curves using this scaling factor. Fits are shown as curves in Fig. 9. If the geometry of the releasing emulsions, is known, the diffusion coefficient can be obtained from these fits.

Typical results obtained from the diffusion box are shown in Fig. 10. The exact solution of Fick's equation for boundary conditions: $x < 0$, $C = C_0$ and $x > 0$, $C = 0$, and the condition of finite concentration at infinite distance from the contact plane is:

$$C = \frac{C_0}{2}\left[1 - \operatorname{erf}\left(\frac{x}{(4Dt)^{1/2}}\right)\right] \tag{1}$$

where erf is the error function.

The lines in Fig. 10 correspond to best fits of Eq. (1). If the diffusion time is long, Eq. (1) is no longer valid for our setup because the condition of finite

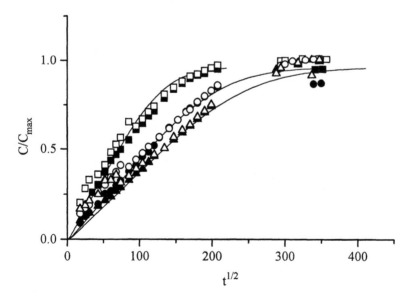

Figure 9 Release of mandelic acid from gel emulsions as obtained from setup 2. The curves are best fits of Fick equations.

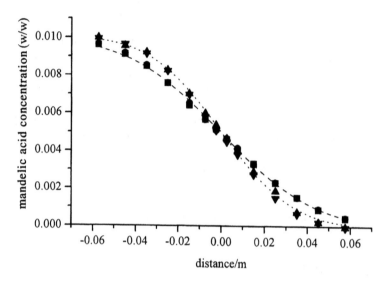

Figure 10 Concentration profiles of mandelic acid for two replicates of the experiment 3 stopped at two different total diffusion times. The curves are best fits of Eq. (1).

concentration at infinite distance is not accomplished. In this case the numerical solution of Fick's equation has to be used with the boundary conditions corresponding to our setup, that is, finite length of diffusion.

Figure 11 shows a set of results for gel emulsions with different volume fraction. From this figure it can be seen that the method is sensitive to differences in diffusion coefficient of about a 10%. Replicas of measurements give diffusion coefficients with standard deviations ranging from 2% up to 10%.

Problems could arise from low stability of the emulsion. If the gel emulsion is completely phase-separated at the end of the experiment, the concentration profile will be a constant value. Simulation at several degrees of phase separation shows that a 10% phase separation does not significantly affect the results. A 20% phase separation would increase the calculated diffusion coefficient still only by a 10%. Larger phase separations increasingly affect the calculated diffusion coefficient, always as a positive increment.

Care has to be taken that the total diffusion time is long enough to show concentration variation in several slices. If the total diffusion time is too short, only the central section will be affected by diffusion and only an upper estimate of the diffusion coefficient can be obtained. This is shown in Fig. 12, where concentration profiles are plotted for several molecules. For citric acid and glucose, only upper estimates of diffusion coefficients can be obtained, in this case 3×10^{-12} m^{2}s^{-1}.

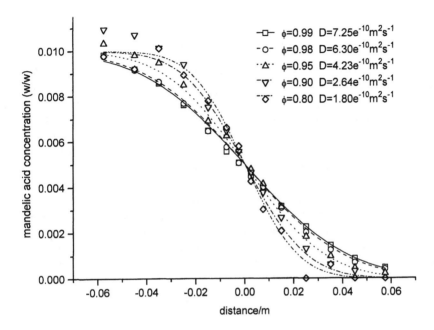

Figure 11 Concentration profiles of mandelic acid as obtained from setup 3 for gel emulsions with different volume fractions. The curves are best fits of Eq. (1).

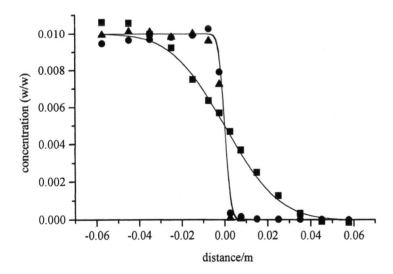

Figure 12 Concentration profiles for several molecules obtained from setup 3. The curves are best fits of Eq. (1).

IV. RESULTS

A. Internal-Phase Volume Fraction Effect

The release of mandelic acid was measured as a function of the total water content of the emulsion. An increase in the apparent diffusion coefficient was observed as the volume fraction was increased. This is shown in Fig. 13 for mandelic acid diffusion coefficients measured from gel emulsions prepared with different surfactants. D depends linearly on the water weight fraction. The diffusion coefficients are higher for the emulsions formed with $C_{10}EO_3$ than for the emulsions formed with $C_{16}EO_4$ while emulsions formed with a mixture of both surfactants show intermediate coefficients. The effect of the internal phase ratio can easily be understood if we consider the system as formed by alternating slices of the internal and external phases. The diffusivity of mandelic acid will be different in both subsystems. For the mandelic acid to diffuse, it has to go across both types of media. If we consider the resistance to diffusion (that is, the inverse of the diffusion coefficient) by analogy to electrical circuits, the total resistance of a collection of resistances in series corresponds to the sum of the resistances. The weight of one and other phases will correspond to their fraction. We find:

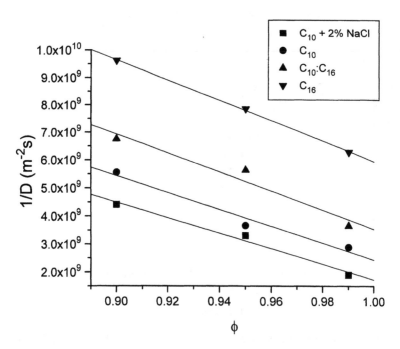

Figure 13 Diffusion coefficients as obtained from setup 1 for gel emulsions prepared with different surfactants as a function of volume fraction. (From Ref. 12.)

$$\frac{1}{D} = \frac{\Phi}{D_1} + \frac{(1 - \Phi)}{D_2} \qquad (2)$$

Which can be rearranged as a linear equation:

$$\frac{1}{D} = \frac{1}{D_2} + \left(\frac{1}{D_1} - \frac{1}{D_2}\right)\Phi \qquad (3)$$

The extrapolation to $\Phi = 0$ corresponds to the resistance of the continuous phase (a W/O microemulsion) and the extrapolation to $\Phi = 1$ corresponds to the contribution of the dispersed phase plus any interfacial contributions. Although, experimentally, one can expect gel emulsion droplet size to grow with volume fraction, this growth is moderate below volume fractions of 0.95. As the presence of interfaces is a minor component of the total resistance, this effect of interfaces reduction is probably within the experimental error. The results of fitting Eq. (3) to the experimental results of Figs. 11 and 12 and others are shown in Table 1.

D_2 is very similar for all the studied systems and close to 3×10^{-11} m^2 s^{-1}. Mandelic acid is very insoluble in pure hydrocarbons (solubility below 100 ppm) while its solubility is greatly increased if a w/o microemulsion is considered. Therefore it is reasonable to consider that mandelic acid transport in the microemulsion is due to reverse micelles transport. The values of D_2 would give an estimation of the microemulsion droplets diffusion. The use of the Stokes-Einstein equation modified for the influence of volume fraction [38] produces reasonable values for the microemulsion droplet radius, around 10 nm [12], in agreement with SAXS measurements [28,39] on similar systems.

As stated before, D_1 corresponds to the contribution of the dispersed phase plus any contribution of the interfaces. The diffusion coefficient of mandelic acid in water is 8.2×10^{-10} m^2 s^{-1} (according to results obtained from Taylor dispersion method [11]), faster than any of the values obtained for D_1. The differences

Table 1 D_1 and D_2 as Obtained from Imposition of Eq. (3) on Several Gel Emulsions

Gel-Emulsion	D_1	D_2
C_{10} + 2% NaCl	5.78e-10	3.40e-11
C_{10}	4.11e-10	3.07e-11
$C_{10}:C_{16}$	2.83e-10	2.66e-11
C_{16}	1.68e-10	2.32e-11
Cremofor	7.4e-10	6e-11
Cremofor	8.0e-10	4e-11
Cremofor	8.1e-10	3e-11

should then be attributed to the different permeability of the interfaces created by the different surfactants. This view is consistent with these results, since a film formed by a surfactant with a hydrophobic tail consisting of 10 carbon atoms would be more easily permeable for a hydrophilic molecule than the one consisting of 16 carbon atoms. Additionally, bigger surfactant molecules result in higher interfacial tension [12] and therefore in a higher energy penalty for hole formation in the film [40].

B. Oil-to-Surfactant Ratio

The effect of oil-to-surfactant ratio on the diffusion coefficients of a hydrophilic molecule from gel emulsions seems to relate mainly to its effect on stability of the emulsion. One could expect the higher the content in surfactant, the higher the diffusion coefficient in the continous phase because of the higher amount of hydrophilic molecule in the oil phase. However, our experimental results show too small differences as a function of this parameter for emulsions that do not suffer any visible modification during the experiment. The sharp increase in diffusion coefficient for oil-to-surfactant ratios higher than 60:40 can be attributed to the increased instability of gel emulsions prepared with higher ratios compared with gel emulsions prepared with lower oil-to-surfactant ratios [24].

C. Temperature

The temperature seems to play an important role in the diffusion of mandelic acid from these emulsions. Some of the increase in diffusion coefficient observed as a function of temperature could be attributed to an increased instability due to the departure from the conditions of maximum stability (see Sec. I.B.3 for details on this point). However, most of the change can be explained in terms of energetic effects on the diffusion. The diffusion coefficient doubles for a 10°C increase of temperature. This agrees well with the expected increment due to energetics of diffusion hindered by a potential barrier that corresponds to an Arrhenius type of equation. From the Stokes-Einstein equation, a similar behavior is obtained, mainly owing to the decrease of viscosity as the temperature rises.

D. Added Electrolytes

Adding electrolytes to a gel emulsion that contains mandelic acid accelerates its release. Diffusion coefficients of mandelic acid as obtained from experiment 2 are shown in Fig. 14 as a function of electrolyte molarity. This increase was found to be related to an increase in the partition coefficient of mandelic acid between the continuous (W/O microemulsion) and dispersed (water) phases of the gel emulsion. The partition coefficient also increases linearly with salt concen-

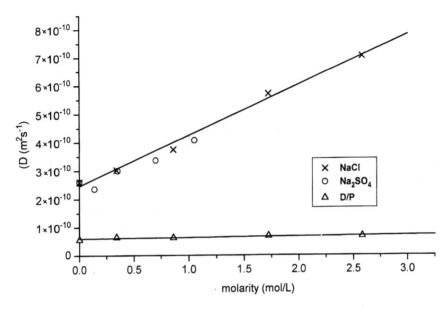

Figure 14 Diffusion coefficients of mandelic acid as obtained from experiment 2 as a function of electrolyte molarity. (From Ref. 12.)

tration in the system; this is probably because of the salting-out effect of the strong electrolyte on the mandelic acid. The most striking feature of this behavior is that dividing the diffusion coefficient by the corresponding partition coefficient produces a constant value. This can be seen with the data displayed in Table 2. The higher the amount of diffusing molecule in the continuous phase, the higher the diffusion coefficient because the activity of the diffusing molecule is in-

Table 2 Partition Coefficient of Mandelic Acid (MA)
Between the Oil and Aqueous Phases and Diffusion Coefficient
over the Partition Coefficient

Surfactant	% NaCl	P	D/P
	0	4.65	5.61e-11
	2	4.71	6.43e-11
	5	5.99	6.26e-11
$C_{10}E_3$	10	8.31	6.01e-11
	15	10.48	5.14e-11

creased in the continuous phase of the gel emulsions; however, the activity in the receptor solution remains the same.

E. Diffusion of Other Molecules

So far we have tested mainly the diffusion of a water-soluble molecule that partitions between the internal and external phases. This partition is mainly due to solubility of the molecule in the microemulsion that constitutes the continuous phase and, especially, to association with the surfactant. Other water-soluble molecules tested are the following: NaCl, citric acid, and fructose. All these molecules diffuse at much lower rates than that of mandelic acid with upper estimates of diffusion coefficients below $10^{-11} m^2 s^{-1}$. The different behavior of these molecules with respect to mandelic acid could be due to a specific association of the aromatic ring with the surfactant ethylene-oxide [41].

V. CONCLUSIONS

The release of water-soluble molecules from W/O highly concentrated emulsions (gel emulsions) has been studied by using several methods. The release methods consist in putting some emulsion in a dialysis bag with a molecular weight cutoff big enough not to affect the diffusion of the desired molecule: Alternatively, for gel emulsions that show a suitable consistency, the emulsion can be directly put in contact with the receptor solution without any barrier. The release can be fitted to Fick's equation with the appropriate boundary conditions; hence the diffusion coefficient within the emulsion is obtained. A different approach to measure the diffusion coefficient of molecules in gel emulsions consists of contacting two emulsions, which differ only in the concentration of the diffusing molecule. Diffusion coefficients can be measured from the concentration profile. The two approaches give essentially the same information and the diffusion coefficients agree fairly well. The effect of several system variables have been studied. The diffusion coefficients are mainly affected by the partition coefficient of the diffusing molecules and the effect of several variables on the diffusion coefficient can be explained by their influence on the partition coefficient. The study of the influence of volume fraction allows to separate a contribution of the interfaces and that of the internal and external phases.

Highly concentrated W/O emulsions (gel emulsions) could be adequate vehicles for drug and cosmetic delivery. These emulsions can be formulated with a range of compositions that allow for a good control of the release in the experimental conditions tested in this work.

ACKNOWLEDGMENTS

Financial support from CICYT (QUI 96-0454) and Generalitat de Catalunya (1995SGR-00498) is gratefully acknowledged. Gabriela Calderó acknowledges CIRIT for financial support.

REFERENCES

1. Becher P. Emulsions: Theory and Practice, 2nd ed. New York: Reinhold, 1965.
2. Block LH. "Pharmaceutical Emulsions and Microemulsions." In: Lieberman HA, Reiger MM, Banker GS, eds. Pharmaceutical Dosage Forms: Disperse Systems, 2d ed, vol 2. New York: Marcel Dekker, 1996:47–109.
3. Davis SS, Hadgraft J, Palin KJ. "Medical and Pharmaceutical Applications of Emulsions." In: Becher P, ed. Encyclopedia of Emulsion Technology, vol 2. New York: Marcel Dekker, 1985:159–238.
4. Knowlton J, Pearce S. Handbook of Cosmetic Science and Technology. Oxford, England: Elsevier Advanced Technology, 1993:95.
5. Breuer MM. "Cosmetic Emulsions." In: Becher P, ed. Encyclopedia of Emulsion Technology, vol 2. New York: Marcel Dekker, 1985:385–424.
6. Flynn GL. "Topical Drug Absorption and Topical Pharmaceutical Systems." In: Banker GS, Rhodes CT, eds. Modern Pharmaceutics, 2d ed. New York: Marcel Dekker, 1990:263–326.
7. Idson B. "Pharmaceutical Emulsions." In: Lieberman HA, Rieger MM, Banker GS, eds. Pharmaceutical Dosage Forms: Disperse Systems, New York: Marcel Dekker, 1996:199–245.
8. Barry BW. Dermatological Formulations. New York: Marcel Dekker, 1983.
9. Junginger HE. In: Kreuter J. ed. Colloidal Drug Delivery Systems. New York: Marcel Dekker, 1994:1–31.
10. Miner PE. In: Laba D, ed. Rheological Properties of Cosmetic and Toiletries. New York: Marcel Dekker, 1993:313.
11. Pons R, Calderó G, García MJ et al. Prog Coll Pol Sci 1996; 100:132–136.
12. Calderó G, Garcia-Celma MJ, Solans C, et al. Langmuir 1997; 13:385–390.
13. Solans C, Comelles F, Azemar N, et al. J Com Esp Deterg 1986; 17:109.
14. Kunieda H, Solans C, Shida N, Parra JL. Colloids Surf 1987; 24:225
15. Solans C, Azemar N Parra JL. Prog Coll Pol Sci 1988; 76:224.
16. Solans C, Dominguez JG, Parra JL, et al. Coll Pol Sci 1988; 266:570.
17. Kunieda H, Yano N, Solans C, Coll Surf 1989; 36:313.
18. Shinoda K, Saito H. J Coll Interface Sci 1968; 26:70.
19. Kunieda H, Shinoda K. J. Disp Sci Tech 1982; 3:233.
20. Kunieda H, Shinoda K. J. Coll Interface Sci 1985; 107:107.
21. Princen HM, Aronson MP, Moser JC. J Coll Interface Sci. 1980; 75:246.
22. Vold RD, Groot RC. J Phys Chem 1964; 68:3477.
23. Bibette J, Roux D, Nallet F, Phys Rev Lett. 1990; 65:2470.

24. Pons R, Carrera I, Erra P, et al: Coll Surf A 1994; 91:259.
25. Kunieda H, Fukui Y, Uchiyama H, Solans C. Langmuir 1986; 12:2136.
26. Ostwald W, Kolloid Z. 1910; 6:103, 7:64.
27. Lissant KJ. J Coll Interface Sci 1966; 22:462.
28. Pons R, Ravey JC, Sauvage S, et al. Colloids Surf 1993;76:171.
29. Kunieda H, Rajagopalan V, Kimura E, Solans C. Langmuir 1994; 10:2570.
30. Solans C, Pons R, Zhu S, et al. Langmuir 1993; 9:1479.
31. Solans C, Carrera I, Pons R, et al. Cosmet Toilet 1993; 108:61–64.
32. Azemar N, Carrera I, Solans C. J. Dispers Sci Technol 1993; 14(6):645.
33. Pons R, Erra P, Solans C, Ravey JC and Stebe MJ. J Phys Chem 1993; 97:12320.
34. Lissant KJ, Mayhan KG. J Colloid Interface Sci 1973; 42:201.
35. Princen HM. J Coll Interface Sci 1979; 71:55.
36. Princen, HM. J Coll Interface Sci 1983; 91:160.
37. Princen HM, Kiss AD. J Coll Interface Sci 1986; 112:427.
38. Batchelor GK. J Fluid Mech 1976; 56:375.
39. Plaza M, Pons R. Unpublished results.
40. Kabalnov A, Wennerström H. Langmuir 1996; 12:276.
41. Christenson H, Friberg SE. J Coll Interface Sci 1980; 75:276.

10
Fluorocarbon Gels

Marie Pierre Krafft
Institut Charles Sadron, CNRS, Strasbourg, France

I. INTRODUCTION

Fluorocarbons [or perfluorocarbons (PFCs)] have unique properties, including outstanding chemical and biological inertness, extreme hydrophobicity, and lipophobicity, as well as exceptional gas-dissolving capacities, very low surface tensions, high fluidity and spreading characteristics, nonadherence and antifriction properties, high density, absence of protons, and magnetic susceptibilities comparable to those of water. These properties are the basis for a range of innovative medical applications [1–3].

The best-known of the fluorocarbon-based products under development are the so-called "blood substitutes," more accurately termed *oxygen carriers*, which consist in injectable submicronic fluorocarbon-in-water emulsions [2,4]. One such emulsion, Oxygent (Alliance Pharmaceutical Corp., San Diego, CA) is being evaluated for use in conjunction with acute normovolemic hemodilution during surgery, with the objective of improving patient safety and avoiding donor blood transfusion [2,5–9]. Recently completed Phase II clinical trials demonstrated that Oxygent reversed transfusion triggers more effectively and for a longer period of time than fresh blood [9]. Treatment of acute respiratory failure by liquid ventilation with a neat fluorocarbon is being investigated in humans [10,11]. A range of PFC-based or stabilized injectable micron-sized gas bubbles are being evaluated as contrast agents for ultrasound imaging [12,13]. Neat fluorocarbons are utilized as a tamponade during repair of giant retinal tears [14]. Fluid PFC emulsions are also in clinical trials for priming bypass circuits during cardiopulmonary surgery [15]. Further applications under investigation include prehospital treatment of trauma patients, prevention and reduction of myocardial infarction and stroke, sensitization of tumor cells to radiation and chemotherapy, organ preservation, and drug delivery [2,16].

A range of pure and well-defined fluorosurfactants have recently been synthesized with a large variety of neutral or charged polar heads derived from polyols, sugars, amino acids, phosphocholine, and other hydrophilic groups [17–19]. Their modular structure allows incremental variation of size, shape, hydrophilic, lipophilic, and fluorophilic characters.

Fluorosurfactants allowed the elaboration of a variety of very stable colloidal systems, including fluorinated vesicles, tubules, and other fibers [1,20,21]. Novel multiple-phase dispersions include diverse direct fluorocarbon-in-water [22], reverse water-in-fluorocarbon [23,24], and hydrocarbon-in-fluorocarbon emulsions [25]; multiple emulsions, including emulsions with three distinct phases, a fluorocarbon, a hydrocarbon and water; microemulsions [26–28]; and gels [29,30]. These preparations are primarily investigated for their potential as drug delivery systems [1,3,20,21,31].

This chapter focuses specifically on the elaboration, properties, and applications of fluorocarbon-based gels. The difficulty in gelifying fluorocarbons arises a priori from the facts that (1) PFCs are very fluid and mobile liquids with very weak intermolecular cohesive interactions and also that (2) the standard gelifying agents are not soluble in these liquids. Fluorocarbon gels have potential in the cosmetic industry because of their permeability to oxygen and capacity to dissolve this gas, their capacity to repel any form of contamination, and their antifriction and spreading properties, which confer to them a unique smooth touch.

II. DISTINCTIVE PROPERTIES OF HIGHLY FLUORINATED MATERIALS

A. Fluorocarbons

The two principal characteristics that are the basis of the applications of fluorocarbons in medicine are their unique gas-dissolving capacities [32] and their exceptional chemical and biological inertness. The first of these characteristics results from the weakness of the intermolecular forces that exist between molecules in liquid fluorocarbons, which facilitates the formation of holes that can host the gas molecules. The second reflects the strength of the intramolecular chemical bonds. The C-F bond is the strongest single bond encountered with carbon in an organic molecule (about 485 kJ mol^{-1}, compared to about 425 kJ mol^{-1} for a standard C-H bond) [33]. The fluorine atom, with an estimated van der Waals radius of 1.47 Å, is significantly larger than the hydrogen atom (1.20 Å) [34]. As a consequence, fluorinated chains are significantly bulkier than hydrocarbon chains, with cross sections of about 30 versus 20 Å^2, respectively [35,36]. Another consequence of the larger size of the fluorine atom is the greater stiffness of fluorocarbon chains, which is related to the loss of gauche/trans conformational freedom [37]. The fluorine atom is also highly electronegative, which provides effective protection of the molecule's backbone [38]. PFCs are metabolically inert and there is no report of enzymatic cleavage of a PFC.

Biocompatibility of pure fluorocarbon is illustrated by the absence of hemolytic activity and innocuity towards cell cultures (fluorocarbons have actually been used to improve cell cultures [39]; cell cultures have also been utilized to monitor the purification and control the purity of fluorocarbons [40]. It is also illustrated by approval for oral use (with up to 1-L doses) as a contrast agent for magnetic resonance imaging in humans [41]. No deleterious effects due to the fluorocarbon were seen in organ preservation, experimental animal work, or human clinical trials with properly engineered emulsions of appropriately selected PFCs [4,7,42].

B. Fluorocarbon-Hydrocarbon Diblocks

Semifluorinated n-alkanes made of a linear perfluorinated chain attached to a hydrocarbon chain ($C_nF_{2n+1}C_mH_{2m+1}$, **1**, FnHm) are essentially apolar except for the CF_2—CH_2 bond, which contributes a small dipolar moment to the compound. Such FnHm diblocks behave as amphiphiles at a fluorocarbon/hydrocarbon interface; they were termed "primitive" amphiphiles [43]. Numerous investigations have been carried out on the structure and properties of FnHm diblocks in the solid state [44–47]. In the liquid state, they form micelles in either hydrogenated or fluorinated solvents, depending on the length of the Fn and Hm segments [43]. (For a recent review on FnHm diblocks, see also Ref. 48.)

Fluorocarbon-hydrocarbon diblocks were shown to be excellent stabilizers of phospholipid-based fluorocarbon emulsions [4,49] and to a lesser extent when poloxamers were used as the emulsifier [50]. The diblocks appear to have a high affinity for the fluorocarbon/phospholipids interface and have therefore been called molecular dowels. The hydrocarbon segment is expected indeed to have a strong affinity for the acyl chain of the phospholipids, while the fluorinated segment is expected to anchor in the fluorocarbon [51]. FnHm diblocks allow the preparation of highly stable fluorocarbon emulsions with predetermined average particle sizes [52]. The FnHm diblocks exhibit lower organ half-lives than would be expected based on their water solubility [53]. The half retention time of perfluorohexyldecane (F6H10) in the liver, after intravascular injection to female rats at a 3.6 g/kg body weight dose of the product in the form of an emulsion containing 25% w/v of F6H10, was measured by ^{19}F NMR to be 25 ± 5 days [22]. The FnHm molecules did not perturb the growth of mice after intraperitoneal administration of 30 g of the diblock per kilogram of body weight. Incubation for 4 days of FnHm with Namalva lymphoblastoid cell cultures did not affect their growth or viability [40]. Fluorocarbon emulsions stabilized by FnHm diblocks showed no detrimental effect toward endothelial cells. Such emulsions are presently being investigated for organ preservation [54,55]. Their effect on intestinal motility has also been studied [56].

C. Fluorinated Surfactants

Fluorinated surfactants (fluorosurfactants) also have unique characteristics [17,57–60]. They reduce surface tensions to a level that cannot be reached with their hydrocarbon analogs. They are also much more efficient: their critical micellar concentrations are typically two orders of magnitude lower, and the chemical inertness of the fluorocarbon chains allows their use in conditions where standard hydrocarbon surfactants would be degraded. Fluorosurfactants have also a particularly strong tendency to self-assemble into well-ordered films, phases, and colloidal systems [31,61,62].

Despite their enhanced surface activity, preliminary biocompatibility data indicate that the introduction of a fluorocarbon chain into a surfactant does not increase its acute toxiciy [17,18,58]. Intravenous LD_{50} values in mice, for example, attain about 8 g/kg and 4 g/kg for certain fluorinated phospholipids (**2**) and trishydroxymethyl aminomethane telomers (**3**), respectively. They are higher than 2 g/kg for certain dimorpholinophosphate-based amphiphiles (**4**) (Table 1). Intraperitoneal administration of the dimorpholinophosphate compound $C_8F_{17}(CH_2)_{11}OP(O)[N(CH_2CH_2)_2O]_2$, dispersed in perfluorooctyl bromide yielded an ip LD_{50} of about 4 g/kg body weight in mice. Hemolytic activity of fluorosurfactants is strongly reduced and often suppressed in spite of their large surface activity [63]. Moreover, hemolysis is seen to decrease with increasing degree of fluorination and consequent increase in surface activity.

III. ASPECTS OF THE COLLOIDAL CHEMISTRY OF PERFLUOROCARBONS AND PERFLUOROALKYLATED AMPHIPHILES

Fluorinated chains are simultaneously hydrophobic and lipophobic. Fluorosurfactants were shown to display a rich, complex, and original colloidal behavior, which is exemplified here.

A. Two-Dimensional Arrangements of Fluorinated Surfactants (Monolayers, Black Lipid Membranes)

Owing to their extreme hydrophobicity and their rigidity, fluorinated chains promote the formation of Langmuir monolayers [64]. The difference in energy between gauche and trans conformation is significantly higher for fluorinated than

Table 1 Examples of Fluorinated Amphiphiles to Be Used as Components of Nanostructures (Vesicles, Tubules, Fibers) and Gels of Various Structures

$C_nF_{2n+1}C_mH_{2m+1}$ (FnHm)   **1**

Fluorocarbon-hydrocarbon diblocks

$C_nF_{2n+1}(CH_2)_mCOO$
$C_nF_{2n+1}(CH_2)_mCOO$ —$OP(O_2)^-OCH_2CH_2N^+(CH_3)_3$

Phospholipids   **2**

$C_{10}F_{21}C_2H_4S[CH_2CHC(O)NHC(CH_2OH)_3]_nH$

Trishydroxymethyl   **3**
aminomethane-telomers

$C_nF_{2n+1}(CH_2)_mOP(O)(N\_\_O)_2$ **4**

Dimorpholinophosphates

$C_nF_{2n+1}(CH_2)_2$
C_mH_{2m+1} CHOP(O)$_2^-$O—
Na$^+$ **5**
OH   OH
HO
OH

Glucophospholipids

$C_nF_{2n+1}C(O)NHCH(CH_2)_3N(CH_3)_2O$ **6**

Amine oxides

$C_nF_{2n+1}C_2H_4SC_2H_4(OC_2H_4)_2OH$ **7**

Polyethoxylated alcohols

Fluorocarbon/hydrocarbon glutamate derivatives
8

$C_nF_{2n+1}(CH_2)_2OC(O)CH\,NHC(O)(CH_2)_7CH=CH(CH_2)_7CH_3$
$\quad\quad\quad\quad\quad\quad CH_2$
$C_nF_{2n+1}(CH_2)_2OC(O)CH$

for hydrogenated chains; as a consequence, fluorinated chains present a lower number of gauche defects [64]. Fluorinated chains have a strong tendency to orient themselves perpendicularly to a water surface due to exceedingly low chain-water interactions. This is illustrated by the fact that a nonpolar molecule, the perfluoro n-eicosane, $C_{20}F_{42}$, can form ordered and stable monolayers at the surface of water [65]. Studies on monolayers made from fluorinated amphiphiles include those from Elbert et al. [61], Barton *et al*. [66] and Jacquemain *et al*. [67]. More recently we have shown on a series of double-tailed fluorinated monomorpholinophosphates, $[C_nF_{2n+1}(CH_2)_mO]_2P(O)N(CH_2CH_2)_2O$, that increasing the Fn/Cm ratio (i.e., the degree of fluorination of the tail), while maintaining the total length constant, results in increased ordering of the monolayer [68]. It was also observed that the hydrocarbon spacer that is inserted between the fluorinated termination and the polar head has a disorganizing effect on the monolayer [69].

Bilayer membranes (BLMs) with a perfluorinated internal film have been produced from combinations of phospholipids and fluorocarbon/hydrocarbon diblocks (**1**) [70]. These fluorinated BLMs are exceptionally stable and sturdy. Their capacitances are two to three times larger (depending on diblock) than in the absence of diblock and are among the highest reported so far. The thickness of the internal fluorinated core of the membrane was, in the case of F6H10, estimated to 20 Å, indicating that the fluorinated chains are not interdigitated.

B.   Self-Assemblies of Fluorosurfactants: Vesicles and Tubules

It is remarkable that even short, single-chain fluorinated derivatives of phosphocholine can yield extremely stable vesicles [71] in the absence of any supplementary associative interaction (electrostatic, hydrogen bonding, etc.), indispensable in the case of hydrogenated surfactants. Some nonchiral fluorinated single-chain surfactants with dimorpholinophosphate polar heads (**4**) form very stable bilayer-based hollow microtubules [72,73] (Fig. 1). These tubules transform reversibly into vesicles depending on temperature. Well-shaped nanotubules were also obtained from fluorinated glucophospholipids (**5**) [74]. The presence of the negative charge causes the formation of tubules to be pH-dependent, higher pH favoring their formation. These findings support the view that hydrogen bonds between polar heads, and hydration play a major role in tubule formation. Lesser hydration of the sugar-derived head groups is related to increased intermolecular hydrogen bonds, which favors membrane crystallization and consequent tubule formation. The structure of fluorinated tubules has been studied by electron microscopy, small-angle x-ray and neutron scattering experiments [75]. Fluorinated microtubules have potential as microcontainers for active agents, and as templates for elaborating new materials.

Bilayers made from fluorinated amphiphiles have a highly ordered, highly hydrophobic internal fluorinated film that has a strong impact on their stability, permeability and in vivo behavior [21,76].

C.   Emulsions with a Fluorocarbon Phase

The recently synthesized, modular fluorinated surfactants allowed the elaboration of a variety of stable direct, reverse or totally non-polar fluorocarbon emulsions.

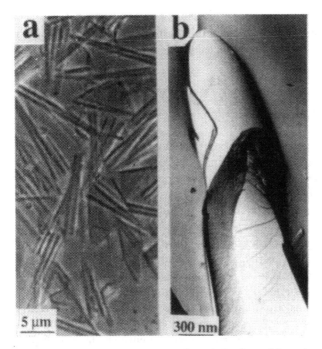

Figure 1   Hollow bilayer-based microtubules formed from the nonchiral fluorinated single-chain dimorpholinophosphates (**4**).

By using the fluorinated surfactant $C_8F_{17}(CH_2)_{11}OP(O)[N(CH_2CH_2)_2O]_2$ (**4**) stable and fluid, reverse water-in-fluorocarbon emulsions have, for example, been obtained in spite of intrinsically unfavorable solubility and diffusivity parameters, interfacial tension and density [23,77]. Fine, fluid, and narrowly dispersed reverse emulsions containing from 1–30% v/v of water have been prepared with perfluorooctyl bromide or perfluorooctylethane as the PFC. A variety of drugs, including antibacterials, vasoactive agents, bronchodilators, mucolytic agents, glucocorticoids, or anticancer agents were incorporated without compromising the emulsion's stability. These emulsions have potential for pulmonary drug delivery, as they should allow uniform and reproducible distribution of the drug throughout the lungs. The release of 5,6-carboxyfluorescein encapsulated in the internal phase of the emulsion was significantly slower than for a reverse water-in-hydrocarbon emulsion, indicating that PFCs can act as a physical barrier to diffusion of encapsulated hydrophilic compounds and can contribute to controlling their release [24,78]. Such reverse emulsions can also be embedded in fluorocarbon gels to allow the incorporation of hydrophilic drugs.

Appropriate combinations of stabilized interfaces also gave access to multiple emulsions with three nonmiscible phases: a fluorocarbon, a hydrocarbon, and water [77]. Water-in-hydrocarbon (or hydrocarbon-in-water)-in-fluorocarbon tri-ple emulsions are of interest for achieving efficient controlled release of both hydrophilic and lipophilic drugs encapsulated in the internal phases of the emul-sion.

IV. FLUOROCARBON GELS

Highly viscous and elastic systems can be obtained by mixing an oil, water and appropriate surfactants under given experimental conditions. The term *gel* covers actually a range of materials with very diverse structures. These gels, whether continuous or bicontinuous, can be conveniently classified in two categories, de-pending on their physical structure. The two types of gels have significantly dif-ferent properties, which can be relevant to their use in cosmetics or pharmacy.

The first class of gels includes those that have a compartmentalized struc-ture. These gels constitute a particular type of emulsion in which the volume fraction of the dispersed phase is so large that their aspect and consistency is that of gels; they are often termed *high-internal-phase-ratio emulsions* (HIPRE) [79], or liquid polyaphrons (from the Greek *aphros*, meaning foam) to reflect the biliquid foam-like polyhedral disperse-phase domain they contain [80]. This class of gels also includes very concentrated viscous emulsions that can display a gel-like consistency, although the packing of their droplets do not reach that of a true polyaphron. Gels can as well be obtained from fluid emulsions by adding a thickener that increases the viscosity.

The second class of gels is constituted by those composed of two inter-locking continuous phases, one of which is liquid and the other consisting gener-ally of long chains of solid or liquid-crystalline materials. Both direct and inverted micelles were reported to form cylindrical micelles: as the surfactant concentra-tion is increased, they start to entangle and form a dynamic network similar to semidilute polymer solutions [81]. This class of gels also includes the so-called hydrogels, which are well known in the biomedical field. Hydrogels are water-swollen tridimensional networks of hydrophilic polymers or copolymers. Bridges between chains are formed by covalent or ionic bonds or by weaker bonds such as hydrogen or van der Waals interactions. Examples of such hydrogels include cellulose, silicates, and interpenetrated polymeric networks.

Certain lyotropic phases can also display viscoelastic properties. This is the case, for example, of lamellar phases when the surfactant chains are in a frozen crystalline or liquid-crystalline state. Another example concerns cubic phases. The latter materials are, however, beyond the scope of this chapter; inter-ested readers can, for example, refer to Refs. 82 to 84.

This section focuses essentially on fluorocarbon gels with potential in medi-cine and cosmetics. The first part is devoted to what we called *gel emulsions* with a PFC phase. It concerns both fluorocarbon-rich systems with a continuous aqueous phase (fluorocarbon-in-water HIPRE and concentrated fluorocarbon emulsions of various types) and water-rich systems (water-in-fluorocarbon HI-PRE) with a continuous PFC phase. A second part describes fluorocarbon gels with a fibrous matrix. Finally, a third short part is devoted to other fluorocarbon gels and uses.

A. Gel Emulsions with an Aqueous Continuous Phase

1. Fluorocarbon-in-Water HIPRE

Despite the unfavorable premises, it was possible to obtain very stable gels with very high concentrations of fluorocarbons, remarkably low amounts of a fluorinated surfactant, and water [29,85]. The fluorinated surfactants used were primarily the neutral amine oxide $C_7F_{15}C(O)NH(CH_2)_3N(O)(CH_3)_2$ (**6**). A large variety of linear and cyclic, light and heavy PFCs were gelified, including trichlorotrifluoroethane (freon 113), perfluorooctyl bromide (perflubron), perfluorooctylethane, perfluorodecalin, a mixture of perfluoro-n-butyldecalin (15–20%) and perfluoroperhydrophenanthrene (80–85%) (Air Product's APF-215), perfluorodiisopropyldecalin (APF-240), and a mixture of perfluorodixylylmethane (60%) and perfluorodixylylethane (40%) (APF-260).

These gels were prepared under nitrogen at room temperature by slowly adding the fluorocarbon to a foamy aqueous solution of the fluorinated amphiphile. Little energy is required. The gels can contain up to 99% v/v of the PFC and as little as 0.2% of the fluorosurfactant. Each preparation can be conveniently characterized by its fluorocarbon/water (FC/W) and surfactant/water (S/W) ratios. For a given amount of surfactant, the larger the FC/W ratio, the more viscous and elastic the preparation (as determined by high yield points); a situation similar to that observed for gels consisting of hydrocarbon HIPREs [86]. A S/W ratio of 1/3 to 1/5 appears to be optimal for obtaining nonflowing fluorocarbon gel consistency. Transparency increases with the boiling point of the PFC.

A freeze-fracture electron micrograph of a concentrated (98% v/v) fluorocarbon gel of this type (Fig. 2) shows that its structure consists of a matrix of micron-sized polyhedral domains of PFC surrounded by thin shells of hydrated surfactant—i.e., of a polyaphron structure [87].

These gels are stable enough to withstand standard heat-sterilization conditions (121°C, 15 min, 10^5 Nm^{-2}). They are also shelf-stable. No change in macroscopic structure was observed after 1 year at room temperature. Neither crystallization of the surfactant nor phase separation was observed.

It is noteworthy that a single water-soluble surfactant suffices to produce such gels when the dispersed phase is a fluorocarbon because, in the case of hydrocarbon polyaphrons, an oil-soluble surfactant was generally considered as

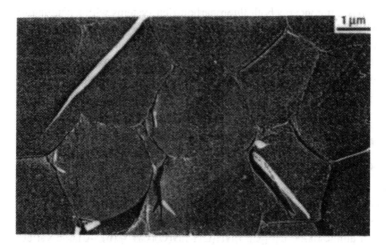

Figure 2 Freeze fracture electron micrograph of a concentrated fluorocarbon-in-water HIPRE (PFC 98.5%) prepared with the fluorinated amine oxide (**6**). The length of the bar represents 220 nm. (From Ref. 29.)

indispensable to allow the oil to spread on the surface of water. In one case, however, hydrocarbon polyaphrons were reported, which were obtained with only sodium dodecyl sulfate (SDS) as the surfactant [88]. The ease of formation of PFC gels with the above-mentioned surfactants can be attributed to the good spreading characteristics of PFCs as well as to the strong hydrophobic interactions that stabilize the interfacial films formed by fluorinated surfactants.

These new fluorocarbon gels have potential for topical applications. They could serve to protect the skin or a wound while remaining permeable to gases. Water-soluble drugs (such as antibiotics), nutrients, and other substances of therapeutic value could be incorporated in the aqueous film, which can also serve as a moisturizer. Preliminary experiments showed no irritation when such fluorocarbon gels were spread on scarified or nonscarified rabbit skin.

2. Concentrated Fluorocarbon Emulsion Obtained by Centrifugation

Centrifugation of a fluorocarbon-in-water emulsion, not unexpectedly, yielded a PFC-rich gelatinous phase and a supernatant liquid phase [89]. Little information was given on the structure, characteristics and properties of the gel phase, which was claimed to be useful as an ointment for treating skin irritations and wounds.

3. Fluorocarbon Emulsions with a Thickener in the Aqueous Phase

A few fluorocarbon gels were reported in which gelification was achieved by adding an appropriate thickener to the aqueous phase of a dilute or moderately concentrated fluorocarbon-in-water emulsion.

An approximately 55 v/v % concentrated emulsion of fluorocarbons (primarily perfluorodecalin) gelified by 1,2-polypropyleneglycol, called Fluorogel, has been prepared with a large amount of a poloxamer as the surfactant, and investigated in rats for accumulation in the skin and for therapeutic efficacy in healing surgical wounds and burns [90]. After a single application of the preparation, the fluorocarbon content of the skin did not exceed 0.04 mg/g of skin. A few PFC particles were found in the deep derma. No accumulation was seen in the skin after 30 successive daily applications of the product, and only minute amounts of the PFC (0.04 mg/g of tissue) were found in the spleen and liver of the animals. Treatment of surgical wounds indicated a roughly 20% acceleration in wound healing. The treatment of burns with Fluorogel was sensitive to active oxygen aeration (76% O_2, 2–3 h/day for 3 days), while non-PFC-containing preparations were not. The mechanism of action of the product was not established.

Another gel formulation has been obtained from fluorocarbon emulsions, in the aqueous phase of which collagen was dispersed [91,92]. Because collagen is the main constituent of skin, it was early recognized to play a significant role in the process of wound healing [93]. Collagen has therefore been proposed for burn dressings. In this application, collagen first acts as a hemostat to coagulate blood, then forms a substrate for cell growth. Collagen solutions enriched with active agents such as antibiotics or nutrients have been found to enhance wound healing and reduce scar formation in fields such as tooth implants, burns, excision wounds, and tendon repair. The authors propose a stable lyophilized collagen product comprising acid-soluble, purified native collagen in combination with platelet-derived growth factors. A series of guinea pigs were wounded and treated with the collagen-platelet mixture to which 0.25 mL/mL of a PFC compound (FC-45) was added after having been emulsified with lecithin (concentration of the dispersed phase was 20–60%). The collagen/platelet/fluorocarbon emulsion mixture was applied to the open wounds. The percent of wound area closed after 5 days was 34.0, 44.6, 32.6, 31.7, 58.0, and 78.2% in the case of controls, collagen only, platelet only, PFC only, collagen plus platelets, and collagen plus platelets plus PFC, respectively.

4. Water-in-Fluorocarbon HIPRE

Viscoelastic and transparent water-in-fluorocarbon gels that can accomodate high water contents have been obtained using non-ionic fluorinated surfactants. Water concentrations ranged from 50–98% [30,94]. Several series of fluorinated amphiphiles allowed the formation of such gels, in particular polyethoxylated alcohols (**7**). The molecule that was most investigated was $C_6F_{13}C_2H_4SC_2H_4(OC_2H_4)_2OH$. As a rule, for water-in-hydrocarbon HIPREs, the larger the water content, the more viscous and more elastic the system. The rheological behavior of these systems is that of a plastic flow with a yield value.

It was observed these fluorinated gels are more transparent than hydrocarbon gels. This difference in optical appearance probably results from closer matching of the refractive index of the oil-surfactant mixture with that of water when the oil is a fluorocarbon and the surfactant a fluorosurfactant. It is noteworthy that the most transparent gels also appeared to be the most stable ones. Since the refractive indexes are related to the Hamaker constant of the materials (water phase and oil phase), their matching would lead to some equalization of this constant in the two phases [95]. Consequently, the attractive forces between water globules could be significantly reduced, preventing coalescence from occurring.

Although these systems are very stable, they cannot be prepared by simple mixing of the ingredients, at least for high water contents; they can therefore not be considered as spontaneously formed and thermodynamically stable. However, their stability is remarkable and could result from some particular organization of the surfactant. It was proposed that the water domains of such gels could be coated by a membrane made of a lamellar liquid crystalline phase or stabilized by the formation of a three-dimensional network of ''plane'' multilayers separating the water domains [96,97].

The structure of the above-mentioned water-in-fluorocarbon gel-emulsions has been investigated by x-ray scattering and small-angle neutron scattering (SANS) using the contrast variation method. It was shown that the presence of a three dimensional network of thick multilayered liquid crystals could be excluded [30,95]. The experimental spectra could not account either for the presence of thin multilayered membranes around the water droplets. The results actually suggest that the continuous phase of the emulsion-gels consists of an inverted water-in-fluorocarbon micellar phase (or microemulsion). In this phase the largest aggregation number of the swollen micelles is in the 200–800 range; the maximum size of the water core is typically 30–50 Å. These gels may thus be described as water-in-(water-in-oil microemulsion) emulsions. The existence of such a water-in-fluorocarbon microemulsion as the continuous phase was also established in the case of certain hydrocarbon gel-emulsions [99]. More details on the structure and rheology of these fluorocarbon gel emulsions are given in refs [100,101].

A study of the binary aqueous phase diagram of the surfactant of interest (**7**) shows that, in order to form such gel-emulsions, the surfactant must practically be ''water insoluble'' [30]: there is no L_1 isotropic phase (Fig. 3). The phase behavior evidences the tendency of the surfactant to form lamellar aggregates: bilayers in L_3, disordered stacks of multilayers in L_2, and organized multilayers in L_α. The

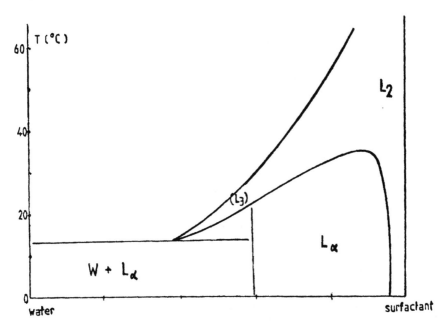

Figure 3 Binary phase diagram of $C_6F_{13}C_2H_4SC_2H_4(OC_2H_4)_2OH$ (**7**), the fluorinated polyethoxylated alcohol used to prepare water-in-fluorocarbon HIPREs. (From Ref. 30.)

tertiary diagram of the surfactant-perfluorodecalin-water system (Fig. 4) shows that stable gels can be obtained only for a very narrow range of oil/surfactant ratios [30].

Potential applications of the above gel-emulsions include their use as microreactors for chemical reactions, as controlled drug release systems and as ointments in pharmacy and ophthalmology.

B. Fluorocarbon Gels with a Fibrous Matrix

1. Gels in Organic (Fluorocarbon or Hydrocarbon) Media, and Involving FnHm Diblocks or Fluorinated Surfactants

FnHm (**1**) can be dissolved in various fluorocarbon and hydrocarbon solvents. In some cases, when the FnHm solutions are cooled below the transition temperature T_g (measured by DSC), they form swollen gels that entrap molecules of solvent to a large extent [102]. Under crossed polarizers, these gels exhibit strong birefringence that disappears above the gel-liquid transition temperature. Under high magnification, the gels appear to be made of microfibrils. It was suggested

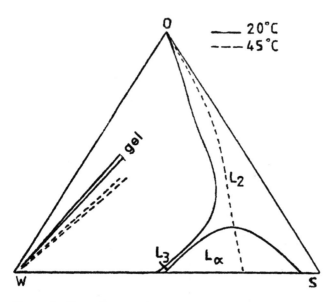

Figure 4   Phase diagram of the ternary system composed of water, perfluorodecalin and fluorinated polyethoxylated alcohol (**7**). The domains of existence of stable and transparent gels are represented at 20 and 45°C. (From Ref. 30.)

that these fibrils could be constituted by the arrangement of the diblock molecules in a lamellar structure that would form very long networked fibers (Fig. 5). The diblock lamellar structures preferentially grow in the direction perpendicular to the fiber's surface [45,103] because of the contrast between the rigidity of the fluorinated moiety and the liquid-like state of the hydrogenated moiety. The solvent is entrapped between the fibrils and is promptly released when the sample is heated or even when a gentle mechanical pressure is applied on the gel. The gel-liquid transition is completely reversible and the transition temperature depends on the solvent and on the chain lengths of the two segments of the diblock. In general fluorocarbon solvents increase the T_g value as compared to hydrocarbons. Microfiber gels were also shown to form from binary mixtures of carbon dioxide and FnHm diblocks [104].

A different type of fluorocarbon gels was also obtained using mixtures of a fluorocarbon, phospholipids, a semifluorinated alkane, and water [105]. Phospholipids were first dispersed in the mixture of fluorocarbon and FnHm. The FnHm diblock was found to be indispensable to disperse phospholipids in a fluorocarbon medium. The mixture was homogenous and fluid. Gelification took place upon addition of small amounts of water. No gels could be obtained in the absence of water. These gels were yellowish and transparent in certain cases. It

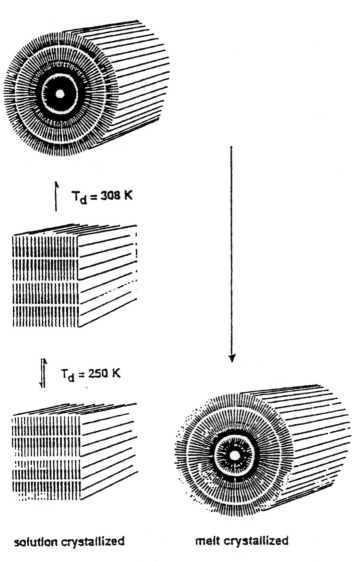

$T_d = 308$ K

$T_d = 250$ K

solution crystallized   **melt crystallized**

Figure 5 Structural model that depicts the molecular packing of a semifluorinated alkane (**1**) (F12H20) in the solid state as long cylinders. (From Ref. 45.)

is proposed that, in analogy with the lecithin-in-organic solvent microgels reported by Luisi et al. [106–108], spherical reverse micelles of phospholipids are initially formed in the PFC/FnHm phase. Upon addition of water, these micelles transform into giant rod-like micelles which rapidly develop into a network.

Amphiphiles with two fluorocarbon chains and one hydrogenated chain grafted on the chiral L-glutamate residue (8), were reported to form very viscous, turbid dispersions in benzene and 2-butanone at room temperature [109]. Optical and electron microscopies revealed the presence of fibers (Fig. 6). Differential scanning calorimetry, NMR and circular dichroism studies showed that these fibers were based on bilayer membrane [110]. This shows that it is possible to obtain stable bilayers in organic media when using amphiphilic molecules with appropriate molecular design [111].

2. Gels in a Continuous Water Medium

It was recently reported that certain fluorinated amphiphiles form dispersions of discrete microtubules—i.e., hollow cylindrical bilayer-based microstructures [72–74]. It was also possible to obtain viscous fibrous dispersions from anionic glucophospholipids, (5), in which a double-tailed hydrophobe is grafted through a phosphate linkage to the O-6 position of a polar glucose head group [74]. These compounds were shown to self-assemble into stable, hollow tubular microstructures when dispersed in water and cooled below the crystal-to-liquid crystal phase transition temperature. When heated, the tubules were observed to convert rapidly into giant vesicles and to simultaneously loose their gel consistency; the tubules spontaneously form again upon cooling.

Gels were also observed to form in dilute dispersions of fluorinated single-chain phosphocholine and maltoside derivatives [71,112]. Freeze fracture electron microscopy indicates that these gels consist in a network of small bilayer membranes of the sponge phase L_3 phase type.

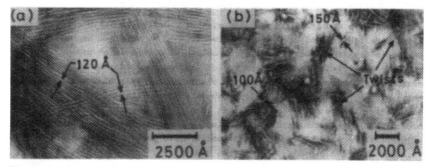

Figure 6 Electron micrographs of fibrous gels of the mixed fluorocarbon-hydrocarbon glutamate derivatives (8) in organic solvents. (a) $n = 8$ in benzene, (b) $n = 10$ in 2-butane. (From Ref. 109.)

C. Miscellaneous Gels

Certain gel ointments containing liquid fluorocarbons were claimed to be useful for treatment of decubitus ulcers and to be better tolerated than fluorinated silicone gels [113]. Flowable aqueous gel-like networks made from polyethylene glycols terminated by perfluoroalkyl tails were used to electrophoretically sequence DNA [114]. The introduction of a fluorocarbon gel or emulsion in an articular space was proposed for the treatment of disorders such as arthritis [115]. Bioactive agents incuding corticosteroids can be present in the preparation.

An "artificial seed" has been patented that contains a plant embryo encapsulated in a hydrated gel containing an oxygenated fluorocarbon together with nutrients, growth hormones, etc. [116]. Embryos of various conifers in alginate capsules containing oxygenated perfluorobutyltetrahydrofurane (FC-77), were claimed to show better germination and less abnormal development than in the absence of the fluorocarbon.

A few other examples of gels involving fluorinated components are briefly cited here, although no use in cosmetics or pharmacy are mentioned. Water-resistant, thermally reversible polymer gels based on fluoroelastomers have been prepared that are useful as protective and consolidating agents for sandstone, marble, bricks, concrete, etc., as waterproofing agents for fibrous materials, and as insulating films [117]. Fluorosilicones were part of compositions whose rheology was claimed to be responsive to electric fields [118].

Certain fluorinated gels were involved in the process of making novel materials. Gels, including a mixed fluorocarbon-hydrocarbon diblock (perfluorododecyl)dodecane, were, for example, involved in the preparation of nanoporous membranes [119]. This process is based on the gelation of solutions containing a methacrylate monomer and the fluorocompound. Curing led to the formation of an interpenetrating polymer network, from which the PFC was leached, leaving the nanoporous material.

Several reports also concern PFC gels to be used as a stationary phase in permeation chromatography. In one example the gel was based on a 4-(2-heptafluoropropoxy-1,2,2-trifluoroethoxy)styrene/divinylbenzene copolymer [120]. The preparation and use of highly fluorinated stationary phases for the analysis of polyfluorinated solutes by reversed-phase high-performance liquid chromatography was also reported [121].

V. USE OF FLUOROCARBON-BASED PRODUCTS IN COSMETICS: OTHER POTENTIAL APPLICATIONS

An emulsion of a fluorocarbon in an aqueous solution of a film-forming hydrosoluble polymer (for example a derivative of keratin, chitin, or cellulose) was reported to provide compositions useful in a number of cosmetic products [122].

Such fluorocarbon-containing compositions are claimed to be particularly smooth and pleasant when applied on the face or on the hair. Fluorocarbons were claimed to be useful in a variety of cosmetic gels—for example, for the retention of makeup [123].

A preparation containing 1–50% w/v of a fluorocarbon emulsion has been patented [124]; the fluorocarbon content of the emulsion is in the 40–100% w/v range. The emulsifier utilized (1–8% w/v) is a nonionic surfactant selected from polyoxymethylene-polyoxypropylene copolymers, ethoxylated sorbitan fatty acid esters, ethoxylated propylene glycols, perfluorinated iminobis(polyoxyalkylenes), or ethoxylated surfactants. This preparation is claimed to be useful for controlling the supply of oxygen to the skin—e.g., for day and night creams, body lotions, cleansing milks, facial and eye makeup, sun lotions, shower gels, shampoos, hair-care rinses, roll-on deodorants, after shave balms, and compact powders.

Another patent concerns a composition destined for cosmetic and topical use, which is made of phospholipids (0.5–20% w/v), fluorocarbons (0.2–100% w/v), and pharmacologically active agents (0.5–12 g/100 mL fluorocarbon) [125,126]. This composition contains particles 50–1000 nm in size composed of an internal core of fluorocarbon surrounded by an uneven number of layers of phospholipid molecules in which the phospholipids' fatty acid chains interact with the fluorocarbon. These particles were quite improperly called "asymmetric lamellar phospholipid aggregates." They consist actually of fluorocarbon emulsion droplets surrounded by phospholipids. For high phospholipid/fluorocarbon ratios, these emulsion droplets coexist with liposomes in the membrane of which small amounts of fluorocarbon may be solubilized, a situation which is typical of fluorocarbon emulsions when phospholipids are utilized as the emulsifier [4]. The active agents listed include dermatological agents (such as rosmaric acid or another virustatic/virucidal agents of vegetal origin), systemic agents (nonsteroidal analgesic/antirheumatics, opiate receptor antagonists or agonists, heparins, histamine antagonists, regulatory peptides, sedatives, or hypnotics), cytostatic, carcinostatic, immunomodular, antibiotic, corticoid, anti-infective or antiacne agents, vaccines, local anaesthetics, antiphlogistics, antihistamine, or antipsoriasis agents. The preparation is claimed to enable the delivery of these active agents to the skin or the eye for local or systemic use and to raise the oxygen concentration in the skin, thus activating certain metabolic processes. The small size of the particles may allow deep penetration into the skin and subcutaneous transport.

Since 1982, when White proposed to use neat fluorocarbons for the treatment of wounds and burns [127], it was shown that the healing rate of diverse types of wounds appears to increase when the concentration of oxygen made available to the viable cells present around the damaged tissues is increased. Fluorocarbon gels and other concentrated fluorocarbon formulations can act as dressings capable of maintaining high oxygen levels in contact with the wound. These preparations will also repel all forms of dust and contaminants and augment the gas-transfer capacity of the wound dressing, thus improving its therapeutic efficacy, while simultaneously providing effective protection of the tissues that have been coated.

REFERENCES

1. Riess JG, Krafft MP. Advanced fluorocarbon-based systems for oxygen and drug delivery, and diagnosis, Art Cells, Blood Subst, Immob Biotech 1997; 25:43.
2. Riess JG. Fluorocarbon-based oxygen delivery: basic principles and product development. In: Chang TMS, ed. Blood Substitutes. New York: Landes Karger, 1997.
3. Riess JG. Highly fluorinated systems for oxygen transport, diagnosis and drug delivery. Colloids Surf 1994; 84:33.
4. Krafft MP, Riess JG, Weers JG. The design and engineering of oxygen-delivering fluorocarbon emulsions. In: Benita S, ed. Submicronic Emulsions in Drug Targeting and Delivery. Amsterdam: Harwood. Academic Publ, 1998:235.
5. Faithfull NS. The role of perfluorochemicals in surgery and the ITU. In: Vincent JL, ed. Yearbook of Intensive Care and Emergency Medicine. Berlin: Springer-Verlag, 1994:237.
6. Zuck TF, Riess JG. Current status of injectable oxygen carriers, Crit Rev Clin Lab Sci 1994; 31:295.
7. Keipert PE, Faithfull NS, Roth DJ, et al. Supporting tissue oxygenation during acute surgical bleeding using a perfluorochemical-based oxygen carrier. In: Ince C, Kesecioglu J, Telci L, Akpir K, eds. Oxygen Transport to Tissue XVII. New York: Plenum Press, 1996:603.
8. Tremper K, Wahr JA. Blood use and non-use: designing blood substitutes. In: Parker MM, Shapiro MJ, eds. Critical Care, State of the Art. Society of Critical Care Medicine, 1995:143.
9. Keipert PE. Perfluorocarbon emulsion to compensate for acute surgical anemia. Proc VIIth Int Symp on Blood Substitutes, Tokyo, Japan, 1997.
10. Hirschl RB, Pranikoff T, Gauger P, et al. Liquid ventilation in adults, children and full-term neonates. Lancet 1995; 346:1201.
11. Leach CL, Greenspan JS, Rubenstein SD, et al. Partial liquid ventilation with per-flubron in premature infants with severe respiratory distress syndrome. N Engl J Med 1996; 335:761.
12. Quay SC. Ultrasound contrast agents development: phase shift colloids. J Ultrasound Med 1994; 13:S9.
13. Schutt EG, Pelura TJ, Hopkins RM. Osmotically stabilized microbubbles sonographic contrast agents, Acad Radiol 1996; S188:190.
14. Chang S, Sun JK. Perfluorocarbon liquids in vitroretinol surgery. In: Banks RE, Lowe KC, eds. Fluorine in Medicine in the 21st Century. Shawbury, UK: Rapra Technologies, 1994, chap. 24.
15. Holman WL, Spruell RD, Ferguson ER, et al. Tissue oxygenation with graded dissolved oxygen delivery during cardiopulmonary bypass, J Thorac Cardiovasc Surg 1995; 119:774.

16. Riess JG. Update on perfluorocarbon-based oxygen delivery systems. In: Tsuchida E, ed. Blood Substitutes. Amsterdam: Elsevier, In press.

17. Riess JG, Krafft MP. Highly fluorinated materials for in vivo oxygen transport (blood substitutes) and drug delivery. Biomaterials 1998;31.

18. Greiner J, Riess JG, Vierling P. Fluorinated surfactants intended for biomedical uses. In: Filler R, Kobayashi Y, Yagupolski LM, eds. Organofluorine Compounds in Medicinal Chemistry and Biomedical Applications. Amsterdam: Elsevier, 1993: 339.

19. Riess JG, Greiner J. Perfluoroalkylated sugar derivatives as potent surfactants for biomedical uses. In: Descotes G, ed. Carbohydrates as Organic Raw Materials II. Weinheim, Germany: VCH, 1993:209.

20. Riess JG, Frézard F, Greiner J, et al. In: Barenholz Y, Lasic DD, eds. Handbook of Nonmedical Applications of Liposomes, Vol. III. Boca Raton, FL: CRC Press, 1996:98.

21. Krafft MP, Riess JG. Elaboration and specific properties of fluorinated liposomes and related supramolecular systems. Cell Mol Biol Lett 1996; 1:459.

22. Riess JG, Cornélus C, Follana R, et al. Novel fluorocarbon-based injectable oxygen-carrying formulations with long-term room-temperature storage stability. Adv Exp Med Biol 1994; 345:227.

23. Sadtler V, Krafft MP, Riess JG. Achieving stable reverse water-in-fluorocarbon emulsions. Angew Chem Intl Ed Engl 1996; 35:1976.

24. Sadtler VM, Krafft MP, Riess JG. Reverse fluorocarbon emulsions for pulmonary drug administration. Proc Symp Control Rel Soc, Stockholm, Sweden, 1997.

25. Krafft MP, Dellamare L, Tarara T, et al. Hydrocarbon oil/fluorochemical preparations and methods of use. PCT WO 97/21425 (June 19, 1997).

26. Mathis GP, Leempoel P, Ravey JC, et al. A novel class of nonionic microemulsions: fluorocarbons in aqueous solutions of fluorinated poly(oxyethylene) surfactants. J Am Chem Soc 1984; 106:6162.

27. Lattes A, Rico-Lattes I. Microemulsions of perfluorinated and semi-fluorinated compounds. Art Cells, Blood Subst, Immob Biotech 1994; 22:1007.

28. Schubert KV, Kaler EW. Microemulsifying fluorinated oils with mixtures of fluorinated and hydrogenated surfactants. Colloids Surf 1994; 84:97.

29. Krafft MP, Riess JG. Stable highly concentrated fluorocarbon gels. Angew Chem Int Ed Engl 1994; 33:1100.

30. Ravey JC, Stébé MJ. Structure of inverse micelles and emulsion-gels with fluorinated nonionic surfactants: a small-angle neutron scattering study. Prog Colloid Polym Sci 1990; 82:218.

31. Riess JG, Krafft MP. Amphiphiles and fluorinated colloidal systems. Biochimie 1998;80.

32. Riess JG, Le Blanc M. Solubility and transport phenomena in perfluorochemicals relevant to blood substitution and other biomedical applications. Pure Appl Chem 1982; 54:2383.

33. Banks RE, Tatlow JC. A guide to modern organofluorine chemistry. J Fluorine Chem 1986; 33:227.

34. Bondi A. Van der Waals volumes and radii. J Phys Chem 1964; 58:441.

35. Tiddy GJT. Concentrated surfactant systems. In Eike E, ed. Modern Trends of

Colloidal Science in Chemistry and Biology. Basel: Birkäuser Verlag, 1985: 148.

36. Rosen MJ. Surfactants and Interfacial Phenomena. New York: Wiley, 1978.
37. Hoffmann H, Kalus J, Thurn H. Small angle neutron scattering measurements on micellar solutions of perfluor detergents. Colloid Polym Sci 1983; 261:1043.
38. Riess JG. Perfluorochemical emulsions for intravascular use: principles, materials and methods. In: Banks RE, Lowe KC, eds. Fluorine in Medicine in the 21st Century. Shawbury, UK: Rapra Technol Ltd., 1994, chap. 20.
39. Lowe KC. Perfluorochemicals in medicine and cell biotechnology. In: Banks RE, Lowe KC, eds. Fluorine in Medicine in the 21st Century. Shawbury, UK; Rapra Technologies, Ltd., 1994, chap. 19.
40. Le Blanc M, Riess JG, Poggi D, Follana R. Use of lymphoblastoid Namalva cell cultures in a toxicity test: application to the monitoring of detoxification procedures for fluorocarbons to be used as intravascular oxygen carrier. Pharm Res 1985; 5: 195.
41. Mattrey RF. The potential role of perfluorochemicals (PFCs) in diagnostic imaging. Art Cells, Blood Subst, Immob Biotech 1994; 22:295.
42. Flaim SF. Pharmacokinetics and side effects of perflubron-based blood substitutes. Art Cells, Blood Subst, Immob. Biotech 1994; 22:1043.
43. Turberg MP, Brady JE. Semifluorinated hydrocarbons: primitive surfactant molecules. J Am Chem Soc 1988; 110:7797.
44. Viney C, Russell TP, Depero LE, Twieg RJ. Transitions to liquid crystalline phases in a semifluorinated alkane. Mol Cryst Liq Cryst 1989; 168:63.
45. Höpken J, Möller M. On the morphology of (perfluoroalkyl)alkanes. Macromolecules 1992; 25:2482.
46. Russell TP, Rabolt JF, Twieg RJ, et al. Structural characterization of semi-fluorinated alkanes: II. Solid-solid transition behavior. Macromolecules 1986; 19:1135.
47. Hoyle CE, Kang D, Jariwala C, Griffin AC. Efficient polymerization of a semi-fluorinated liquid crystalline methacrylate. Polymer 1993; 34:3070.
48. Lo Nostro P. Phase separation properties of fluorocarbons, hydrocarbons and their copolymers. Adv Colloid Interf Sci 1995; 56:245.
49. Riess JG, Sole-Violan L, Postel M. A new concept in the stabilization of injectable fluorocarbon emulsions: the use of mixed fluorocarbon-hydrocarbon dowels. J Dispers Sci Technol 1992; 13:349.
50. Meinert H, Fackler R, Knoblich A, et al. On the perfluorocarbon emulsions of second generation. Biomat, Art Cells, Immob Biotech 1992; 20:805.
51. Cornélus C, Krafft MP, Riess JG. About the mechanism of stabilization of fluorocarbon emulsions by mixed fluorocarbon/hydrocarbon additives. J Colloid Interf Sci 1994; 163:391.
52. Cornélus C, Krafft MP, Riess JG. Improved control over particle sizes and stability of concentrated fluorocarbon emulsions by using mixed fluorocarbon/hydrocarbon dowels. Art Cells, Blood Subst, Immob Biotech 1994; 22:1183.
53. Weers JG, Arlauskas RA, Harris M, Otto S. The use of lipophilic secondary fluorocarbons to solve the emulsion stability/organ retention dilemma in blood substitutes. Art Cells, Blood Subst, Immob Biotech 1996; 24:458.
54. Voiglio EJ, Zarif L, Gorry F. et al. Aerobic preservation of organs using a new

perflubron/lecithin emulsion stabilized by molecular dowels. J Surg Res 1994; 63: 439.

55. Mathy-Hartert M, Krafft MP, Deby C, et al. Effects of perfluorocarbon emulsions on cultured human endothelial cells. Art Cells, Blood Subst, Immob Biotech 1997; 25:563.

56. Bouley L, Krafft MP, Dutoit P, et al. Viability of the rat ileum perfused with various oxygen carriers. In preparation. Proc VII Int Symp on Blood Substitutes, Tokyo, Japan, 1997 (Abstract)

57. Kissa E. Fluorinated Surfactants: Synthesis, Properties, Applications. New York: Marcel Dekker, 1994.

58. Riess JG, Krafft MP, Naon RN. Biological aspects and toxicity of fluorosurfactants and other highly fluorinated materials. In: Florence AT, Walters KA, Attwood K, eds. Surfactants in Biological Systems. New York: Marcel Dekker. In preparation.

59. Kunieda H, Shinoda K. Krafft points, critical micelle concentrations, surface tension, and solubilizing power of aqueous solutions of fluorinated surfactants. J Phys Chem 1976; 80:2468.

60. Shinoda K, Hato M, Hayashi T. The physicochemical properties of aqueous solutions of fluorinated surfactants. J Phys Chem 1972; 76:909.

61. Elbert R, Folda T, Ringsdorf H. Saturated and polymerizable amphiphiles with fluorocarbon chains. Investigation in monolayers and liposomes. J Am Chem Soc 1984; 106:7687.

62. Kunitake T. Synthetic bilayer membranes: molecular design, self-organization and applications. Angew Chem Int Ed Engl 1992; 31:709.

63. Riess JG, Pace S, Zarif L. Highly effective surfactants with low hemolytic activity. Adv Mater 1991; 3:249.

64. Goldmann M, Nassoy P, Rondelez F. Search for perfectly ordered dense monolayers. Physica A 1993; 200:688.

65. Li M, Acero AA, Huang Z, Rice SA. Formation of an ordered Langmuir monolayer by a non-polar chain molecule. Nature 1994; 367:151.

66. Barton SW, Goudot A, Bouloussa O, et al. Structural transitions in a monolayer of fluorinated amphiphile molecules. J Chem Phys 1992; 96:1343.

67. Jacquemain D, Grayer Wolf S, Leveiller F, et al. Dynamics of two-dimensional self-aggregation: pressure and pH-induced structural changes in a fluorocarbon amphiphile at liquid-air interfaces. An x-ray synchrotron study. J Am Chem Soc 1990; 112:7724.

68. Jeanneaux F, Giulieri F, Krafft MP. In preparation.

69. Krafft MP. Colloidal systems made from highly fluorinated amphiphiles. 15th Int Symp Fluorine Chem, Vancouver, 1997.

70. Krafft MP. Cohen J. Fluorinated black lipid membranes. In preparation.

71. Krafft MP, Giulieri F, Riess JG. Can single-chain perfluoroalkylated amphiphiles alone form vesicles and other organized supramolecular systems. Angew Chem Int Ed Engl 1994; 32:741.

72. Giulieri F, Krafft MP. Riess JG. Stable fluorinated fibers and rigid tubules from single-chain perfluoroalkylated amphiphiles. Angew Chem Int Ed Engl 1996; 34: 1514.

73. Giulieri F. Krafft MP. Self-organisation of single-chain fluorinated amphiphiles with fluorinated alcohols. Thin Solid Films 1996; 284–285:195.

74. Giulieri F, Guillod F, Greiner J, et al. Glucophospholipids—a new family of anionic tubule-forming amphiphiles. Eur Chem J 1996; 2:239.

75. Krafft MP, Giulieri F, Imae T. Small-angle scattering and electron microscopy investigations of microtubules made from perfluoroalkylated glucophospholipids and their hydrogenated analogs. In preparation.

76. Riess JG, Krafft MP. Fluorinated phosphocholine-based amphiphiles as components of fluorocarbon emulsions and fluorinated vesicles. Chem Phys Lipids 1995; 75:1.

77. Riess JG, Krafft MP. Reverse fluorocarbon emulsions and their use for drug delivery via the pulmonary administration and their use for the obtention of multiple emulsions. Fr Pat Appl 94/07068 (1994).

78. Sadtler VM, Krafft MP, Riess JG. Reverse water-in-fluorocarbon emulsions as a drug delivery system: an in vitro study. Proc 2nd World Congress on Emulsion, Bordeaux, France, 1997.

79. Lissant KJ. The geometry of high-internal-phase-ratio emulsions. J Colloid Interf Sci 1966; 22:462.

80. Sebba F. Foams and Biliquid Foams—Aphrons. Chichester: Wiley, 1987.

81. Luisi PL, Scartazzini R, Haering G, Schurtenberger P. Organogels from water-in-oil microemulsions. Coll Polym Sci 1990; 268:356.

82. Ravey JC, Stébé MJ. Phase behaviour of fluorinated nonionic surfactant systems. Prog Coll Polym Sci 1987; 73:127.

83. Boden N, Jolley KW, Smith MH. Phase diagram of the cesium pentadecafluorooctanoate (CsPFO)/H_2O system as determined by [133]Cs NMR: comparison with the CsPFO/D_2O system. J Phys Chem 1993; 97:7678.

84. Chittofrati A, Boselli V, Visca M, Friberg SE. Perfluoropolyether ammonium carboxylates: SAXRD study of lamellar liquid-crystals in water and formamide. J Dispers Sci Technol 1994; 15:711.

85. Krafft MP, Riess JG. Viscoelastic compositions of fluorinated organic compounds. PCT WO 95/09606 (April 13, 1995).

86. Pons R, Erra P, Solans C, et al. Viscoelastic properties of gel-emulsions: their relationship with structure and equilibrium properties. J Phys Chem 1993; 97:12320.

87. Sebba F. Foams and Biliquid Foams-Aphrons. Chichester:Wiley, 1987.

88. Eberts G, Platz G, Rehage H. Elastic and rheological properties of hydrocarbon gels. Ber Bunsenges Phys Chem 1988; 92:1158.

89. Moore RE. Preparation of a gel having gas transporting capability. US Pat 4,569,784 (Feb. 11, 1986).

90. Oxynoid OE, Sydliarov DP, Aprosin YD, Obraztsov VV. Application of fluorocarbon emulsions as components of cosmetics and medical ointments. Art Cells, Blood Subst, Immob Biotech 1994; 22:1331.

91. Magdassi S, Roys M, Shoshan S. Interactions between collagen and perfluorocarbon emulsions. Int J Pharm 1992; 88:171.

92. Shoshan S, Michaeli D Magdassi S. Processes for the preparation of storage stable collagen products. US Pat. 5,073,378 (Dec. 17, 1991).

93. Shoshan S. Wound Healing. Int Rev Connective Tissue Res 1981; 9:1.

94. Delpuechl JJ, Matos L, Moumni EM, et al. Matériaux viscoélastiques isotropes à base d'eau, de tensioactifs fluorés et d'huiles fluorées, leur procédé d'obtention et leurs applications dans divers domaines, tels que ceux de l'optique, de la pharmacologie et de l'électrodynamique. Fr Pat Appl 2 630 347 (April 20, 1988).

95. Ravey JC, Stébé MJ. Small angle neutron scattering studies of aqueous gels with fluorinated nonionic surfactants. Physica B 1989; 156–157:394.

96. Barry BW. Viscoelastic properties of concentrated emulsions. Adv Colloid Interf Sci 1975; 5:37.

97. Friberg SE, Mandell L, Fontell K. Acta Chem Scand 1969; 23:1055.

98. Groves MJ. Anomalous densities of dilute emulsions prepared by self-emulsification. J Colloid Interf Sci 1978; 64:90.

99. Solans C, Dominguez JG, Parra JL, et al. Gelled emulsions with a high water content. Colloid Polym Sci 1988; 266:570.

100. Ravey JC, Stébé MJ, Sauvage S. Water in fluorocarbon gel emulsions: structures and rheology. Colloids Surf 1994; 91:237.

101. Ravey JC, Stébé MJ, Sauvage S. Fluorinated gel-emulsions. J Chim Phys Phys - Chim Biol 1994; 91:259.

102. Twieg RJ, Russell TP, Siemens R, Rabolt JF. Observations of a "gel" phase in binary mixtures of semifluorinated n-alkanes with hydrocarbon liquids. Macromolecules 1985; 18:1361.

103. Höpken J, Pugh C, Richtering W, Möller M. Melting, crystallization and solution behavior of chain molecules with hydrocarbon and fluorocarbon segments. Makromol Chem 1988; 189:911.

104. Wikramanayake R. Enick R. Turberg M. The phase behavior and gel formation of binary mixtures of carbon dioxide and semifluorinated alkanes. Fluid Phase Equilibria 1991; 70:107.

105. Krafft MP, Riess JG. Reverse gels with a continuous fluorocarbon phase. Fr Pat Appl 2,737,135 (1995).

106. Willemann H, Walde P, Luisi PL, et al. Lecithin organogel as matrix for transdermal transport of drugs. J Pharm Sci 1992; 81:871.

107. Scartazzini R. Luisi PL. Organogels from lecithin. J Phys Chem 1988; 92:829.

108. Scartazzini R, Luisi PL. Reactivity in an optically transparent lecithin-gel matrix. Biocalalysis 1990; 3:377.

109. Ishikawa Y, Kuwahara H, Kunitake T. Self-organisation of solvophobic, double-chained fluorocarbon derivatives in organic media. Chem Lett.: 1989; 1737.

110. Kuwahara H, Hamada M, Ishikawa Y, Kunitake T. Self-organization of bilayer assemblies in a fluorocarbon medium. J Am Chem Soc 1993; 115:3002.

111. Ishikawa Y, Kuwahara H, Kunitake T. Self-assembly of bilayers from double-chain fluorocarbon amphiphiles in aprotic organic solvents: thermodynamic origin and generalization of the bilayer assembly. J Am Chem Soc 1994; 116:5579.

112. F. Giulieri, M.P. Krafft, in preparation.

113. Maeda T, Yoshikawa M, Arai J. Cosmetics containing liquid fluorocarbons with high oxygen permeability and water- and oil-repellency. Jpn Kokai Tokkyo Koho JP 05 70,319 (March 23, 1993).

114. Menchen F, Johnson B, Winnick MA, Xu B. Gel-like networks used to sequence DNA. Electrophoresis 1996; 17:1451.

115. Walters MA, Hopkins RM, Klein DH. Fluorocarbon for diagnosis and treatment of articular disorders, PCT Appl WO 9733563 (Sept. 18, 1997).

116. Carlson WC, Hartle JE, Bower BK. Oxygenated artificial botanic seed. US Pat 5,236,469 (Aug. 17, 1993).

117. Mascia L, Moggi G. Reversible polymer gels based on fluoro elastomers for coatings. Eur Pat Appl EP 481,283 (April 22, 1992).

118. Stangroom JE. Electro-rheological fluids/electric field responsive fluids. Eur Pat Appl EP 284,268 (Sept. 28, 1988).

119. Gankema H, Hempenius MA. Gel template leaching: an approach to functional nanoporous membranes. Macromol Symp 1996; 102:381.

120. Juhl HJ, Heitz W. Fluorinated gels for gel permeation chromatography. Makromol Chem 194:963 1993.

121. Krafft MP, Jeanneaux F. Le Blanc M, Riess JG. Highly fluorinated stationary phases for analysis of polyfluorinated solutes by reversed-phase high-performance liquid chromatography. Anal Chem 1988; 60:1969.

122. Arnaud P, Mellul M. Cosmetic or dermatologic oil in water dispersion composition, for composite film—comprises organo fluorohydrocarbon phase and liquid phase dispersed in aqueous solution of water soluble film forming polymer free of hydroxyethyl, for aqueous nail varnish. Fr Pat Appl 2 688 006 & 2 687 932 (Feb. 27, 1992).

123. Ito T, Momose S. Cosmetic gels for improving retentivity of makeup cosmetics. Jpn. Kokai Tokkyo Koho JP 07 196,446 (Aug. 1, 1995).

124. Roeding J, Stanzl K, Zastrow L. Carrier for controlling supply of oxygen to skin—comprises oxygen-carrying fluorocarbon and non-ionic surfactant containing aqueous emulsion. US Pat Appl 94/256180 (1994).

125. Gross U, Roeding J, Stanzl K, Zastrow L. Topical formulation containing dermatological or systemic drug in asymmetric lamellar agregates containing phospholipid and fluorocarbon, giving deep skin or subcutaneous penetration, DE Pat 4221256 (Feb. 1994).

126. Gross U, Roeding J, Stanzl K, Zastrow L. Cosmetic for raising oxygen concentration in skin—contains asymmetric lamellar aggregates of phospholipid(s) and fluorocarbon loaded with oxygen in cosmetic carrier. DE Pat 4221255 (Feb. 1994).

127. White D. Use of perfluorocarbons as wound treatment. US Pat 4,366,169 (Dec. 28, 1982).

11

Vapor Pressures of Fragrances After Application: Some Fundamental Factors

Stig E. Friberg
Clarkson University, Potsdam, New York

I. INTRODUCTION

The olfactory perception of fragrance compounds is certainly the decisive factor for consumer selection of perfumes, eau de colognes, and after-shaves; the development of these formulations has a long and fascinating history [1], beginning in Mesopotamia [2] and Egypt some 5000 years ago. It may be mentioned that the name Arabia (''land of fragrance'') illustrates the attitude of the ancient Greeks toward the oriental countries with respect to their fragrance art. Formulation perfumery dates back a few hundred years; Jean Paul Feminis is credited with initiating the intentional formulation of perfumes in 1695 with his combination of floral and citric notes with herbs [3] to create "Eau Admirable." A later turning point was the introduction of synthetic fragrances in 1921, when the well-known Chanel No. 5 was introduced [4].

The art of composing perfume blends is well established, with emphasis on balance between top, middle, and bottom notes to prevent the volatile elements from being lost too early. In the hands of skilled artists in the area, this balancing has developed into a most beautiful work of genius that may well be compared with the great musical creations.

Recently, the entire situation has changed because of government regulations severely restricting the use of alcohol as a solvent and, with the exception of some attempts to use long-chain alcohols or glycols [5], the formulations have become geared to the use of surfactants to achieve the necessary compatibility between water and the fragrances, which are strongly hydrophobic organic compounds. Microemulsions [6] are an attractive alternative, as are liposomal solutions [7,8]. This new approach to formulation has inspired a number of investigations to clarify the different phases and structures in water-surfactant-fragrance systems [9–20]. In addition to the information about phase regions in such systems, it is essential to obtain knowledge about vapor pressure.

II. FRAGRANCE VAPOR PRESSURES

Early investigations on the variation of fragrance vapor pressure with time as compared with traditional alcohol-based formulations [21] showed that the vapor pressure followed an exponential function:

$$\frac{p}{p_0} = A \times e^{-kt} \tag{1}$$

in which p is the vapor pressure, p_0 is the vapor pressure of the pure fragrance, A is the relative vapor pressure at the beginning of the process, and k is the constant monitoring the decline of vapor pressure with time (Fig. 1).

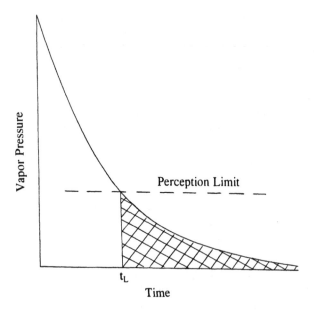

Figure 1 The principal vapor pressure variation with time of a fragrance compound after application.

A. Time Dependence

The consequence of the exponential function is obvious: all the added fragrance represented by the shaded area in the figure is wasted—an economic factor of importance.

By choosing p_0 equal to the pressure at $t = 0$, Eq. (1) is simplified to

$$p = e^{-kt} \tag{2}$$

Assuming the lowest level for perception equal to fraction L of p_0, the time to reach this value is

$$t_L = \frac{-(\ln L)}{k} \tag{3}$$

The fraction of originally added fragrance wasted is

$$F_L = \int_{-(\ln L)/k}^{\infty} e^{-kt}dt \bigg/ \int_0^{\infty} e^{-kt}dt$$

$$F_L = L \tag{4}$$

showing a very simple relationship.

Hence, ideally, the vapor pressure should initially be at a low level but remain as constant as possible with time to drop to zero suddenly at the end. The economic gain of such a modification is significant. With the vapor pressure constant at a level $L(1 + \partial)$ until complete removal of the fragrance, the time for perception limit is $t_{L+\partial}$.

$$t_{L+\partial}L \times (1 + \partial) = \int_0^{\infty} e^{-kt}dt = \frac{1}{k} \tag{5}$$

$$t_{L+\partial} = \frac{1}{L(1+\partial)k} \tag{6}$$

and the increase time of perception may be calculated as follows:

$$r = \frac{t_{L+\partial}}{t_L} = \frac{1}{L(1+\partial)\ln L} \tag{7}$$

Reasonable values to evaluate the economic importance of the reduced vapor pressure is $L = 0.5$ and $\partial = 0.05$. For such a case, the increased time for perception becomes 2.6 times the original time.

B. Surfactant-Association Structures

These evaluations are valid for alcohol-based formulations; the conditions in more complex systems such as microemulsions or liposomal solutions remain to be evaluated. The vapor pressures from organic solvents solubilized in micelles has been extensively investigated by Christian et al. [22–26], who established the thermodynamic basis for such phenomena.

The results are reported in the form

$$K_m = \frac{X_{\mathrm{mic}}}{C_s} \tag{8}$$

and the activity coefficient is defined as

$$\gamma_{\mathrm{mic}} = \frac{f_A}{f_A^0 X_{\mathrm{mic}}}$$

where f_A is the fugacity of the solute and f_A^0 is the fugacity of the solvent in the pure state. The trends in the activity coefficients were established [27].

These results leave a large fund of very exact knowledge of the fundamental thermodynamics of micellization. They will provide an excellent basis for future efforts to relate the thermodynamics to structural details of the micelles and its changes during solubilization.

Vapor pressure determinations in microemulsions are comparatively few and mostly concerned with water pressures in W/O microemulsions, with some results concerned with the fundamental thermodynamics [28–31].

These results are of interest for the relations between fragrances and different personal care products, but it must be borne in mind that a product changes composition rapidly after application, because more than 90% of the water is lost within half an hour. Hence, the vapor pressures in the water-poor part of the system is of pronounced interest and vapor pressures in the entire phase diagram must be clarified.

III. PHASE DIAGRAMS

A combination of a complete phase diagram with vapor pressure measurements first appeared very recently [32]. The diagram, Fig. 2, is useful because the surfactant laureth 4 (Brij 30) behaves as a single well-defined compound in spite of its commercial quality.

The phase behavior, Fig. 2, is described in detail, because a significant part of the conclusions about vapor pressures depends on details. The solubility of phenethyl alcohol in water is small, 2.1% by weight, and the solubility of the surfactant is even smaller, <1%. Water is soluble in phenethyl alcohol to 7.5%, while the phenethyl alcohol and the surfactant are mutually completely soluble. With increased amount of surfactant, the water becomes solubilized into inverse micelles at higher surfactant amounts, reaching a maximum value of 46% water by weight at a phenethyl alcohol/surfactant ratio of 3/7. The liquid surfactant

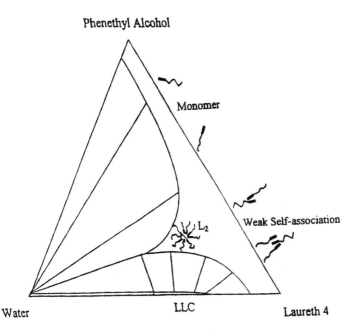

Figure 2 Phase diagram of water, laureth 4, and phenethyl alcohol (see text).

dissolved water to 12% and to a lamellar liquid crystal between 27 and 50% water. The liquid crystal solubilized phenethyl alcohol to 2% by weight.

The resulting two-phase region between the phenethyl alcohol (PEA)/surfactant solution with solubilized water and the lamellar liquid crystal covered the lower part of the phenethyl alcohol/surfactant solution from zero phenethyl alcohol to a phenethyl alcohol/surfactant ratio of 3/7, at which point maximum water was solubilized. This last point was in equilibrium with an aqueous solution with 1.3% phenethyl alcohol and with the lamellar liquid crystal of high water content 49.5% and small amounts of phenethyl alcohol (1.3%), leaving a small two-phase region along the water/surfactant axis between the water and the liquid crystal.

The second two-phase region consists of the aqueous solution with phenethyl alcohol dissolved in the range 1.3–2.1% and the part of the phenethyl alcohol/surfactant solution with phenethyl alcohol/surfactant ratios in excess of 3/7 with water solubilized to the limit. It should be observed that the tie lines in this two-phase region are determined exactly from the vapor pressure data.

A. Vapor Pressures, Association Structures

The vapor pressure of phenethyl alcohol in its interaction with water is described in Fig. 3. The right part of the figure shows the reduction of phenethyl alcohol

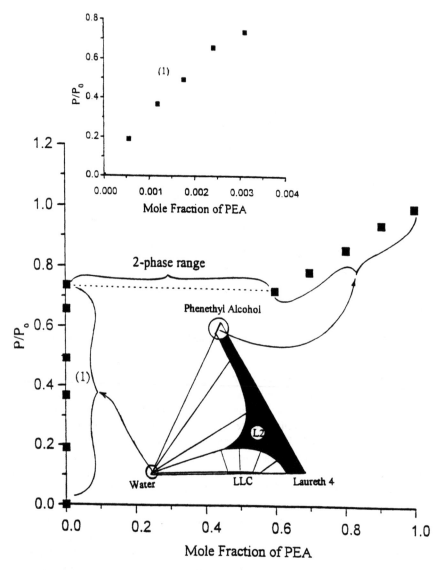

Figure 3   Vapor pressures in water/phenethyl alcohol solutions.

vapor pressure with added water, slightly in excess of those for an ideal solution. When phenethyl alcohol is added to water, the vapor pressures are much higher than these in an ideal solution.

The pressure variation when surfactant is added to phenethyl alcohol is characterized by an initial part of slightly lower values than these for an ideal solution (dashed curve), while in the surfactant-rich part, vapor pressures slightly higher than those in an ideal solution were found (Fig. 4).

The vapor pressures for compositions along the limit of water solubility in the phenethyl alcohol/surfactant region (Fig. 5) show the regular variation with phenethyl alcohol mole fraction calculated on the two amphiphiles only. The values follow those for an ideal solution except for the range of high water content, in which they are significantly higher.

B. Thermodynamic Consequences

These results provide complementary information to the extensive studies on micellar solubilization about the importance of the colloidal state per se for the

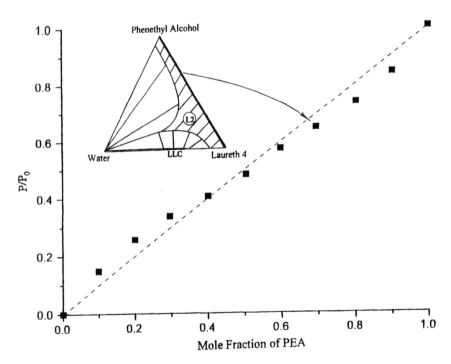

Figure 4 Vapor pressures in laureth 4/phenethyl alcohol solutions.

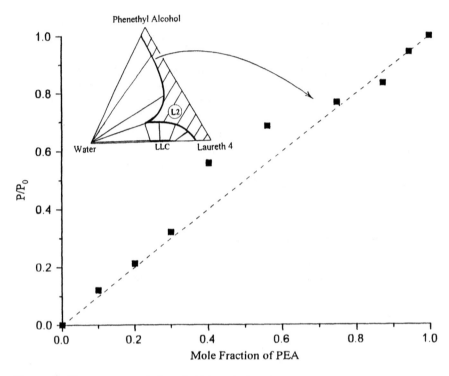

Figure 5 Vapor pressures in laureth 4/phenethyl alcohol solutions with maximum water content.

activity of a solubilized organic compound. It should be observed that the natural logarithm of the activity gives a measure of the free energy change for transferring the compound from its pure state to its solubilized one.

$$\mu = \mu^0 + RT\ell n a_i \tag{9}$$

in which the standard state $a_i = 1$ is chosen as the pure compound i. In the present treatment the conditions in the gaseous state are considered ideal from zero pressure to p_0 (10^{-3}–10^{-4}). Hence, an activity constant, γ, is defined by

$$\mu = \mu^0 + RT\ell n(\gamma x_i) \tag{10}$$

in which x_i is the mole fraction of solubilized material. It is easily realized that for an ideal condensed solution $\gamma = 1$, and that $RT\ln\gamma$ is the free energy change when one mole of solubilizate is transferred from an ideal solution with mole fraction x_i to a real solution also of mole fraction x_i, but an activity coefficient of γ_i.

With this background, the free energy change becomes $2.5\ell n\gamma_i$ kJmol^{-1}.

It is instructive to compare the values of γ and the free energy change at dilute solutions when changing the solvent from surfactant to water. This change means transversing media of different hydrophobic/hydrophilic character from the liquid surfactant with its weak association of polar groups over a gradual change to an inverse micellar solution with increased water content, to a phase change to a liquid crystal with water content varying from 27 wt% to 50 wt% and another phase change to water with extremely low concentration of surfactant and no micelles.

The solution in the nonionic surfactant, Table 1, shows $\gamma = 1.5$ or $\mu-\mu_{ideal}$ $= 1.0$ kJmol^{-1}. The difference from ideality is small; obviously the interaction between the phenethyl alcohol and the surfactant is not significantly influenced by the weak association of the latter.

Adding water to the surfactant to its maximum solubility 12 wt%, $x_w = 0.61$, Table 1, increases the chemical potential difference to 2.7 kJmol^{-1}; still a small value. The phase change to lamellar liquid crystal results in an increase to 10.2 kJmol^{-1}, reflecting the constraint by the ordered structure on the location of the phenethyl alcohol (Fig. 6). Maximum water in the lamellar liquid crystal gives a smaller increase to 12.5 kJmol^{-1}, illustrating the influence both of the

Table 1 The Activity Coefficient and Chemical Potential Change for Phenethyl Alcohol (PEA) and Phenethyl Acetate (PEAc) in Water/Laureth 4 (L4) Combinations and in Selected Organic Solvents

Water/L4 wt% ratio	Structure	γPEA	γPEAc	$\mu - \mu_{ideal}$ PEA	$\mu - \mu_{ideal}$ PEAc
100/0	Liquid	3.4×10^2	5.2×10^3	14.6	21.2
50/50	Lamellar liquid crystal	$\approx 1.5 \times 10^2$	47.3	12.5	9.6
27/73	Lamellar liquid crystal	≈ 60	12.3	10.2	6.2
12/88	Liquid	2.9	4.0	2.7	3.5
0/100	Liquid	1.5	1.3	1.0	0.65
Decane	Liquid	—	8.4	—	5.3
Tetraethylene glycol	Liquid	—	3.1	—	2.8
Benzene	Liquid	—	2.4	—	2.2

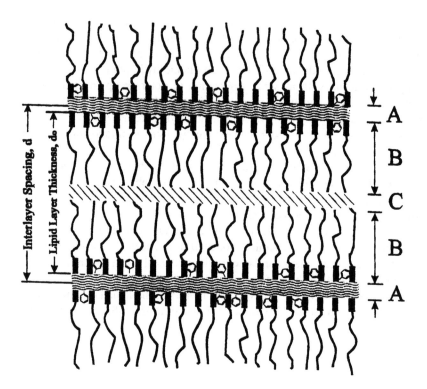

Figure 6   The structure of a lamellar liquid crystal may be envisioned as consisting of three layers: (A) aqueous layer and polar groups, (B) amphiphile hydrocarbon chains, and (C) solubilized compared between the B layers. The main location of a solubilized aromatic compound is indicated.

high water content and the location of the aromatic alcohol close to the aqueous part (Fig. 6). The value is now approaching the value for water 14.6 kJmol^{-1}.

The solubility of the PEA in water (Fig. 2), is extremely small; γ is at a level of 10^2–10^3 and the free energy difference from an ideal solution is large. This reflects the predominantly hydrophobic nature of the phenethyl alcohol modified only by the hydroxide group.

Hence, the hydrogen bond interaction plays only a small role for the vapor pressures, but extremely weak associations such as those of a polyethyleneglycol alkyl ether in an aromatic hydrocarbon [33] are of some consequence for the vapor pressure of an aromatic compound such as phenethyl alcohol. The vapor pressure of phenethyl alcohol in the nonionic surfactant shows values slightly less than those in an ideal solution for mole fractions of surfactant less than 0.5

(Fig. 4). Under these conditions the surfactant exists in monomeric state as shown by Christenson [33], and all parts of the molecule interact with the solute. The vapor pressure is now entirely monitored by the molecular interactions. Mole fractions in excess of this value leads to a weak association of the surfactant polar parts [33], and their interaction with the solute is now to some degree restricted enhancing the frequency of interaction between the nonpolar part of the surfactant and the phenethyl alcohol. The small excess vapor pressure compared with an ideal solution for surfactant mole fractions greater than 0.5 is most probably due to this limitation of molecular interactions. This conclusion is supported by the comparatively high value for decane solutions [32].

The values of vapor pressures should be useful to estimate its variation during evaporation of a formulation after application. The results of the first investigation [34] showed the vapor pressure of phenethyl alcohol during evaporation to agree in a satisfactory manner with values predicted from vapor pressure measurements of the phases involved.

IV. CONCLUSION

It is obvious that such investigations as these are only a very approximate introduction to a complex field, but the results are encouraging.

REFERENCES

1. Ballantyne K. Parfüm Kosmet 1995; 76:716.
2. Lyons AS, Petrocelli RJ. Die Geschichte der Medizin im Spiegel der Kunst. Berlin: Dumont Press, 1980.
3. Rimmel E. The Book of Perfumes. London: Illstein Press, 1988.
4. Sell CS. *Organic Chemistry in the Perfume Industry,* vol. 24, no. 8. Chemistry in Britain. 1988:791. Royal Society of Chem, Burlington House, London.
5. Ebrsama T, Teshima Y. Eur. Pat. Appl. 687,460 1995.
6. Siciliano M. U.S.Pat. 5,490,982 1996.
7. Chung SL, Tan CT, Tuhill IM, Scharpf LG. U.S. Pat 5,283,056 1994.
8. Juszynski M, Azoury R, Rafaeloff R. SÖFW 1992; 118:811.
9. Labows JN, Brahms JC, Cagan RH. In: Rhein L, Rieger M, eds. Surfactants in Cosmetics. 2d ed. New York: Marcel Dekker. In press.
10. Herman SJ. Cosmet Toilet 1994; 109:71.
11. Uchiyama H, Christian SD, Scamehorn JF, et al. Langmuir 1991; 7:95.
12. Tokuoka Y, Uchiyama H, Abe M, Ogino K. J. Coll Interface Sci. 1992; 152:402.
13. Suhaimi H, Rose LC, Ahmad FBH. Pertanika J Sci Technol 1995; 3(1):141.
14. Hamdan S, Ahmed FBH, Laili CR, Fauziah H. Oriental J Chem 1995; 11(3):220.
15. Tokuoka Y, Uchiyama H, Abe M. Coll Polym Sci. 1993; 272:317.

16. Abe M, Mizuguchi K, Kondo Y, et al. J Coll Interface Sci 1993; 160:16.
17. Friberg SE, Vona SAJr, Brin AJ. Coll Surf. In press.
18. Yang J, Rong G, Friberg SE, Aikens PA. Int J Cosmet Sci 1996; 18:43.
19. Friberg SE, Vona SAJ. Soap Cosmet Chem Spec, August 1994, page 32.
20. Friberg SE, Yang J, Huang T. Ind Eng Chem Res 1996; 35:2856.
21. Sorrentino F. PhD thesis, University of Bourgogne, Dijon, 1984.
22. Christian SD, Tucker EE, Land EH. J Coll Interface Sci 1981; 84:423.
23. Tucker EE, Christian SD. Faraday Symp Chem Soc 1982; 17:11.
24. Christian SD, Tucker EA, Smith GA, Bushong DS. J Coll Interface Sci 1986; 113: 439.
25. Smith GA, Christian SD, Tucker EE, Scamehorn JF. J Coll Interface Sci 1989; 130: 254.
26. Mahmoud FZ, Higazy WS, Christian SD, et al. J Coll Interface Sci. 1989; 131:96.
27. Tucker EE. In: Christian SD, Scamehorn JF, eds. Solubilization in Surfactant Aggregates. New York: Marcel Dekker, chap 13.
28. Chew CH, Wong MK. J Disp Sci Technol. 1991; 12:495.
29. Ueda M, Schelly ZA. J Coll Interface Sci 1988; 124:673.
30. Cavallo JL, Rosano HL. J Phys Chem 1986; 90:6817.
31. Zulauf, M, Eicke, H.F. J Phys Chem 1979; 83:486.
32. Friberg SE, Huang T, Fei L, et al. Progr Coll Lett 1996; 15:1172.
33. Christenson H, Friberg SE, Larsen DW. J Phys Chem 1980; 84:3633.
34. Friberg SE, Szymula M, Fei L, et al., Int J Cosmet Sci 1997; 19:259–270.

12

Liposomes

Klaus Stanzl
Dragoco Gerberding & Co. AG, Holzminden, Germany

I. INTRODUCTION

In 1994, the total value of the cosmetic and toiletry market, including personal care stood at approximately $25 billion, a 4–5% increase over 1993. This figure includes skin-care products ($4.5 billion), color cosmetics ($4.22 billion), and sun-care products ($1.40 billion). The strong growth in sales of treatment products is probably the reason for the growth of the market (Table 1).

The projected worldwide retail sales growth indicates a strong increase for skin-care products until the year 2000 to 6.3 billion, estimated on a 6.3% average annual growth rate (Table 2).

Sales of new modern treatment products fueled a large portion of the increase, while other traditional cosmetic and personal care products suffered a loss in momentum. Skin-care products were somewhat immune to 1994s' sluggish increase, maintaining a steady 8% growth. New lines of high-performance skin-care products are merging to capture consumers, whose tastes are changing to favor multifunctional products that offer a healthful appearance.

Since its discovery in 1965 by Bangham and coworkers [1], the liposome has not ceased to spark interest in many different fields of application. It took over 20 years until the liposome found its way into cosmetic products. In 1986 the first cosmetic products containing liposomes were launched by L'Oreal (Niosomes) and Christian Dior (Capture).

Today, liposomes are in widespread use in cosmetic products. Primarily, they are to be found in products intended to make particularly valuable cosmetic substances bioavailable or to transport such substances into deeper skin

Table 1 Worldwide Cosmetic
and Cosmeceutical Industry
Market Segments, 1995

Skin care	38%
Color cosmetics	36%
Sun care	12%
Antiaging	8%
Antiacne	2%
Pigmentation	2%
Depigmentation	1%
Hair loss	1%

layers. Liposomes are also used without any added ingredient to improve skin condition.

Liposomes are lipid vesicles composed of one or more lipid bilayers surrounding an equal number of aqueous compartments. Traditionally, the membrane components of liposomes have been phospholipids, particularly phosphatidyl cholines, partly because they are the building blocks that nature itself uses to form membranes. However, bilayer membrane vesicles have been constructed using single-chain amphiphiles or nonionic surfactants in which the principles of formation and physical properties are so similar to those of conventional liposomes that there is no reason why they should not be taken into the fold and treated as subsets of the general case.

Table 2 Projected Worldwide Retail Sales Growth of
Selected Cosmetic Industry Segments ($ Millions)

Market segment	1995	2000	Average annual growth (%)
Skin care	4500	6122	6.3
Color cosmetics	4220	4569	1.6
Sun care	1400	1636	3.2
Antiaging	1000	1464	7.9
Antiacne	225	247	1.9
Pigmentation	210	307	7.9
Depigmentation	150	204	6.3
Hair loss	150	183	4.1

Source: Leading Edge Report.

When phospholipids are equilibrated with excess water or aqueous salt solutions, they form liquid crystals of the smectic mesophase type related multilamellar vesicular structures that are composed of bimolecular lipid membranes separated by an aqueous layer. Furthermore, as such multilamellar vesicles are disrupted by means of an ultrasonic disintegrator, the vesicles become very much smaller, and most of them appear to be composed of unilamellar liposomes of either spheroidal or crescent-shape. These lipid vesicles were named *liposomes* by Kinsky et al. [2]

II. COMPOSITION

A wide range of amphiphiles with strongly contrasting structures both lipid and polar portions are capable of forming themselves in sheets when mixed under low shear conditions with water. The molecules align side by side in like orientation, "heads" up and "tails" down. These sheets then join tails-to-tails to form a bilayer membrane that enclosed some of the water in phospholipid sphere. Typically, these spheres create a multilamellar structure of concentric spheres separated by layers of water. A large variety of ingredients can be entrapped within liposomes. Polar ingredients are sequestered into the internal aqueous phase, while nonpolar ingredients partition between the aqueous compartments and the lipid bilayer. This property makes liposomes valuable vehicles for transporting cosmetic or pharmaceutical ingredients to various targets.

Membranes composed wholly of unsaturated lipids are fluid, highly permeable, prone to oxidation and easily disrupted by contact with serum.

Saturated lipids form stable liposomes if incubated above the phase transition temperature.

Using high-shear processing instead of mixing the amphiphiles with low shear can lead to unilamellar liposomes.

Cholesterol reduces the fluidity of membranes above the phase transition temperature; therefore the addition is often necessary to unsaturated membranes to achieve sufficient stability.

In contrast cholesterol increases the fluidity of membranes below the phase transition temperature, so that its addition in saturated membranes, which are usually in the lower, gel phase at room temperature reduces the stability.

A. Phospholipids

Natural phospholipids (phosphatides) are triglycerides with two long chain fatty acids (R_1, R_2) and a phosphate containing head group (Fig. 1).

The fatty acids can be of different chain length and grade of saturation and may have different hydrophilic groups linked to the phosphate.

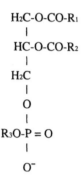

H₂C-O-CO-R₁
 |
HC-O-CO-R₂
 |
H₂C
 |
O
 |
R₃O-P = O
 |
O⁻

Figure 1 Chemical structure of phospholipids.

In mammals, the fatty acids of membrane phospholipids have even numbered chains of 16 or greater, up to 24. The fatty acids may be either saturated or unsaturated with single unsaturations being located usually in the middle of the chain in the 9 position. The double bonds are usually all in the *cis*-configuration. Multiple unsaturations are spaced along the chain every three carbon so that none of the double bonds is conjugated with each other.

Plant lipids have lower proportions of saturated fatty acids than those of egg yolk or mammalian sources.

The most important phospholipids are phosphatidyl ethanolamine (PE) (Fig. 2) and phosphatidyl choline (PC) (Fig. 3).

PC is the predominant phospholipid found in natural membranes.

PE has a smaller head group than PC, and the presence of hydrogen directly attached to the nitrogen of ethanolamine permits interactions of adjacent molecules in the membrane by hydrogen bonding.

Both product categories consist of a mixture of phosphatides linked with various saturated and unsaturated fatty acids, like palmitic-, stearic-, oleic-, linoleic-, or linoleic acid. The fatty acid composition of phosphatidyl cholin from soybean is shown in Table 3.

In sphingolipids (Fig. 4), the lipid sphingosine (Fig. 5) forms one of the hydrocarbon chains and is linked directly to the phosphate.

Sphingolipids can be isolated from brain, spleen, or spinal cord.

O-CH₂-CH₂-N⊕H₃
 |
-P=O
 |
O⁻

Figure 2 Chemical structure of the phosphatidyl ethanolamine part of phospholipids.

O-CH₂-CH₂-N⊕(CH₃)₃

$$O\text{-}CH_2\text{-}CH_2\text{-}N^{\oplus}(CH_3)_3$$
$$|$$
$$\text{-}PO_2\text{-}O^-$$

Figure 3 Chemical structure of the phosphatidyl choline part of phospholipids.

Table 3 Composition of Fatty Acids of Phosphatidyl Choline from Soya Bean

Fatty acid	Fatty acid R_1 (%)	Fatty acid R_2 (%)	Total (%)
C 16:0 Palmitic acid	24.0	1.7	12.9
C 18:0 Stearic acid	7.9	1.0	4.4
C 18:1 Oleic acid	10.9	10.0	10.5
C 18:2 Linoleic acid	52.4	80.6	66.5
C 18:3 Linolenic acid	4.7	6.7	5.7

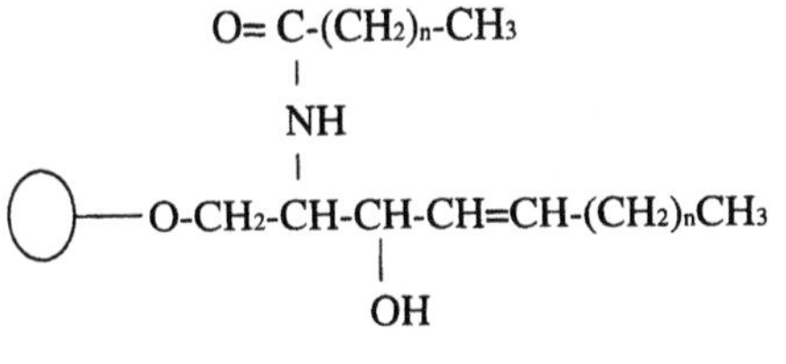

$$O= C\text{-}(CH_2)_n\text{-}CH_3$$
$$|$$
$$NH$$
$$|$$
$$\bigcirc\!\!-\!\!O\text{-}CH_2\text{-}CH\text{-}CH\text{-}CH=CH\text{-}(CH_2)_nCH_3$$
$$|$$
$$OH$$

Figure 4 Chemical structure of sphingolipids.

$$NH_2$$
$$|$$
$$HOCH_2\text{-}CH\text{-}CH\text{-}CH=CH\text{-}(CH_2)_n\text{-}CH_3$$
$$|$$
$$OH$$

Figure 5 Chemical structure of sphingosin.

NH-CO-R^1
|
HOCH$_2$-CH-CH-CH=CH-(CH$_2$)$_n$-CH$_3$
|
OH

Figure 6 Chemical structure of ceramide.

B. Glycolipids

If a fatty acid is linked to the amino group of sphingosin, ceramides are yielded (Fig. 6). If, in addition; to the terminal alcohol group of sphingosin, a sugar molecule is linked, cerebrosides are received (Fig. 7).

C. Nonphospholipids

Hadjani et al. [3–9] have made lipid vesicles called *niosomes* from single or two-tailed ether or ester derivatives of polyglycerol or polyoxyethylene. These liposomes are being produced on a large scale by L'Oreal in France for use in cosmetic products. Different combinations of lipid moieties and polar head groups joined by an ether or ester have been tested for niosome stability and permeability [10]. There exist a large number of amphiphilic non-phospholipid molecules, which are able to form bilayers. The hydrocarbon chain can vary from simple fatty acids to more complex alkyd types with several unsaturated carbon-carbon bonds in the chain. The head group can be as simple as —OH and can also be a very complex molecule. The bonding between the head group and the hydrocarbon chain can be of the ether or ester type, or it can be an amid bond. Philippot published a very good summary [11]. An extract is given in Table 4.

In 1984 Sebag et al. [12] presented at the IFSCC Congress the development of a group of nonionic lipids composed of two saturated hydrocarbon chains of different length and a polygycerol polar head with an ether link (Figure 8).

The varied lengths of these lipid chains modulate the transition tempera-

NH-CO-R^1
|
Sugar-O-CH$_2$-CH-CH-CH=CH-(CH$_2$)$_n$-CH$_3$
|
OH

Figure 7 Chemical structure of cerebroside.

Table 4 Principal Wall-Forming Materials

Classification	Hydrocarbon chain	Bond	Head group
Fatty alcohol	C_{12}–C_{20} (0–1 unsaturation)		—OH
Fatty acid	C_{12}–C_{20} (0–1 unsaturation)		—COOH
Ethoxyl. fatty alcohol	C_{12}–C_{20} (0–1 unsaturation)	—O	—$(CH_2CH_2O)_{2-5}H$
Ethoxyl. fatty acid	C_{12}–C_{20} (0–1 unsaturation)	—CO—O	—$(CH_2CH_2O)_{2-5}H$
Fatty acid diethanolamide	C_{12}–C_{20} (0–2 unsaturation)	—CO—N	—$(CH_2CH_2OH)_2$
Fatty acid sarcosinate	C_{12}–C_{18} (0 unsaturation)	—CO—N—CH_3	—CH_2—COOH

Source: From Ref. 11.

ture—that is, the fluidity of the vesicles that manage these properties. There is a significant difference of fluidity as a function of the hydrocarbon chain length. The hexadecyl derivative ($C_{16}G_2$) exhibits a sharp gel-liquid crystal phase transition at 40°C. The dodecyl derivative ($C_{12}G_2$) exhibits a gel lamellar transition at a temperature lower than 5°C [13]. For the dialkyl chain derivative $C_{12}/C_{16}G_7$ (Fig. 8), the transition temperature occurs between 5 and 25°C. These results can be compared to those obtained with Phosphatidyl choline, where the transition temperature is around 40°C and egg or soybean lecithin with a transition temperature below 5°C. Due to its transition temperature below the room temperature, the compound $C_{12}/C_{16}G_7$ is ideal to manufacture vesicles at a low temperature which is favorable for the encapsulation of thermosensitive ingredients. The addition of a slight percentage of charged lipids like dimyristyl phosphate to the $C_{12}/C_{16}G_7$ increases the zeta potential of vesicular dispersions, which increases stability.

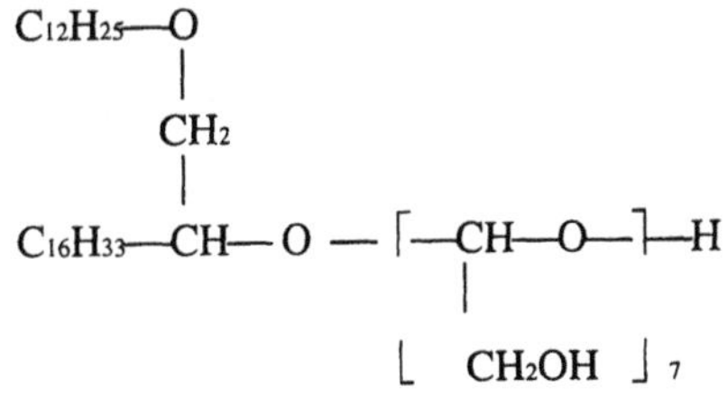

Figure 8 Nonionic lipid $C_{12}C_{16}G_7$.

D. Types of Liposomes

Liposomes can be classified according to their size, morphology, or method of preparation.

1. Morphology

Unilamellar vesicles (UL). Unilamellar vesicles can be liposomes of any size that are composed with one bilayer membrane only.

Multilamellar vesicles (MLV). MLVs are liposomes of any size with more than one bilayer membrane. Since even a liposome of just two bilayers is at least twice the size of an SUV, MLVs are easy to distinguish from SUVs by size. MLVs are the type of liposome formed by gentle manual shaking of dry phospholipids in water. The lamellarity of these MLVs depends on lipid composition among other factors, but it usually varies among 5 and 20 bilayers. Liposomes with lower numbers of lamellae are sometimes referred as oligolamellar.

2. Size

Small unilamellar vesicles (SUV). This term refers to liposomes of the smallest possible size that the shape of bilayer membranes will permit from a steric point of view. Ultrasonication, together with certain high-pressure extrusion techniques, is the method that is capable to give liposomes in this small size range of 50–100 nm.

Large unilamellar vesicles (LUV). Introduction of ether solutions of PC into hot aqueous buffer forms large planar sheets of bilayer membrane that folded in on themselves. Liposomes produced according to this method where of the order of 500 nm in size.

Intermediate-sized unilamellar vesicles (IUV). This is a term that has not been widely used. It refers to liposomes within the 100- to 200-nm region. Liposomes of this size can be prepared by high-pressure extrusion or by detergent dialysis.

3. Method of Preparation

Many methods of liposome preparation have been developed [14,15]. Amselem et al. describe several methods for large scale preparation of liposomes [16]. A summary is given in Table 5.

III. PROPERTIES

A. Phospholipids

Phospholipids are amphiphilic molecules. They are soluble only in organic solvents. In water, they aggregate spontaneously. The structure of the aggregates

Table 5 Preparation Possibilities for Liposomes

Method	Liposomes received
Mechanical shaking	MLV
Organic solvent(s) replacement	MLV, OLV, UV
Removal of detergent from mixed micelles	OLV, UV, SUV
Ultrasonic irradiation	SUV
High-pressure extrusion (above 5000 psi)	UV
Medium-pressure extrusion (up to 2000 psi)	UV, MLV

depends on the chemical nature of the head group, the length and degree of saturation of the fatty acyl chain, and the pH and ionic strength of the aqueous phase.

The hydrated lamellar phases formed by most phospholipids are also a basic feature of biological membranes. The hydrated lamellar phases can occur in at least two different states, the so-called gel phase, in which the lipid molecules are packed in a quasicrystalline two-dimensional lattice, and the liquid crystalline phase, in which lipids retain their lamellar arrangement but are now able to diffuse rapidly in the plane of the lipid bilayer [17]. The transition temperature of the gel to liquid crystalline phase depends on the nature of the head group and the length and degree of unsaturation of the chain. It can vary from −60°C for dimyristoleyl PC to 80°C for ditetracosanoyl PC. Plant and egg PCs have a transition temperature below 0°C, while mammalian PCs are considerably higher. Phospholipids with longer fatty acid chain have higher transition temperatures.

Unsaturation and branching lead to lower transition temperatures. Head groups which are able to form intermolecular hydrogen bonds, like PE, have higher transition temperatures [18].

The physical properties of membranes, depend on their phase state. The permeability for liquid-crystalline bilayers is higher than for gel-phase bilayers. At the phase transition sometimes a sudden jump of the permeability can be observed [19].

With this in mind, the physicochemical behavior of liposomes can be tailored to one's need by modifying the structure of the chains and the head group.

Water diffuses rapidly through phospholipid bilayers and biological membranes. To obtain a clearer understanding of this diffusion process, Blume et al. [20,21] measured the water permeability of several lipid classes as a function of chain length and head group structure. Their findings can be summarized as follows:

For PC, the length of the saturated fatty acyl chain has no marked influence on the water permeability. When the chain length is kept constant and the structure of the head group is varied, one observes a reduction in permeability only for phosphatic acid, probably due to the intermolecular hydrogen bonds between the head groups. Much larger effects are observed when the linkage or the structure of the chains is altered. The diether analogue of PC, dihexadecyl-phosphatidyl choline (DHPC), and the compound 13cyPC, a phosphatidyl choline with two ω-cyclohexyl fatty acyl chains, show much lower water permeability than the corresponding PC. DHCP forms interdigitated gel state bilayers [22]. This leads to a reduction in bilayer thickness. If the water permeability depends on the bilayer thickness, an increase in permeability would be expected. This is not observed and confirms results for ester that water permeability does not depend on bilayer thickness. The low activation energy for water permeation, the sudden jump in permeability at the phase transition temperature, and the absolute values of the permeability coefficients support the view that water diffusion probably continues through fluctuating pores or defects [23,24].

Synthetic phospholipids can be synthesized with the phospholipase D reaction. Phospholipids with an ethylenglycol head group can then be further modified by linking to the residual hydroxyl group other molecules. Polymerizable moieties have been linked to the fatty acyl chain also to the lipid head groups [25,26].

The major advantage of polymerized phospholipids is the greater long-term stability of these systems compared with monomeric lipids.

B. Glycolipids

Ceramides, cholesterol, and fatty acids are the most abundant lipids in the stratum corneum. (Table 6). The fatty acids found in stratum corneum consist of a series of straight chain saturated species from 16–28 carbon in length [27]. These lipids

Table 6 Composition of Stratum Corneum Lipids

Cholesterolesters	10.0%
Triglycerides	0.0%
Fatty acids	9.1%
Cholesterol	26.9%
Ceramide 1	3.2%
Ceramide 2	8.9%
Ceramide 3	4.9%
Ceramide 4	6.1%
Ceramide 5	5.7%
Ceramide 6	12.3%
Cholesterol sulfate	1.9%
Others	11.1%

form the multiple intercellular lipid bilayer in mammalian stratum corneum, which constitute the epidermal water barrier. However, unlike all other biological membranes, the stratum corneum does not contain phospholipids, as a result the ability of the stratum corneum lipid mixture to form bilayers has been questioned.

Ceramides alone or in combination with cholesterol do not form liposomes. However, when ceramides combined with cholesterol and either fatty acids or cholesterol sulfate were sonicated at temperatures above 80°C, stable, small unilamellar vesicles ranging from 20–150 nm in diameter were obtained [28].

C. Nonphospholipid Molecules

Like in phospholipid or glycolipid liposomes, cholesterol can modify the properties of nonphospholipid vesicles. Cholesterol can intercalate between amphiphile hydrocarbon chains in bilayers, and thereby allows the intermixing of different acyl chains without phase segregation and broadens the range of temperature within which the crystalline liquid-crystalline transition occurs [29]. Phytosterols and ethoxylated sterols impart similar effects.

Ionogenic membrane amphiphiles can be used in conjunction with any of the nonionics, ranging from 0.05 to 10 mol%. Of the anionic modules, the fatty acids must be used at a pH level of less than 7. Diacylphosphates can be used over a broad pH range. Spacer modules can be employed primarily to modulate the accessibility of the external bilayer. Ethoxylated amphiphiles and propoxylated amphiphiles allow the placement, on top of the outermost bilayer surface.

IV. MANUFACTURING

The production of liposomes is a four-step process (Fig. 9, Fig. 10). Natural phospholipid fractions are available as a plastic, viscous material. The first step

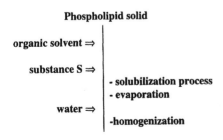

Phospholipid solid

organic solvent ⇒

substance S ⇒

- solubilization process
- evaporation

water ⇒

-homogenization

Liposomes loaded with S

Figure 9 Preparation of loaded liposomes from phospholipids.

is the mixing of amphiphilic molecules in organic solvents. In case of the preparation of loaded Liposomes with hydrophobic ingredients, the organic solvent must simultaneously be a solvent for the added substance. The next step is the separation from the solvent followed by the addition of water. The addition of water causes the detachment of phospholipid bilayers. The last step is the homogenization. The mechanical energy input at this stage determines the particle size and homogeneity of the liposomes formed. For the manufacturing of loaded liposomes, several well-described methods are currently available [30].

An alternative liposome concept was introduced by Röding [31] (Fig. 11). The principle of this system is extraordinarily simple. Natipide II consists of densely packed liposomes. They are so tightly packed that they penetrate one another completely. The system is semisolid and transparent, a gel composed exclusively of liposomes. If a substance can be dissolved in this vesicular gel, then it must be in the enclosed phase.

This method is particularly suitable for the preparation of liposomal systems with hydrophilic and amphiphilic substances. However lipophilic terpenoids such as bisabolol can be encapsulated up to 3% (w/w).

The first evidence that vegetable oils associate with liposomes in significant amounts was described in U.S. 4,874,553 [32], and in 1989 the electron microscopic proof of stabilization in "propeller liposomes" was presented [33].

Several techniques exist for the elimination of organic solvents. Solvent evaporation [34] is the classical one. A second technique is lyophilization. A final technique is the spray-drying method [35].

Detergents can be eliminated with the dialysis techniques [36].

The hydration step seems simple. Depending on whether the lipid mixture obtained is a film or a powder, the quantity of aqueous phase or buffer can vary by 50%. The hydration time for a film is much longer, up to 20 h, compared to a powder, where the hydration time is short (1 to 2 h).

The homogenization is the ultimate step to achieve liposomes of uniform size. There are many commercial machines available, like the Ultra Turrax (Janke

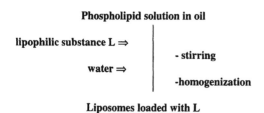

Figure 10   Preparation of phospholipids loaded with lipophilic substances.

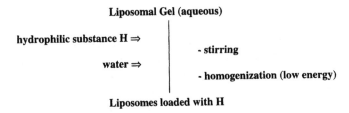

Figure 11 Preparation of loaded liposomes starting from a liposomal gel.

und Kunkel, Germany), the Microfluidizer (Microfluidics Corp., USA), the Nano-jet (Nanojet Engineering, Germany), or the Brogli (Italy).

Depending on the stirring time and the equipment used, microvesicles of different particle sizes are obtained [37].

A totally different Liposome was prepared by Riess [38] et al. He described a system composed of perfluorocarbons and phospholipids, where he found a monolayer of lipids around the perfluorocarbon containing core followed by the regular bilayer system. Therefore he got always an uneven number of lipid layers around the core. Stanzl et al. called this system the asymmetric oxygen carrier system (A*O*C*S) [39]. As perfluorocarbons are excellent carriers for oxygen, they used this system to transport oxygen to the skin.

V. STABILITY

The stability of unloaded liposomes depends on the lipid composition, the manu-facturing process, and the bacterial contamination.

Liposomes composed of egg lecithin show the lowest stability compared to egg lecithin cholesterol mixtures (5:1) or egg lecithin α-tocopherol mixtures (5:1). The time it takes that 50% of the liposomes are deteriorated was only 13 days at room temperature and 100 days at 4°C. Cholesterol increased this time to 50 days and 180 days, respectively. The best effect of stabilization was ob-tained with liposomes containing α-tocopherol [40].

Without cholesterol, a rapid increase in the size of liposomes can be ob-served. The oxidation of unsaturated lipids may occur as well as their hydrolysis. The addition of α-tocopherol improves the stability. However, when disodium EDTA was added to chelate heavy metals and block their catalytic effect, the oxidation process was suppressed.

The long-term stability of liposomes prepared from hydrogenated phospha-tidyl choline was investigated by Arakane et al. [41]. When incompletely hydrogenated soy phosphatidyl choline liposomes were stored at 40°C, pH decrease and the concomitant change of liposomal appearance were observed. It was found that this pH decrease was induced by the peroxidation of the residual unsaturated fatty acyl chains in the soy PC. On the contrary, the liposomes composed of completely hydrogenated soy phosphatidyl choline, did not change their appearance, pH and particle size at a wide range of thermal conditions. If dicetylphosphate, a negative charged lipid was incorporated into the liposomes the PC hydrolyzes drastically.

Chemical or enzymatic hydrolysis can cut off a fatty acid chain to give single chain amphiphiles. In high concentrations these lysophospholipids can disrupt membranes, and, indeed, can be highly toxic for cells and whole organisms.

Hydrogenated soybean phosphatidyl choline vesicles were prepared by microfluidization, ultra sonication and detergent dialysis [42]. In this system only liposomes produced by detergent dialysis are physically stable.

This study shows that the proper choice of the manufacturing process is crucial for the stability of a certain liposome.

When the stability of loaded liposomes is observed, besides the issues mentioned before, the stability of the encapsulated ingredient also has to be taken into account.

Liposomes, especially of natural origin, are prone to bacterial contamination. If the liposome dispersion is not produced and stored under sterile condition, the liposome dispersion has to be preserved. Preservatives can migrate into the water compartment or in the hydrophobic part of the liposome, which makes them unavailable for protecting the product. The choice of the preservative system is essential for its efficacy. The efficacy of the preservative system also has to be checked over time, because it takes some time before the equilibrium between available preservatives and encapsulated preservatives is achieved.

VI. CHARACTERIZATION OF LIPOSOMES

A. The Freeze-Fracture Technique

This technique allows the detection of encapsulated lipophiles and nonencapsulated lipophils. The samples are placed between two copper plates, cryofixed by being plunged into liquid ethane, broken and etched under high vacuum and low temperature and vapor-coated with Pt and C. The replicas are assessed under the transmission electron microscope. Liposomes are mostly fractured in the bilayer, oil droplets randomly [43,44].

B. With Cryoelectron Microscopy

This technique [45] makes it possible to determine the essential physical parameters like particle size, size distribution, lamellarity and its distribution and simultaneously to detect the presence of nonvesicular constituents such as oil droplets. In order to determine these parameters, the liposomes must be imaged in their native state. Only cryofixation and subsequent cryoelectron microscopy of the particles in amorphous ice make it possible to recognize the structure and form of the vesicles and other particles.

VII. EFFICACY OF PREPARATIONS CONTAINING LIPOSOME FOR TOPICAL APPLICATION

A. The Skin

Stratum corneum is composed of interlocking, vertical columns of polyhedric, protein enriched corneocyte cells without nuclei but with thickened membranes embedded in a lipid enriched matrix [46]. The corneocyte cells originate in the basal layer as keratinocytes. As they migrate upwards they flatten and gradually lose their nuclei, becoming enriched in protein.

The extracelluar lipids of the stratum corneum exist in layers called intercellular lamellae. They appear to arise as ellipsoidal organelles in the first suprabasal cell layer (stratum spinosum), and they continue to accumulate in the stratum granulosum until they account for up to 25% of the volume of the cytosol.

The ceramides, and the sphingolipids of which they are the major species, are presumed to be of great importance for the epidermal barrier. Elias [47] has shown that mammalian stratum corneum contains multiple intercellular lipid bilayers that constitute the epidermal water barrier.

The epidermis functions as a permeability barrier that prevents excessive loss of the body fluids required for life. This barrier is mainly located in the stratum corneum, which is organized into a two compartment system of protein-enriched cells embedded in lipid-enriched intercellular membrane bilayers. The lipids are a mixture of Sphingolipids, cholesterol, and free fatty acids (Table 7), which form intercellular membrane bilayers. Lipid synthesis occurs in the keratinocytes, and the newly synthesized lipids are delivered by lamella bodies to the interstices of the stratum corneum during epidermal differentiation. Proksch published this very interesting review of Elias's research [48].

Disruption of the barrier layer by topical acetone treatment brings increased synthesis of free fatty acids, Sphingolipids and cholesterol in the living layers of the epidermis, leading to barrier repair. Cholesterol and sphingolipid synthesis are regulated by the rate-limiting enzymes HMG CoA reductase and serine palmitoyl transferase, respectively. Acute barrier disruption leads to increased levels of both enzymes. After acute and chronic disturbances, not only lipid but DNA synthesis is stimulated. Topical application of enzyme inhibitors after acetone treatment delays barrier repair. It is interesting that artificial barrier by latex occlusion prevents an increase in lipid and DNA synthesis. In addition, increased

Table 7 Variation in Lipid Composition During Epidermal
Differentiation and Cornification

Lipid	Basal layer	Granular layer	Cornified
Phospholipids	44.5 ± 3.4	25.3 ± 2.6	6.6 ± 2.2
Cholesterolsulfate	2.6 ± 3.4	5.5 ± 1.3	2.0 ± 0.3
Neutral lipids	51.0 ± 4.5	56.5 ± 2.8	66.9 ± 4.8
Free sterols	11.2 ± 1.7	11.5 ± 1.1	18.9 ± 1.5
Free fatty acids	7.0 ± 2.1	9.2 ± 1.5	26.0 ± 5.0
Triglycerides	12.4 ± 2.9	24.7 ± 4.0	Variable
Stero wax/esters	5.3 ± 1.3	4.7 ± 0.7	7.9 ± 1.2
Squalene	4.9 ± 1.1	4.6 ± 1.0	6.5 ± 2.7
n-Alkane	3.9 ± 0.3	3.8 ± 0.8	8.2 ± 3.5
Sphingolipids	7.3 ± 1.0	11.7 ± 2.7	24.4 ± 3.8
Glucosylceramides	3.5 ± 0.3	5.8 ± 0.2	trace
Ceramides	3.8 ± 0.2	8.8 ± 0.2	24.4 ± 3.8

epidermal lipid and DNA synthesis in essential fatty acid deficiency can be reversed by topical application of the n-6 unsaturated fatty acids.

Epidermal ceramide synthesis following skin damage was examined by Holleran [49]. The investigator reports that barrier recovery is rapidly repaired (up to 75%) within 6 hours after injury. However, the initial lipid synthesis is limited to cholesterol and fatty acid, and de novo sphingolipid synthesis does not begin until more than 6 h after injury. Their study of β-chloro-L-alanine, an inhibitor of sphingolipid synthease, led them to conclude that the early barrier repair (shortly after injury) depends on a preformed pool of lipids within the epidermis. If this supposition is confirmed, any skin treatment which supplements this lipid pool would represent a true skin protectant against future damage.

An article by Imokawa et al. [50] provides information about the function of the stratum corneum lipids as skin moisturizing agents. In previous publications, Imokawa et al. [51,52] reported that the stratum corneum lipids which form the lamellar structure serve as water modulator in the stratum corneum. This water holding function of the stratum corneum is thought to be completely different from the water permeability barrier. Recent evidence suggests that ceramides are a main determinant for both functions. Ceramides with short, nonbranched, saturated alkyl chains are mainly associated with the water-holding capacity of the stratum corneum, whereas acylceramides with linoleic acid, or ceramides with long-chain alkyl groups serve as the permeability barrier.

This review of the structure of skin clearly shows that phospholipid or ceramide vesicles should interfere with the lipid barrier of the skin.

B. Penetration of Liposomes into Human Stratum Corneum

One of the key questions which always arise concerning liposome is: Are intact liposomes able to penetrate into the stratum corneum or they able to penetrate even into deeper layers of the skin?

Mezei et al. [53] and later Cevc [54] found evidence for the penetration of intact liposomes into deeper layers of the skin. Other studies [55,56] showed controversy results. In a study to investigate the interaction between liposomes and the skin Junginger et al. [57] used freeze-fracture electron microscopy (FFEM) and small x-ray scattering (SAXS). The investigators treated human stratum corneum in vitro with three types of liposomes. The liposomes had comparable sizes and were mainly unilamellar. The main difference between the different types of liposomes was their hydrophilicity and the charge of the head groups of the lipids. The results reported by the authors indicate that the degree of interaction between vesicular dispersion and the skin mainly depends on the physicochemical properties of the compounds of which the liposomes are composed. The intersection is most likely to be dependent on the size, mean number of bilayers, head groups, and alkyl chain length of the lipids. Although the different types of liposomes had comparable size lamellarity, and alkyl chain length, the compositional differences between the vesicles were very complex. Therefore it was very difficult to draw conclusions on the effect of each of the phospholipids on the interaction between vesicles and stratum corneum from these experiments. The interactions can be limited to only the interface, but the changes in lipid structure can also extend to deeper layers of the stratum corneum.

Another study conducted by Artmann and Röding [58] proves the uptake by the skin of phospholipids of liposome applied to pig skin. In detail, the authors found that 3 h after application, the concentration of PL in the horny layer is 200 times higher than in the first section of living tissue. Another significant concentration drop was found from the stratum corneum to the subcutaneous fat by a factor of 10,000. From 1 mg PL in the horny layer only 0.1 µg arrives in the deep-lying skin layers. This effect was shown of a one-time application of liposomes without occlusion. In other experiments the authors have proven the transport of macromolecules and drugs by using the occlusive technique [59,60].

The presented data show that the maximum useful dosage is 1 mg PL/cm^2 skin. Thus only a limited amount of phospholipid can be taken up by the horny layer. The distribution of the phospholipid concentrations in the skin tissues shows that liposomes have a particularly high affinity to the horny layer (Table 8).

This study shows that 99.5% of the total measured phospholipid in skin is accumulated in the horny layer.

Gehring et al. [61] reported the influence of various topical preparations with and without active ingredients on the cutaneous blood flow. His results indi-

Table 8 Phospholipid Concentration
in Skin 3 h After Application of 1 mg
PL/cm^2 Skin

Skin section	μg PL/g tissue
Horny Layer	100 000
Epidermis	500
Dermis	20
Subcutis	8
Subcutis—4 mm	8
Subcutis—5 mm	12

cate an influence of liposome preparations on the blood circulation in the corium. This lead to the conclusion that the liposomes must have penetrated the stratum corneum and the epidermis at least fragmentarily. Consistently, all three empty liposomes caused a decrease in the blood supply; no comparable effect could be detected with the application of the base cream DAC under identical test conditions. The decrease reached a maximum after 90 min. In this context, the formulation of the liposomes seems to be of relatively little importance, since he observed no significant differences between three types of preparation.

Korting [62] topical applied liposomes of the large, unilamellar vesicle type and found a dose-dependent alteration in the morphology of both the stratum corneum and the living part of the epidermis. Of particular interest were shrunken lipid droplets found between the corneocytes and keratinocytes.

Corneocytes throughout the SC and keratinocytes in the upper most layers of the living epidermis exhibited particularly osmophilic membranes, indicating lipid transfer. Liposomes, both intact and in remnants, were sometimes seen between corneocytes of the upper stratum corneum.

The possible presence of liposomal lipids within the SC was supported by the detection of gold particles, which the researchers used as internal markers in the liposomes. However, the authors found no evidence of an uptake of intact liposomes by the living epidermis or of their passage through this compartment of the skin. It appears that the soy lecithin phospholipids used in the formation of the liposomes are taken up by the keratinocyte cell membranes. Liposomes are then disrupted, and the contents spill out where this occurs.

None of these studies could prove that liposomes stay intact after application and can penetrate to deeper layers of the skin, or if they are metabolized and therefore destroyed over time. For cosmetic application this really does not matter. As long as the applied products are safe only the effect on the skin is important.

C. The Influence of Liposomes on the Moisture Content of Human Skin

At a lecture at the Nattermann Phospholipid Workshop, Ghyczy [63] presented the results of a study the aim of which was to evaluate how liposomes with similar physical properties but different chemical composition influence skin moisture content after a single topical application. The composition of the phospholipids evaluated is given in Table 9.

The liposomes (A) that contain a high content of phosphatidyl choline increased the skin humidity by a maximum value of 40% after 30 min, followed by a decrease during the following 2.5 h. On the contrary, the other preparation (B) showed only weak effects, and the treatment with preparation (C) led to a reduction in the skin moisture content by 15%, probably due to the high content of phospholipids with hydrophilic head groups.

Another study over 4 weeks with liposomes was conducted by the same author [64]. The purpose of this study was whether a long-term hydration of the skin could be achieved by using phosphatidyl choline enriched liposomes which turned out to be the only effective preparation in his first study. Three different liposome preparations were applied to the volar side of the forearm twice daily over 4 weeks, and the hydration state of the skin was measured 30 min later with the corneometer CM 820 (Courage and Khazaka, Köln). He observed that all preparations tested increased the skin moisture content with maximum values reached on day 7. Ghyczy concluded that liposomal phospholipids are able to influence skin moisture content, the kind and extent of this effect depends on the phospholipid head groups. While phospholipids with highly negatively and/or hydrophilic phospholipids did not show the desired effects. Those with a high content in zwitterionic and the relatively lipophilic phosphatidyl choline increased skin humidity significantly, the maximum value being reached after 1 week of treatment followed by a steady state.

Table 9 Phospholipid Composition of the Applied Liposomes

	Choline	Ethanolamine	Phosphatidyl inositol (PI)	Phosphatidic acid (PA)	N-Acetyl PE
A	80	4	11[a]	8	
B	28	2		11[a]	22
C	10	25	20	20	4

[a] PI and PA together.

D. The Influence of Liposome Morphology on Skin Moisture

Garcia-Anton et al. [65] recently succeeded in developing a new series of liposomes. They used high pressure microfluidizing techniques to gain plurilamellar multivesicular Liposomes (PML). They got liposomes entrapping liposomes, coated with electrically charged colloidal polymeric matrix that prevents aggregation of the vesicles and makes them stable in the presence of surfactants.

The authors compared different formulations containing PML, MLV (multilamellar vesicles), and DRV (dehydration–rehydration vesicles) in different excipients: acrylic polymers, polysaccharide polymers, O/W, W/O emulsions, and Hispagel 200 for their hydration properties of human skin.

They observed that when the different liposomes are dispersed in Hispagel 200, the maximum increase in cutaneous hydration occurred at $t = 1$ h. The hydration level remains practically constant for 4 h. PML increased the skin moisturization status better than DRV and MLV.

When the liposomes are dispersed in a mixture of acrylic polymers (carbomer), PML and MLV showed maximum hydration at $t = 1$ h whereas DRV showed the maximum at $t = 0.5$ h. The hydration level remains practically constant for 4 h. PML gave higher hydration values. PML suspended in a standard O/W emulsion gave higher and sustained hydration level for 4 h compared to DRV and MLV. On the contrary when a standard W/O emulsion was used as excipient, maximum hydration occurs after 0.5 h, with a significant decrease throughout the 4 h of the experiment.

E. Liposomes and Skin Roughness

Ghyczy [66] reported on the effects of phospholipids on skin roughness. He compared the following products concerning their effects in reduction of wrinkles:

1. A mixture of PC 76%, PA 8%, PE 4% and other lipids
2. A mixture of PC 24.3%, Ethanol 16%, water dem. 56.5%, NaOH (10%) 1.36% and no other phospholipids
3. DAC cream o/w emulsion without phospholipids

The treatment was stopped after day 28 (Fig. 12). In contrast to DAC cream, the results achieved with liposomes described above under 1 above were marked on day 3 and lasted up to day 7, followed by a continuous decrease of the effects up to day 14, when a kind of steady state can be observed. Two days after cessation of the treatment, the skin-smoothing effect was no longer detectable. The liposomes prepared according to formulation 3 show a different course of activity. Up to day 7, a continuous and impressive decrease in skin roughness can be observed, followed by a kind of steady state starting as early as day 7 and lasting

the whole treatment period, whereas on day 3 the observed effect is not different to that of DAC cream and much less pronounced than the skin smoothness achieved with the mixture described in item 1 above. This indicates a different mode of action. The result obtained with mixture 2 might be explained by a continuous uptake and accumulation of the phosphatidyl choline liposomes in the horny layer, followed by a strong interaction with the intercellular lipid barrier. Ghyczy speculated that this might lead to an increased fluidity of the lipid sheets caused by the supply of phospholipids with a high content of unsaturated fatty acids and highly hydrated polar head groups, and to a swelling of corneocytes, both effects being contributing to the long lasting and impressive smoothing effect.

F.   A*O*C*S Liposomes Carry Oxygen into the Skin

The partial pressure of oxygen in the skin changes with the oxygen loading of the blood. In addition to parameters which influence oxygen availability in the capillaries, such as smoking and other microangiopathic risk factors, increasing age leads to a decrease of oxygen in the skin and represent a measuring parameter for the experimental monitoring of skin aging [67,68]. A study conducted by Stanzl et al. [69] has proven that the oxygen partial pressure of the skin can be increased by using liposome-like systems. These vesicles used by the authors are composed of a perfluorocarbon core surrounded by phospholipids. They used perfluorodecalin and phospholipids with a high amount of phosphatidyl choline. This results in vesicle, where the core is surrounded by only a monolayer of phospholipids, followed by the regular bilayer systems known from the liposomes. That was the reason why this vesicle was called asymmetric oxygen carrier system (A*O*C*S), because it is possible to solubilize large amounts of oxygen in the perfluorodecalin and it was used to transport molecular oxygen into the skin.

In a follow-up study [70], Stanzl et al. proved that the oxygen-carrying

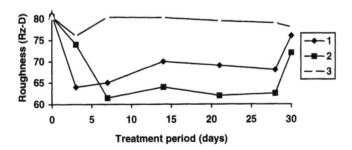

Figure 12   Skin roughness of two liposomal (1,2) preparations and a control (3).

vesicles improved the skin moisture content, reduced the number of wrinkles, and reduced the depth of wrinkles compared with the same product without per-fluorodecalin (Table 10). They also found that the oxygen partial pressure was significantly increased by over 100%.

In the same publication, they showed a dose-dependent improvement of skin surface parameters. They tested five products containing various amounts of the oxygen carrier. This study showed that the skin moisturizing effect is based on the liposomes, while the reduction in wrinkle depth could be correlated to the amount of oxygen in the formula, with a correlation coefficient of 0.9855.

G. The Influence of Encapsulation of Vitamin A Derivatives and the Effect on the Skin

The penetration of liposome-encapsulated *trans*-retinoic acid through different state of the skin was investigated by Imbertz [71] using in vitro diffusion experiments on human cadaver skin. Simple phospholipid liposomes and a more complex pH sensitive preparation containing 0.05% t-retinoic acid were applied to isolated human stratum corneum, isolated human epidermis and dermatomed human skin, along with one of two nonliposomal controls. The control solution were either 0.05% t-RA in ethanol or saturated solutions of t-RA in Transcutol/water mixtures. Skin-specific interactions were differentiated from solution thermodynamic effects by replacing some experiments using a silicone-rubber membrane. The influence of dose, volume and occlusion was investigated. The results showed no evidence of enhanced epidermal penetration, nor of decreased percutaneous absorption from liposomal t-RA versus unencapsulated controls under realistic dose conditions.

However, under one set of exaggerated dose conditions (consisting of a large, nonoccluded dose of t-RA in PC liposomes), there was a suggestion that diffusion of t-RA across the lower skin layers may have been retarded by the

Table 10   Change in Depth of Wrinkles, Number of Wrinkles, and Moisturization of the Skin[a]

Product	Skin smoothness	Reduction number of wrinkles	Moisturization
Product with PFD	36.2 ± 2.2	10.0 ± 0.6	36.3 ± 1.9
Product without PFD	21.9 ± 1.3	3.3 ± 0.3	22.1 ± 3.2

[a] After 14 days of application of a product containing perfluorodecalin encapsulated in A*O*C*S compared to a product containing liposomes only.

liposomal components. Biodisposition studies revealed that liposomes delivered more drug to the skin and less into the internal organs than the commercially available vitamin A acid creams and gels [72].

Zatz investigated the permeation of retinyl palmitate from several vesicles, including those containing nonionic surfactant vesicles [73]. Both excised hairless mouse and human skin were utilized as model membranes in these permeation experiments. Intact retinyl palmitate was not detected in the receptor. Following application of various nonionic surfactant vesicle preparations to hairless mouse skin, the label accumulated in the receptor at a rate comparable to or greater than from mineral oil. The opposite occurred with excised human skin. Penetration into the receptor was commensurate with accumulation in the viable human skin tissue. Most of the label residing within human skin following application of the NSV preparation was found in the stratum corneum, to an extent significantly greater than for either of the other vehicles. The authors concluded that because of the importance of storage in the stratum corneum hairless mouse skin is not a good model for retinyl palmitate permeation.

Redziniak et al. [74] studied the effects of retinoids free or encapsulated in liposomes on human skin cells in culture. In vitro retinoid acids modulates cell growth and differentiation [75]. The authors studied the activity of different esters of vitamin A (acetate, propionate, palmitate) free or encapsulated in liposomes, compared to the control vitamin A acid (trans retinoic-acid). The activity was evaluated by measuring the growth of the cells. Table 11 gives the composition of the liposomes used.

Nonencapsulated vitamin A propionate was the most active ester. The activity was close to the activity of vitamin A acid. At a concentration of 6 μM, when the free vitamin A propionate and the unloaded liposome do not have any activity, the liposome-ester combination is very active. Moreover it clearly appears that the liposome potentates the vitamin A acid effect by inhibiting the slightly cytotoxic effect of vitamin A overdose. At 4 μM the vitamin A propionate presents an activity similar to the vitamin A acid liposome.

Table 11 Composition of Liposomes

Ingredient	% (w/w)
Dipalmitoylphosphatidyl choline	78.0
Stearic acid	7.0
Beta-sitosterol	7.0
Antioxidants	0.4
Vitamin A derivatives	7.6

H. Encapsulation of Salicylic Acid

Salicylic acid was incorporated into phospholipid vesicles by Zatz [76]. Entrapment of SA and flux through an inert Silastic membrane were a function of pH. From mathematical treatment of data obtained at more than one value of pH, the relative contribution ionized and unionized species were determined. Both entrapment and flux were dominated by the uncharged form of the compound. An increase in lipid concentration resulted in an increase in entrapment and corresponding reduction in the steady-state flux of SA through the membrane. Penetration from liposomes was slower. And the amount crossing the membrane at 24 h was significantly less.

I. Encapsulation of Anti-Inflammatory Ingredients

Korting [77] compared radiolabeled triamcinolone acetonide incorporated into a conventional ointment and in parallel into a liposomal formulation on its effect on patients suffering from atopic eczema and psoriasis vulgaris. Upon topical application, he found the skin-surface levels were about equal, while the dermal levels were markedly higher with the liposomal form. In contrast, the blood levels as well as the levels in the thalmic region were markedly lower upon application of the liposomal product. Correspondingly, identical concentrations of radioactive hydrocortisone incorporated into a liposome and a conventional form were found in the horny layer of excised human skin, while drug concentration were higher upon application of the liposomal form in the viable epidermis as well as the dermis both 30 and 300 min after application [78]. As it is impossible at the moment to discriminate between encapsulated and free drugs in the various compartments, the focus should be more on the efficacy of the various preparations. According to the results obtained, liposome encapsulation can increase the efficacy of a topical glucocorticoid even further as compared with a conventional preparation containing a penetration enhancer. This does not apply to all indications for topical glucocorticoids but seemingly to those where the anti-inflammatory activity rather than the antiproliferative one is stressed.

The anti-inflammatory activity of chamomile on topical application is well known [79,80]. This effect is said to come from the essential oils, azulenes, α-bisabolol, and dicloethers [81,82]. In UVB-induced erythema, Kerscher [83] demonstrated that the anti-inflammatory activity of a conventional chamomile cream preparation was even better that that of two hydrocortisone preparations. As shown by chromametry and visual evaluation of erythema and scaling, the author proved that liposome preparations consisting of phospholipid mixture have a marked activity in inflamed human skin possibly exceeding the one of other types of vehicle preparations. Furthermore, the activity of a 2% chamomile liposome gel preparation in inflamed human skin is markedly higher than a liposome dispersion containing 10% chamomile extract.

The author speculated that instead of using phospholipid-based liposomes, formulations of stable liposomes from mixtures of "skin lipids" may further increase the activity. The use of skin lipid liposomes as effective drug delivery systems has been documented by Egbaria and Weiner [84–86]. They demonstrated that topical application of α interferon encapsulated in skin lipid liposomes showed a greater reduction of lesion scores than interferon entrapped in liposomes prepared from phospholipid mixtures. In in vitro diffusion experiments the application of such liposomes resulted in almost twice the amount of interferon deposited in the deeper layers of the skin from a 15 mg/mL skin lipid liposomal system compared to a 50 mg/mL phospholipid-based system containing the same amount of interferon.

Furthermore, topical application of skin lipid liposomes prepared by the dehydration/rehydration method showed twice the effectiveness of those prepared by the reverse phase evaporation method. Similar differences were observed between LUV and MLV cyclosporine A formulations both with skin lipids and phospholipid-based liposomal formulations in in-vitro human skin studies. These findings indicate that the lamellarity of the liposomes as well as the extent of bilayer-drug interaction play a crucial role in the liposome aided transfer of drugs into deeper skin layers [87].

J. Liposome-Encapsulated Enzymes for DNA Repair

Skin cancer is a widely distributed decease throughout the western hemisphere. It is strongly linked to solar UV exposure. UV radiation can alter the DNA. The most frequent lesions are pyrimidine dimer, formed by a cyclobutane ring between adjacent pyrimidines, and the 6-4 pyrimidine-pyrimidone photoproduct.

DNA photoproducts are efficiently removed by the excision repair process. The first step of recognition and incision near the site of damage is accomplished by two different enzymes. In the second step, the lesion is removed by exonucleases, and in the third step, undamaged DNA is resynthesized using the opposite strand as a template.

DNA repair rates have been significantly increased by increasing the first step of excision repair through the introduction of exogenous DNA repair enzymes [88]. Liposomes offer a practical vehicle to deliver enzymes to skin cells [89]. The optimal liposome composition to encapsulate DNA repair enzymes is pH-sensitive, owing to the presence of phosphatidyl ethanolamine and oleic acid in the membrane [90]. Liposomes encapsulated in T4 endonuclease V are termed *T4N5 liposomes* [91].

T4N5 liposomes in a topical lotion penetrate mouse and human skin and localized in the epidermis, producing little or no local or systemic toxicity. Dimers are removed more quickly from epidermal DNA resulting in dramatic changes during chronic UV exposure, e.g., reduces skin wrinkling and lower skin cancer. The dose-response curve is similar in shape to those of dimer repair.

T4N5 liposomes were applied to mouse skin in vivo and added to cultured murine keratinocytes in vitro [92]. The fate of the liposome membrane was followed using a fluorescent, lipophilic dye, and the fate of the enzyme was traced by immunogold labeling. In vivo, the liposome penetrated the stratum corneum, localized in epidermis and appendages of the skin and were found inside basal keratinocytes. The enzyme was found inside keratinocytes treated in vitro and in the epidermis, hair follicles, and sebaceous glands of topical treated skin. The results demonstrated that liposomes can deliver encapsulated proteins into cells of the skin in vivo.

UV-induced skin cancer as well as wrinkling is caused in part by pyrimidine dimers in DNA. T4N5 liposomes enhance repair in a dose-dependent manner up to a plateau. Topical application of DNA repair enzymes offers the possibility of preventing UV-induced skin damage even after sun exposure.

K. Special Use of Phospholipid Liposomes

Today fragrances are formulated almost exclusively with the help of ethanol. In the United States efforts are being made to replace volatile organic compounds (VOC),like ethanol, in cosmetics. This is the reason why some companies already launched products with high perfume levels based on nonionic surfactants. Recent work has revealed that vesicles produced from Soya phospholipids liposomes are also suitable for the preparation of perfume-containing formulations [93,94]. Ghyzcy [95] presented at the IFSCC Congress in London in 1993 a specific fraction for the preparation of aqueous dispersions of water-insoluble fragrances and aromas. This fraction exhibits the following important properties:

1. Low concentrations of phosphatidyl ethanolamine (since this reacts with aldehyds and ketons to yield Schiff's base).
2. Simple processing in all normal homogenizing apparatus
3. Liquid product form to simplify industrial application

These properties were achieved with a unique composition of phospholipids called Phosal NAT-50-PG. Table 12 shows the limiting parameters of Phosal NAT-50-PG.

The perfume formulations based on this unique combination are milky liquids that do not burn or produce cooling sensation on the skin but rather create a pleasant, satiny skin feel. They are highly excellent compatible with the skin and, owing to the content of phospholipids reduce skin roughness.

Table 12 Characterization of Phosal NAT-50-PG

Phospholipids	
Phosphatidyl choline	>15%
Lyso-phosphatidyl choline	<3%
Phosphatidyl ethanolamine	<3%
N-Acetyl phosphatidylethanolamine	>15%

L. Safety of Liposomes

Zonneveld et al. [96] reviewed the numerous reports from laboratories, who injected liposomes to individual humans either for diagnostic or biodistribution studies. In most studies, the liposomes were injected intraveniously in large concentrations; to date, no major difficulties have been reported.

It is well known that parenteral injection of multilamellar liposomes into animals results in a massive uptake of the liposomes by the macrophages in the liver, spleen, and bone marrow [97].

Detailed acute toxicity studies in mice and beagle dogs failed to demonstrate significant hematological, biochemical, immunological, or histopathological toxicity associated with as many as six injections of liposomes over a 2-week period [98].

Liposomes do not present any special safety problem. Used for topical application their innocuity can be checked by the standard methods used for other types of lipid-water systems. The safety of liposome dispersions depend upon that of the polar lipids used and of the additives.

VIII. CONCLUSION AND FUTURE OUTLOOK FOR LIPOSOMES FOR TOPICAL APPLICATION

Liposomes were first described by Bangham in the early 1960s and shortly thereafter were put to use as models for biological membranes in many branches of biology and medicine.

Liposomes have found widespread use in cosmetic products. Since Capture was launched by Dior in 1986, numerous products followed claiming the use of liposomes. Liposome-based cosmetics have, in fact, contributed to the field of liposomology by demonstrating that stable vesicles can be prepared on an industrial scale.

It is still unclear if liposomes penetrate through the skin. Some studies indicate that their structure seems to be lost in the very first layers of the stratum corneum. However, recent studies by Yarosh have shown that there might be a penetration into the deeper layers of the skin and the active substance could be deposited into a living cell. In the first case, if liposomes improve the activity of a substance, it is because they increase its local concentration in the epidermis and upper dermis. In the second case, liposomes transport the active ingredient directly to the living cell.

Liposomes have been also proposed as therapeutic drug carrier systems for the topical treatment of skin disease. By enhancing drug accumulation at the site of administration, liposomes may improve its activity and reduce serious side effects due to undesirable systemic absorption. To a certain extent, this relates to their seemingly successful use in cosmetics. For a long time however, there has been remarkable doubt that liposome preparations would also meet the requirements of long-term stability and adequate control. Stability and control are no longer issues, and as the toxicological profile, especially for the natural liposomes, is excellent. Liposomes should still have a bright future in cosmetic products.

REFERENCES

1. Bangham AD, Standish MN, Watkins JC. The action of steroids and streptolysin S on the permeability of phospolipic structures to cations. J Mol Biol 1965; 13:138.
2. Kinsky SC, Haxby J, Kinsky RA, Damel RA. Biochim Biophys Acta 1968; 152:174.
3. Vanlerberghe G, Handjani-Vila MR, Ribier A. Les "niosomes," une nouvelle famille de vesicule, et bas d'amphiphile nonioniques. Colloq Nat CNRS 1978; 983:303.
4. Handjani-Vila RM, Ribier A, Vanlerberghe G. L niosomes. In: Pusieux F, Delattre J, eds. Les Liposomes. Paris: Techniques Documentation Lavoisier, 1985.
5. Azmin MN, Florence AT, Hadjani-Vila RM, et al. The effect of nonionic surfactant vesicles (niosome) entrapment on the absorption and distribution of methotrexate in mice. J Pharm Pharmacol 1985; 37:237.
6. Azmin MN, Florence AT, Hadjani-Vila RM, et al. The effect of niosomes and polysorbate 80 on the metabolism and excretion of methotrexate in the mouse. J Microencaps 1985; 3:95.
7. Handjani R, Ribier A, Vanlerberghe G, Handjani-Vila RM. Preparation of more stable niosomes useful in preparation of cosmetic creams, pharmaceutical products, etc., and obtained with non-ionic lipid-phase and an aqueous phase. French Patent 2,597,346 (1987).
8. Vanlerberghe G, Handjani-Vila RM. Aqueous dispersions of lipid spheres and compositions and contents of same. U.S. Patent 4,772,471 (1988).
9. Griat J, Handjani-Vila RM, Ribier A, et al. Cosmetic and pharmaceutical preparations containing niosomes and a water-soluble polyamide, and a process for preparing these compositions; encapsulated aqueous phase in spherule formed from non-

ionic amphiphilic lipid; external phase as aqueous poly-B-alanine. U.S. Patent 4,830,857 (1989).

10. Florence AT. Nonionic surfactant vesicles: preparation and characterization. In: Gregorioadis G, ed. Liposome technology, vol 1. Boca Raton, FL: CRC Press, 1992.

11. Philippot JR, Milhaud PG, Puyal CO, Wallach DFH. Non-phospholipid molecules and modified phospholipid components of liposomes. In: Philippot JR, Schuber F, eds. Liposomes as Tools in Basic Research and Industry. Boca Raton, FL: CRC Press, 1994:41–57.

12. Sebag H, Vanlerberghe G. Preprints XIII Congress of the International Federation of Societies of Cosmetic Chemists (IFSCC), 1984; I:111–127.

13. Vanlerberghe G, Handjani-Vila RM, Berthelot C, Sebag H. Synthése et activité de surface comparée d'une série de nouveaux dérivés non-ioniques. Proceedings of the 6th International Congress on Surface Activity, Zurich, 1974, pp 139–155.

14. Bangham AD. Preparation of liposomes and methods for measuring their permeability. In: Hesketh HL et al, eds. Techniques in Lipid and Membrane Biochemistry: II. Techniques in Life Science. Ireland: Elsevier Biomedical, 1982:1.

15. Lasic DO. The mechanism of vesicle formation. Biochem J 1988; 156:1–11.

16. Amselem S, Barenholz Y. A large scale method for preparation of sterile and nonpyrogenic formulations for clinical use. In: Gregoriadis G, ed. Lipsome Technology, 2d ed. Boca Raton, FL: CRC Press, 1992.

17. Cecec G, Marsh D. Phospolipid Bilayers: Physical Principles and Models. New York: Wiley, 1978.

18. Blume A, Rice DM, Wittebort RJ, Griffin RG. Molecular dynamics and conformation in the gel and liquid-crystalline phases of phospatidyl ethanolamine bilayers. Biochemistry 1982; 21:6220–6230.

19. Kanesish MI, Bong TY. Cluster model of lipid phase transitions with application to passive permeation of molecues and structure relaxation in lipid bilayers. J Am Chem Soc 1978; 100:424–432.

20. Blume A. Phospolipids as basic ingredients. In Braun-Falco O, Korting HC, Maibach HI, eds. Liposome Dermatics—Griesbach Conference. Berlin/Heidelberg/New York: Springer-Verlag, 1992:30.

21. Engelbert HP, Lawaczeck R. The H_2O/D_2O exchange across vesicular lipid bilayers: lecithins and binary mixtures of lecithins. Berlin Buns Phys Chem 1985; 89:754–759.

22. Ruocco MJ, Siminovitch DJ, Griffin RG. Comparative study of the gel phases of ether- and ester-linked phosphatidyl cholines. Biochemistry 1985; 24:2406–2411.

23. Carruthers A, Melchior DL. Studies of relationship between bilayer water permeability and bilayer physical state. Biochemistry 1983; 22:5797–5807.

24. Lawaczeck R. Defect structure membranes: routes for the permeation of small molecues. Berlin Buns Phys Chem 1988; 92:961–963.

25. Blume A. Phase transition of polymerizable phospholipids. Chem Phys Lipias 1991; 57:253–273.

26. Lange A. Synthese und Eigenschaften kopfgruppenpolymerisierbarer Lipide mit verschiedener Spacerlänge. Thesis, University of Kaiserslautern, 1991.

27. Wertz PW, Swartzendruber DC, Madison KC, Downing DT. Composition and morphology of epidermal lipids. J Invest Dermatol 1987; 89:419–425.

28. Wertz PW, Abraham W, Landmann L, Dowing DTR. Preparation of liposomes from stratum corneum lipids. J Invest Dermatol 1986; 87:582–584.
29. Wallach DFH. Receptor Molecular Biology. New York: Marcel Dekker, 1987.
30. Rickwood D, Hames BD, eds. The Practical Approach Series: Liposomes as a Practical Approach. Oxford/New York/Tokyo: IRL Press,
31. Röding J. Natipide II: New Easy Lipsome System. Seifen Öle Fette Wachse 1990; 14:509–512.
32. Hagner J, Ghyczy M, Feyen V, et al. U.S. Patent 4,874,553 (1986).
33. Röding J. Proceedings of the 47th Annual Meeting of the Electron Microscopy Society of America. San Francisco: San Francisco Press, 1989:744.
34. Deamer D, Bangham AD. Biochim Biophys Acta 1976; 443:629.
35. Redziniak, G, Meybeck A. Production of Pulverulent Mixtures of Lipidic and Hydrophobic Constituents. U.S. Patent 4,508,703.
36. Weder HG, Zubuehl O. In: Gregoriadis G, ed. Liposomes Technology, vol 1. New York: Wiley, 1988:855.
37. Ghyczy M, Gareiss J. Industrial production of liposomes for topical application. 17th IFSCC Congress, Yokohama, vol 1. 1992:193–206.
38. Postel M, Chang P, Roland J, et al. Fluorocarbon/lecithine emulsions: identification of EYP coated fluorocarbon droplets and fluorocarbon empty vesicles by freeze fraction electron microscopy. Biochim Biophys Acta 1991; 1086:95–98.
39. Stanzl K, Zastrow L, Röding J, Artmann C. Ein neues kosmetisches Produkt mit molekularem Sauerstoff [A new cosmetic product containing molecular oxygen]. Eur Cosmet 1993; 1:39.
40. Hernandez-Caselles T, Villalain J, Gomez-Fernandez JC. Stability of liposomes on long-term storage. J Pharm Pharmacol 1990; 42:397–400.
41. Arkane K, Hayashi K, Naito N, et al. Liposomes: the long-term stability and the topical effect on the skin. 17th International Congress of the IFSCC, Yokohama, vol I. 1992:172–192.
42. Thoma K, Schmid A. Influence of composition and manufacturing technique on size and homogeneity of small unilamellr liposomes. 5th International Conference on Pharmaceutical Technology, vol III. Paris, May 30–June 1, 1989:15–24.
43. Röding J, Muller T, Lautenschlager H. Electronenmikroskopischer Nachweis von Liposomen in einem Hautpflegegel. Seifen Fette Öle Wachse 1989; 3:88–89.
44. Röding J. Stabilität, physikalischer Eigenschaften und Charakterisierung von Liposomen in flüssigen und halbfesten Zubereitungen. Parfum Kosmet 1990; 2: 80–89.
45. Taylor KA, Gleser RM. Electron diffraction of frozen hydrated protein crystals. Science 1974: 186: 1036–1037.
46. Elias PM, Jass, Cosmet Toilet 1991; 106:47.
47. Elias PM. Epidermal lipids, barrier function, and desquamation. J Invest Dermatol 1983; 80:445–495.
48. Proksch E. Br J Dermatol 1993; 128:473–482.
49. Holleran. Cosmet Toilet 1991; 106:64.
50. Imokawa et al. J Invest Dermatol 1991; 96:845–851.
51. Imokawa et al. J Invest Dermatol 1985; 83:282–284.
52. Imokawa et al. Arch Dermatol Res 1989; 281:45–51.

53. Mezei M, Gulasekharem. Liposomes: A selective drug delivery system for the topical route of administration: I. Lotion dosage forms. Life Sci 1980; 26:1473–1477.

54. Cevc. Private communication, 1993.

55. Ganesan MG, Weiner ND, Flynn GL, Ho HFH. Influence of liposomal drug entrapment on percutaneous absorption. Int J Pharm 1984; 20:139–154.

56. Knepp VM, Szika FC, Guy RH. Controlled drug release from a novel liposomal delivery system: I. Investigation of transdermal potential. J Contr Rel 1990; 12:25–30.

57. Bouwstra JA, Hofland HEJ, Spies F, et al. Changes in the structure of the human stratum corneum induced by liposomes. In: Braun-Falco O, Korting HC, Maibach HI, eds. Liposome Dermatics. Berlin/Heidelberg/New York: Springer-Verlag, 1992: 121–136.

58. Röding J, Artmann C. The fate of lipsomes in animal skin. In: Braun-Falco O, Korting HC, Maibach HI, eds. Liposome Dermatics. Berlin/Heidelberg/New York: Springer-Verlag, 1992:185–194.

59. Artmann C, Röding J, Ghyczy M. Liposomes from soya phospholipids as percutaneous drug carriers. I. Qualitative in vivo investigations with antibody loaded lipsomes. Drug Res 1990; 40(II-12):1363–6365.

60. Artmann C, Röding J, Ghyczy M. Liposomes from soya phospholipids as percutaneous drug carriers: II. Quantitative in vivo investigations by macromolecules and salt loaded liposomes with radioactive labeling. Drug Res 1990; 40(II-12):1365–1368.

61. Gehring W, Klein M, Gloor M. Influence of various topical liposome preparations with and without active ingredients on the cutaneous blood flow. In Braun-Falco O, Korting HC, Maibach HI, eds. Liposome Dermatics. Berlin/Heidelberg/New York: Springer-Verlag, 1992:315–319.

62. Korting et al. Liposomal penetration. Cosmet Toilet 1995; 110:19.

63. Ghyczy M. Control of skin penetration by liposomes (review. Lecture at the Nattermann Phospholipid Worshop on Liposomes at Skin, Paris, Dec. 5, 1990.

64. Röding J, Ghyczy M. Control of skin humidity with liposomes: stabilization of skin care oils and lipophilic active substances with liposomes. Seifen Öle Fette Wachse 1991; 10:372–378.

65. Garcia-Anton JM, Nieto A, Del Pozo A, et al. Plurilamellar mutivesicular liposomes: methodology and cosmetic application. In: Fact and Illusions in Cosmetics—IFSCC Congress. Montreaux, France, 1995:249–255.

66. Ghyczy M. Natipide II and Phosal: the influence on skin roughness of different vesicular systems. Presentation at the In-cosmetics. Frankfurt, 1992.

67. Elstner EF. Der Sauerstoff: Biochemie, Biologie, Medizin. Mannheim, Germany: BI-Will-Verlag, 1990.

68. Artmann C, Röding J, Stanzl K, Zastrow L. Oxygen in the skin—a new parameter of skin aging. Seifen Öle Fette Wachse 1993; 15:942.

69. Stanzl K, Zastrow L, Röding J, Artmann C. A new cosmetic product containing molecular oxygen. Eur Cosmet 1993; 1:39.

70. Stanzl K, Zastrow L, Röding J, Artmann C. The effectiveness of molecular oxygen in cosmetic formulations. Int J Cosmet Sci 1996; 18:137–150.

71. Imbertz D, Kasting G, Wickett R. Influence of liposomal encapsulation on the pene-

tration of retinoic acid through human skin in vitro. J Soc Cosmet Chem 1994; 45: 119–134.

72. Foong WC, Harsanyi BB, Mezei M. Biodisposition and histological evaluation of topically applied retinoic acid in liposomal, cream, and gel dosage form. In Hanin I, Pepeu G, eds. Phospholipids: Biochemical, Pharmaceutical, and Analytical Considerations. New York: Plenum Press, 1990:279–282.

73. Zatz J, Guenin EP. Skin permeation of retinyl palmitate from vesicles. J Soc Cosmet Chem 1995; 46:261–270.

74. Franchi J, Coutadeur MC, Archambault JC, et al. Effects of retinoids free or encapsulated in liposomes on human skin cells in culture. 17th IFSCC Congress, Yokohama, Japan, vol I. 1992:126–139.

75. Jetten AM. Fed Proc 1984; 43:134–139.

76. Zatz JL, Guenin EP. Effect of salicylic acid encapsulation by phospholipid vesicles on transport through an inert membrane. J Soc Cosmet Chem 1995; 46.

77. Korting HC. In: Braun-Falco O, Korting HC, Maibach HI, eds. Liposome Dermatics. Berlin/Heidelberg/New York: Springer-Verlag, 1992:320.

78. Wohlrab W, Lasch J. Penetration kinetics of liposomal hydrocortisone in human skin. Dermatologica 1987; 174:18–22.

79. Aertgeerts J. Erfahrungen mit Kamillosan, einem standardisiertem Kamillenextrakt in der dermatologischen Praxis Arztl Kosmetol 1984; 14:502–504.

80. Aertgeerts J, Albring M, Klaschka F, et al. Vergleichende Prüfung von Kamillosan Creme gegenüber steroidalen (0.25% hydrocortison, 0.75% fluocortinbutylester) und nicht-steroidalen (5% Bufexamac) externa in der Erhaltungstherapie von Ekzemerkrankungen. Zeitschr Hautkr 1985; 60:270–277.

81. Della Loggia R, Tubaro A, Dri P, et al. The role of flavonoids in the antiinflammatory activity of Chamomila recutita. Prog Clin Res 1986; 213:481–484.

82. Kadir R, Barry BW. α-Bisabol, a safe penetration enhancer for dermal and transdermal therapeutics. Int J Pharm 1991; 70:87–94.

83. Kerscher MJ. Influence of liposomal encapsulation on the activity of an herbal nonsteroidal anti-inflammatory drug. In: Braun-Falco O, Korting HC, Maibach HI, eds. Liposome Dermatics. Berlin/Heidelberg/New York: Springer-Verlag, 1992:329–337.

84. Egbaria K, Ramachandran C, Weiner N. Topical delivery of ciclosporin: evaluation of various formulations using in vitro diffusion studies in hariless mouse skin. Skin Pharmacol 1990; 3:21–28.

85. Egbaria K, Ramachandran C, Weiner N. Topical application of liposomally entrapped ciclosporin evaluated by in vitro diffusion studies with human skin. Skin Pharmacol 1991; 4:21–28.

86. Egbaria K, Weiner N. Liposomes as a topical drug delivery system. Adv Drug Del Rev 1990; 5:287–300.

87. Weiner N, Williams N, Birch G, et al. Topical delivery of liposomal encapsulated interferon evaluated in a cutaneous herpes guinea pig model. Antimicrob Agents Chemother 1989; 33:1217.

88. Hejimakers J. Characterization of genes and proteins involved in excision repair of human cells. J Cell Sci 1987; suppl 6:111–125.

89. Yarosh D. Topical application of liposomes. J Photochem Photobiol (B) 1990; 6: 445–449.

90. Yarosh D, Tsimis J, Yee V. Enhancement of DNA repair of UV damage in mouse and human skin by liposomes containing a DNA repair enzyme. J Soc Cosmet Chem 1990; 41:85–92.

91. Yarosh D, Kibitel J, Green L, Spinowitz A. Enhanced unscheduled DNA synthesis in UV irradiated human skin explants treated with T4N5 liposomes. J Invest Dermatol 1990; 97:147–150.

92. Yarosh D, Bucana C, Cox P, et al. Localization of liposomes containing a DNA repair enzyme in murine skin. J Invest Dermatol 1994; 103:461–469.

93. Juszynski M, Azoury R, Rafaeloff R. Perfume loaded liposomes. Seifen Öle Fette Wachse 1991; 117:159–262.

94. Jusynski M, Azoury R, Rafaeloff R. Fragrance loaded lyophilized liposomes. Seifen Öle Fette Wachse 1992; 118:811–815.

95. Ghyczy M. In: Nattermann phospholipids GmbH: A Mirror of Continuous Research. E. Hoff Scientific Brochure No. 6. 1996:45–53.

96. Zonneveld GM, Crommelin DJA. Liposomes: parenteral administration to man. In: Gregoriadis G, ed. Liposomes as Drug Carriers. London: Wiley, 1988:795–817.

97. Senior J. Fate and behavior of liposomes in vivo: a review of controlling factors. CRC Crit Rev Ther Drug Carrier Syst 1987; 3:123–193.

98. Hart IR, Fogler WE, Poste G, Fidler IH. Toxicity studies of liposome-encapsulated immunomodultors administered intravenously to dogs and mice. Cancer Immunol Immunother 1910; 157–166.

13

Liposomes and Follicular Penetration

Linda Margaret Lieb
University of Utah Health Sciences Center, Salt Lake City, Utah

I. INTRODUCTION

The hair-targeting properties of liposomes would greatly benefit the hair cosmetic industry. Topically applied liposomes have properties that can be used in the hair industry to, for example, enhance hair conditioning, replace lost melanin or color in the hair shaft, and cosmetically treat hair disorders such as hair loss.

The data we currently have relate primarily to drug delivery. Our laboratory has shown liposomes to have unique qualities, such as an ability to target the hair shaft or follicle opening. This information tells us much of the behavior of the liposome with the tested agent or drug within the hair follicle. Although liposomes have an affinity for the follicular unit, it is not at all clear why. We may speculate that there is possibly some assimilation between phospholipid liposomal bilayers and components of the inner root sheath, which are also made up of phospholipids.

Therefore, we are not at a point in our study of liposome follicular penetration to ascertain the mechanism. What we can do is to state the formula type and focus on how the liposome acts on the hair follicle. This information can then be applied to the cosmetic realm. We assume that in using cosmetics, our goal is to produce a superficial or localized effect on the hair follicle, whereas with drugs, our goal may be more complicated, meaning that we may wish to penetrate the local follicular circulation.

II. FOLLICULAR PENETRATION OF LIPOSOME BILAYERS

Lieb performed a study to determine the follicle penetration abilities of rhoda-mine phosphatidylethanolamine (RHPE) (Lissamine, Molecular Probes, Eugene, OR) labeled reverse evaporation (REV) liposomes [1]. The purpose of this study was to gain insight into the possible mechanism whereby topically applied lipo-somes facilitate penetration into pilosebaceous units (PSU). Lissamine, a neutral phospholipid analog in which the fluorescent marker, rhodamine B sulfonyl group, is covalently bonded to the amine group of the phosphatidylethanolamine, was chosen because it remains associated with the liposomal bilayers in a biologi-cal milieu. This gave us the added advantage of determining the follicular deposi-tion of liposomal bilayers into PSU. RHPE was added directly to a lipid film of phosphatidylcholine:cholesterol:phosphatidylserine at a mole ratio of 1:0.5:0.1 to form a total lipid concentration of 25 mg/mL. Of this preparation, 40 mL was applied in vitro to hamster ears using a Franz cell apparatus. The presence of the rhodamine-linked fluorescent marker in the pilosebaceous unit was viewed through fluorescent and confocal laser scanning microscopy 24 h after topical application to the hamster ear.

Fluorescence microscopy results show the presence of RHPE, presumably intercalated with liposomal bilayers, inside the follicular opening, as well as ho-mogenously distributed throughout the PSU, centering around the follicular shaft. Confocal laser scanning microscopy ''three-dimensional'' views revealed the presence of RHPE ''coating the entire follicular shaft'' and RHPE deep within the sebaceous glands. Overall interpretation of the microscopic views revealed RHPE, intercalated within liposomal bilayers, selectively deposited in the ham-ster ear pilosebaceous units. A lack of visible marker in regions surrounding the PSU suggests that deposition into PSU is greatly enhanced as compared with deposition into the dermis. The depth of penetration and the areas reached within the PSU of the marker varied greatly, probably owing to differences in diffusion kinetics, rates of drying, or other factors. A likely scenario for deposition of a liposomal application might be dehydration of the liposome followed by a bilayer partitioning and packing of the bilayers into the follicular opening, diffusion down into the hair shaft, mixing with sebum, followed by diffusion into the seba-ceous duct and down into the sebaceous glands.

In summary, it was shown, through confocal and fluorescent microscopic studies, that liposome bilayers were found coated completely around the hair shaft of a Syrian hamster hair follicle, 24 h after topical application [1].

Cevc illustrated the effect of ''transfersomes'' on the delivery into nude mice hair follicles [2]. Transfersomes are defined as a ''self-adaptable and opti-mizing mixed lipid aggregate.'' Standard liposomes are converted to trans-fersomes through the addition of ''suitable edge-activators into the aggregate membrane,'' such as surfactants [2]. Fluorescently labeled transfersomes appear to heavily stain some but not all hair follicle shafts near the skin surface.

III. LIPOSOME-ENHANCED FOLLICULAR DELIVERY OF AGENTS POTENTIALLY USEFUL FOR THE COSMETIC INDUSTRY

Lieb et al. [3] found enhanced penetration with multilamellar phospholipid liposomes of carboxyflourescein, a water-soluble dye, into Syrian hamster ear pilosebaceous units in vitro. Multilamellar phospholipid liposomes demonstrated the highest follicular penetrability over other ''cosmetic-type'' vehicles, such as lauryl sulfate, propylene glycol, and ethanol. Significant amounts were found in follicular openings and sebaceous glands as early as 12 h after topical application. This group also found that phospholipids processed into bilayers were necessary for delivery; i.e., nonprocessed phospholipid suspensions did not work [3]. This has important implications in the cosmetic industry as a method to transfer dye into hair follicles.

Lieb et al. [4] showed that cimetidine (an antiandrogen) could be manipulated liposomally to target hair follicles. The amounts of deposition of cimetidine into skin compartments varied depending on whether phospholipid or nonionic liposomes were used [4]. However, it was also found that binding of a drug to liposomal bilayers may also cause inhibition of drug action in vivo.

Deposition into skin may also vary depending on drug charge only [4]. This was the case also with cimetidine, a small weak base (MW = 252; pKa = 6.8). Liposomes appeared to withhold delivery of un-ionized cimetidine (pH = 8.3) into pilosebaceous units compared to an aqueous solution. This was not the case when cimetidine was in the ionized form (pH = 5.5).

This shows that drug association with liposomal bilayers forms a complex dynamic. This must be taken together when considering using liposome-drug complexes as topical cosmetic or therapeutic agents. Cimetidine would be a useful topical agent in the cosmetic industry, since it has antiandrogenic effects [5]. This effect would be useful in an antiacne and anti-hair-loss agent.

Very little data have been generated on the pharmacokinetics topical drugs complexed with liposomes. Lieb et al. [1] studied the deposition of salicylic acid from phospholipid multilamellar liposomes into the Syrian hamster ear in vitro as a function of time and lipid concentration using Franz diffusion cells. The data demonstrated that the percentage of salicylic acid deposition into pilosebaceous units increased linearly with time within the tested time period (0–24 h). Furthermore, the amount in the pilosebaceous units correlated with the amount in the dermis for the first 12 h. It was also found that the minimum lipid concentration/area of skin necessary for optimum follicular deposition was ~ 1 mg/cm^2. Salicylic acid would also be a useful agent in cosmetics because of its antikeratolytic activities.

IV. COSMETICALLY USEFUL MACROMOLECULES ENHANCED TO HAIR FOLLICLES BY LIPOSOMES

Many cosmetic ingredients are macromolecules (e.g., polymers). The following studies of liposomal penetration into hair follicles using macromolecules can apply to cosmetic situations.

Li et al. [6] found specific delivery with liposomal entrapped melanin in in vitro tissue culture, where aqueous control showed no drug localization. It was found that when the melanin was entrapped in a phosphatidylcholine liposome, it deposited into both the periphery and the follicle cells themselves.

Li and Hoffman have recently shown phosphatidylcholine liposomes that entrap either fluorescent dye calcein or the pigment melanin then deliver these molecules to the hair follicle and hair shafts of mice in vivo when applied topically [7]. Negligible amounts of these molecules entered the dermis, epidermis, or bloodstream, which illustrates the selective delivery into the hair follicles. Phosphatidylcholine liposomes appeared to be key for selective follicle delivery, as naked calcein and melanin were found trapped in the stratum corneum and prevented from entering the follicle. These findings show a ready topical route for agents of biotechnological origin and possible means to treat diseases associated with the hair follicle (i.e., alopecia or hirsutism).

A. Applications Using Proteinaceous Compounds

Yarosh et al. have developed a liposomal formula, called T4N5 liposomes, that contains the DNA repair enzyme T4 endonuclease V, used to enhance the removal of lesions in DNA after ultraviolet irradiation. They were multilamellar, phospholipid-based liposomes at 20 mM concentration [8]. The liposomes were mixed into a 1.5% hydrogel (Carbopol-941, BF Goodrich) in phosphate buffer and the liposomal formula was labeled with a fluorescent lipophilic dye to follow the skin deposition upon topical application [9]. It was found that the fluorescent liposomes had localized in the epidermis as well as in cells surrounding the hair follicles by 1 h. The liposomes remained localized in the epidermis and hair follicles with little penetration in the dermis up to 18 h.

Niemiec et al. assessed the delivery of a hydrophilic protein (alpha interferon) and a hydrophobic protein (cyclosporine A) into pilosebaceous units (PSU) of hamster ears using four types of liposomes [10]. Three of the formulas were very similar in that they all contained polyoxyethylene-10-stearyl ether and cho-

lesterol (refered to as Non-#). Non-1 and Non-2 differed from Non-3 in that Non-1 also contained glyceryl dilaurate and Non-2 instead contained glyceryl distearate. The fourth liposome was a phospholipid-based formulation (PC). It was found that although the Non-1 formula enhanced the delivery of both proteins to all the strata (PSU, dermis, cartilage, dorsal) over PC liposomes, the PC liposomes provided a "more selective" delivery into the PSU, i.e., a higher concentration of the peptide in the PSU relative to the other strata.

Further studies with very high-molecular-weight peptides—i.e., monoclonal antibodies—illustrated the importance of formula optimization when working with phospholipid-based liposomes [11]. Only the charged liposomes led to deposition of the monoclonal antibody into hamster ear sebaceous glands. This was confirmed through confocal microscopy, viewing the follicular deposition from topically applied liposomal formulations of FITC-antibody. The label was found in the stratum corneum, in hair follicle openings, and within the hair follicle. It was also found deep in the follicles at the matrix cell level. It was found from these studies that a charge-charge interaction may be important for transport of antibody into hair follicles.

B. Applications Using Genetic Compounds

Li et al. found specific delivery with liposome-entrapped high-molecular-weight DNA [12] in hair-bearing tissue culture.

Lieb et al. [13] demonstrated localized deposition of antisense oligonucleotides to human hair follicles in vitro using cationic liposomes. They found significant concentrations at germinal components of the follicle (i.e., hair bulb) using topically applied fluorescein-conjugated oligonucleotides (22–25 base; approximately 5000 MW). This has important implications for gene therapy approaches to treat hair, as the hair bulb is the site for hair regrowth.

Weiner et al. [11] recently showed delivery of fluorescently labeled DNA plasmid into the ventral surface of hamster ear hair follicles and perifollicular glands, specific to nonionic/cationic (NC) liposomal formulations. The NC liposomal formulations contained glyceryl dilaurate, cholesterol, polyoxyethylene-10 stearyl ether (POE-10), and 1,2 dioleoyloxy-30(trimethylammino) propane (DOTAP) at a weight percent ratio of 50:15:23:12. Phosphatidylcholine:cationic liposomes were not successful under these conditions.

Liposomes used to mediate topical delivery and skin transfection of β-galactosidase gene expression constructs (LacZ plasmid) have penetrated and effectively expressed the encoded transgene in epidermis, dermis, and hair follicles of murine skin [14,15].

Li and Hoffman topically applied phospholipid liposomes containing entrapped LacZ plasmid on the shaved backs of BALB/c mice [14]. Expression of β-galactosidase gene specifically in the hair follicles (indicated by a dark blue stain) was found 3 days after treatment. Topical application of "naked" LacZ plasmid (in a buffer solution only) resulted in no evidence of expression in murine skin. This has important implications for the gene therapy treatment of hair diseases.

Alexander and Akhurst performed a similar study where, instead of entrapping the plasmid inside MLV vesicles, they used LacZ gene complexed with cationic unilamellar liposomes [15]. In this study, DNA was complexed with a commercial preparation of N-{1-(2,3-dioleoyloxy) propyl}-N,N,N-trimethyl-ammonium-methyl-sulfate (DOTAP) in a ratio of 1:1.6 (w/w). The gene penetrated and expressed its encoded protein not only in the hair follicles but in the epidermis and dermis as well. They were able to quantify their results as well as to determine optimum concentrations (267 μg/mL) and expression time (24–48 h posttreatment). One wonders, when comparing these results to the above study, whether differences in how the drug, in this case, a gene, associates with the liposome affects the eventual route and local of deposition in skin, or whether the type of liposome (neutral MLV versus cationic unilamellar) plays a role.

Lieb et al. [16] recently showed topical delivery and expression of LacZ plasmid (an indication of gene penetration and localization in the hair follicle) in human scalp skin in vivo using a fetal scalp skin graft nude mouse model [17]. Successful delivery and expression of LacZ in cationic liposomes (Lipofectin and Lipofectamine from Gibco BRL Life Technologies, Inc. and DOTAP from Boehringer Mannheim) depended not on cationic liposome formula but on hair type. Gene expression was found only in scalp with terminal follicles (larger follicles in scalp), not in vellus follicles (smaller follicles). Expression of LacZ in terminal haired grafts were found in the following areas: (1) in regions of the epidermis and the dermis surrounding the follicle as well as in the companion layer of the hair follicle and (2) in regions in the sheath surrounding the hair follicle and in subcutaneous tissue near the follicles. Removing hair from follicles allowed some increase in delivery and expression in the scalp skin and deep in the follicles.

In a separate study, multilamellar phospholipid liposomes were compared to DOTAP for topical delivery and expression of genes into human hair follicles. This study is discussed in detail below.

A 25 mg/mL multilamellar phospholipid liposome with a mole ratio of 1:0.25:0.2 phosphatidylcholine:cholesterol:phosphatidylserine was prepared using a rotary evaporator technique [18]. 300 μg/mL of LacZ plasmid was added to the hydration phase. 150 μL/cm² of this formula and a separate 300 μg/mL LacZ plasmid:500 μg/mL DOTAP cationic lipid formula was applied to a human terminal-haired graft at 0 and 6 h. Two 3-mm biopsies were taken 48 h after the last dose from the graft treated with the DOTAP formula. Three 3-mm biopsies were taken from three different areas of the phospholipid liposome-treated graft, one from the edge, one from midcenter and one from the center. The pretreated and treated biopsies were processed for LacZ expression, embedded in paraffin, sliced on edge parallel to the follicle in serial sections, and viewed microscopically. Serial sections allowed for a more detailed account of the ratio of the number of stained follicles to total follicle number.

Results show that, for each of the treated grafts, approximately 85–95% of the entire follicles were stained (from the infundibulum to the root, including the sebaceous glands, inner and outer root sheaths, and the entire root). In those cases where follicles were not stained, it was always in the upper portion of the very center of the biopsy. We suspect that this could be due to a decrease in the bioavailability of the stain. The root region in most of the follicles was intensely stained over the pretreatment graft. Staining intensity was similar between the DOTAP and the phospholipid liposome formula.

The data thus presented would be applicable for cosmetics containing nucleic acids and/or large, negatively charged polymers. In summary, liposomal delivery appeared to be most successful with cationic charged liposomes. With this, it can be hypothesized that the cosmetic agent of interest would need to be complexed with the positively charged lipid bilayers for transfer to the hair follicle.

V. CONCLUSIONS

Liposomal interaction within the hair follicle differs depending on the type and the composition. It may be that the feasibility of liposomes as an aid for specific follicular targeting will depend on its individual design.

The studies of Lieb, Niemiec, Yarosh, and Li and Hoffman suggest the relationship of phospholipid comprising liposomes as an exclusive formula for selective penetration into hair follicles. It is also implied that the liposomal formula must be "custom-tailored" for the agent or cosmetic for delivery to the follicle. These data lay the foundation for future studies, where follicle penetration would be measured as a function of liposomal lipid content and type.

REFERENCES

1. Lieb LM. Formulation Factors Affecting Follicular (Pilosebaceous Route) Drug Delivery as Evaluated with the Hamster Ear Model. PhD thesis, Ann Arbor: University of Michigan, 1994.
2. Cevc G. Transfersomes, liposomes and other lipid suspensions on the skin: permeation enhancement, vesicle penetration, and transdermal drug delivery. Crit. Rev. Ther. Drug Carrier Syst. 1996; 13:257.
3. Lieb LM, Ramachandran C, Egbaria K, Weiner N. Topical delivery enhancement

with multilamellar liposomes into pilosebaceous units: I. In vitro evaluation using fluorescent techniques with the hamster ear model. J. Invest. Dermatol. 1992; 99: 108.

4. Lieb LM, Flynn G, Weiner N. Follicular (pilosebaceous unit) deposition and pharmacological behavior of cimetidine as a function of formulation. Pharm. Res. 1994; 11:1419.

5. Hennessy JV. Comparative antandrogenic potency of spironolactone and cimetidine: assessment by the chicken cockscomb topical bioassay. Proc. Soc. Exp. Biol. Med. 1986; 182:443.

6. Li L, Lishko LV, Hoffman RM. Liposomes can specifically target entrapped melanin to hair follicles in histocultured skin. In Vitro Cell Dev. Biol. 1993; 29A:192.

7. Li L, Hoffman RM. Topical liposome delivery of molecules to hair follicles in mice. J. Dermatol. Sci. 1997; 14:101.

8. Ceccoli J, Rosales N, Tsimis J, Yarosh DB. Encapsulation of the UV-DNA repair enzyme T4 endonuclease V in liposomes and delivery to human cells. J Invest Dermatol. 1989; 93:190.

9. Yarosh D, Bucana C, Cox P, et al. Localization of liposomes containing a DNA repair enzyme in murine skin. J. Invest. Dermatol. 1994; 103:461.

10. Niemiec SM, Ramachandran C, Weiner N. Influence of nonionic liposomal composition on topical delivery of peptide drugs into pilosebaceous units: an in vivo study using the hamster ear model. Pharm. Res. 1995; 12:1184.

11. Weiner N, Lieb LM. Developing uses of topical liposomes: Delivery of biologically active macromolecules. In: Paphadjopoulos, D, Lasic DD, eds. Medical Applications of Liposomes. Amsterdam: Elsevier, 1998.

12. Li L, Lishko V, Hoffman RM. Liposome targeting of high molecular weight DNA to the hair follicles of histocultured skin: a model for gene therapy of the hair growth processes (letter). In Vitro Cell Dev. Biol. Anim. 1993; 29A:258.

13. Lieb LM, Liimatta AP, Bryan RN, et al. Description of the intrafollicular delivery of large molecular weight molecules to follicles of human scalp skin in vitro. J. Pharm. Sci. 1997; 86:1022.

14. Li L, Hoffman RM. The feasibility of targeted selective gene therapy of the hair follicle. Nature Med. 1995; 1:705.

15. Alexander MY, Akhurst RJ. Liposome-mediated gene transfer and expression via the skin. Hum. Mol. Genet. 1995; 4:2279.

16. Lieb LM, Jorgensen CM, Morgan JR, Krueger GG. Delivery of transgenes to follicles of human scalp in vivo. J Invest Dermatol. 1997; 108:594.

17. Lane A, Scott G, Day K. Development of human fetal skin transplanted to the nude mouse. J. Invest. Dermatol. 1989; 93:787.

18. New RRC. Preparation of liposomes. In: New RRC, ed. Liposomes: A Practical Approach. Oxford, England: IRL Press, 1990:33.

14

Cyclodextrins in Cosmetics

Dominique Duchêne
Université Paris-Sud, Châtenay Malabry, France

Denis Wouessidjewe
Université Joseph Fourier, Meylan-Grenoble, France

Marie-Christine Poelman
Université René Descartes, Paris, France

I. INTRODUCTION

Cyclodextrins were described by Villiers in 1891 [1] and studied at the start of the twentieth century by Schardinger [2–4]. However, these works did not result in cyclodextrin development. It is only in the past 20 years that their value was recognized owing to their ability to include guest molecules in their cavity. Now, many industries—such as food, agriculture, tobacco, pharmacy and cosmetics—have patented different cyclodextrin uses [5].

II. CYCLODEXTRINS

A. Main Cyclodextrins

1. Natural Cyclodextrins [6]

Cyclodextrins are obtained by enzymatic degradation of starch. The enzyme (cycloglycosyl transferase) appears in various bacillus broths. According to the nature of the bacillus and the reaction conditions, three main cyclodextrins can be produced, constituted by 6, 7, or 8 glucopyrannose units; they are α-, β-, and γ-cyclodextrins, respectively.

The structure of these molecules is such that the primary hydroxyl groups are located on one side of the ring constituted by the cyclodextrin, and the second-

Table 1 Main Characteristics of Natural Cyclodextrins

	α-Cyclodextrin	β-Cyclodextrin	γ-Cyclodextrin
Number of glucopyranose units	6	7	8
External diameter (Å)	14.6	15.4	17.5
Internal diameter (Å)	4.9	6.2	7.9
Water solubility at 25°C (g/100 mL)	14.50	1.85	23.30

ary hydroxyl groups on the other side. Differences in tensions result in the fact that the side on which the primary hydroxyl groups are located is slightly narrower than that on which the secondary hydroxyl groups are found. The internal cavity bears the glucosidic groups. The consequence of such a structure is that the external part of the ring molecule is hydrophilic while the internal cavity is rather hydrophobic.

The three natural cyclodextrins differ in their diameter, a function of the number of glucopyrannose units, and in their water-solubility (Table 1). It can be seen that the solubility does not vary regularly from α-cyclodextrin to γ-cyclodextrin. In fact, β-cyclodextrin is the least water-soluble of the three natural cyclodextrins. This could be related to the odd number of glucopyrannose units resulting in the formation of dimers solution, with the lowest solubility.

The particular configuration of cyclodextrins gives them the ability to include in their cavity hydrophobic molecules or the apolar part of amphiphilic molecules. The main requirement for the occurrence of an inclusion is compatibility between the steric hindrance of the guest molecule and the cavity of the host cyclodextrin. The cavity must be large enough to allow penetration of the guest molecule, but not too large, in order to permit the appearance of weak interactions (hydrophobic interactions, for example) between the two molecules.

2. Cyclodextrin Derivatives [7]

For many years, γ-cyclodextrin was only a by-product of the preparation of either α-cyclodextrin or β-cyclodextrin, and because of the small size of the α-cyclodextrin cavity, allowing the inclusion of only small molecules, β-cyclodextrin was almost the only cyclodextrin studied and employed. However, owing to its low water solubility, chemists focused on the synthesis of water-soluble derivatives. Presently, new production technologies have been developped for γ-cyclodextrin which is, nowadays, easily available in form of the natural product or of its derivatives [8].

(a) Methylated Cyclodextrins. Methylated cyclodextrins are among the first cyclodextrin derivatives to have been studied with a view to their pharmaceutical use [9,10].

Various methyl ether derivatives are proposed on the market: dimethyl and trimethyl cyclodextrins and partially methylated and randomly methylated β-cyclodextrins. Dimethyl cyclodextrins are substituted mostly on C2 and C6, while trimethyl cyclodextrins are substituted on C3 as well. Surprisingly, the substitution of hydroxyl groups by methyl groups leads to a remarkable increase in water-solubility, since in the case of β-cyclodextrin derivatives, the water-solubility at room temperature (1.85 g/100 mL for β-cyclodextrin itself), is 57 g/100 mL and 31 g/100 mL for dimethyl and trimethyl β-cyclodextrin, respectively. However, a major drawback of these derivatives is an exothermic dissolution phenomenon resulting in a decrease in solubility with an increase in temperature [10].

(b) Hydroxypropyl Cyclodextrins. Hydroxypropylation of cyclodextrins occurs in a completely random manner, resulting in products that are not well defined with respect to the position of the hydroxypropyl groups. Products marketed are a mixture of various hydroxypropyl ethers of cyclodextrins only characterized by their substitution degree and the exact nature of the hydroxypropyl group. They cannot crystallize and are amorphous. All the hydroxypropyl cyclodextrins are highly water-soluble: over 50 g/100 mL at room temperature; the water-solubility increases with an increase in temperature [11].

(c) Other Derivatives. Glucosyl and maltosyl cyclodextrins, more or less substituted, are marketed. They are highly water-soluble. For example, the diglucosyl β-cyclodextrin has a water solubility of 140 g/100 mL, compared with only 1.85 g/100 mL for the original β-cyclodextrin [12,13]

Hydroxyethyl cyclodextrins are presented as very closely resembling the hydroxypropyl cyclodextrins, especially with respect to their water-solubility. However, they are not easily available on the market [14].

The new sulfobutyl ether derivatives are proposed for their very high solubilizing effect and thus are very often compared with hydroxypropyl derivatives [15].

Finally, cyclodextrin polymers are water-soluble for the low molecular weights, whereas high-molecular-weight products are water-insoluble but capable of swelling [16].

B. Inclusion in Cyclodextrins

1. Inclusion Formation [17]

The formation of an inclusion compound generally occurs in aqueous solution or at least in the presence of a small quantity of water. Roughly, the inclusion process mechanism takes place as follows. When a cyclodextrin is dissolved, the water molecules inside the cyclodextrin cavity are in an unfavorable hydrophobic environment and cannot create hydrogen bonds with the surrounding glucosidic groups. For this reason, they are more or less attracted by the water molecules from the bulk, and—from a very simple thermodynamic standpoint—they are more often outside the cyclodextrin cavity than inside. Obviously, the cyclodextrin cavity cannot remain empty, and any other molecule present in the medium will be trapped into the cavity. The supermolecule obtained can be separated in solid form, either by spontaneous precipitation or by evaporation of the aqueous phase.

In fact, this inclusion mechanism is a succession of equilibria:

Solid guest ↔ dissolved guest

Dissolved guest + dissolved cyclodextrin ↔ dissolved inclusion compound

Dissolved inclusion compound ↔ solid inclusion compound

Each of these equilibria is dependent on a constant. In the case of the equilibrium resulting in the formation of an inclusion compound, the constant K is the affinity constant (affinity of the guest molecule for the cyclodextrin cavity), also called the stability constant (stability of the inclusion compound under the form of a supermolecule and not in dissociated form). A low-stability constant means that in solution, the inclusion compound will be highly dissociated. On the other hand, a high-stability constant means that the inclusion will be present mainly in undissociated form in solution. Knowledge of this constant is very useful, because it governs the yield of inclusion formation as well as the behavior of the inclusion in liquid medium. This constant depends not only on the nature of the guest molecule and of the cyclodextrin but also on the nature of the surrounding medium.

2. Use of Inclusions

The inclusion of a guest molecule in a cyclodextrin cavity constitutes a molecular encapsulation, resulting in modification of the physicochemical characteristics of the guest molecule with respect to its state, stability, solubility, undesirable side effects, etc.

(a) Modification of Physical State.   Gas or liquids can be included in cyclodextrins, leading to solid inclusion compounds that, after dissolution, will release the included product with its former characteristics. This has been applied to various essential oils [18], but it seems highly improbable that the very numerous constituents of such products can all be included with the same yield and released at the same speed, because, due to their different chemical natures, their stability constant cannot be the same.

(b) Stabilization.   One can consider that inclusion of an essential oil is a stabilization because it reduces its volatility [19]. Similarly, the inclusion of solid compounds capable of sublimation can reduce their volatility [20].

However, chemical stabilization can be obtained for oxidizable products when they are presented as solid inclusion compounds. In fact, the molecular encapsulation of a guest by a cyclodextrin can be considered as a molecular coating resulting in a protection against oxygen, air, and light. Examples of such stabilization are numerous [21] and particularly concern liposoluble vitamins [22].

It is not at all the same when the inclusion compound is in solution, because then dissociation takes place according to the stability constant, and the free guest can be normally decomposed by the liquid medium [21]. As a consequence, the only thing that can be obtained in solution is a slowdown of decomposition kinetics, related to the stability constant of the inclusion compound in the liquid medium.

However, and despite the fact that stabilization cannot, by inclusion in a cyclodextrin, be total, an increase in stability of tixocortol 17-butyrate 21-propionate (a dermocorticoid) has been demonstrated in petroleum jelly or emulsion-based ointment after inclusion in β-cyclodextrin [23].

(c) Decrease in Undesirable Side Effects.   Likewise, the molecular encapsulation of a guest by a cyclodextrin is a molecular coating protecting the guest against external factors: it can protect various mucosae against undesirable effects of the guest. This can concern unpleasant odors of volatile products [24] or a bitter (or other unpleasant) taste of some components [25]. It can also concern the irritating power of an active ingredient applied to the skin and has been demonstrated for indomethacin [26] or tretinoin [27] included in β-cyclodextrin or hydroxypropyl β-cylodextrin.

(d) Solubilization.   The increase in water-solubility obtained by inclusion of a poorly water-soluble guest in a cyclodextrin is in fact an increase in apparent solubility. What is to be considered is the whole: dissolved free guest plus solubilized guest included in the cyclodextrin.

It must not be forgotten that, to have its complete normal activity, the guest must be released from the cyclodextrin. This can be accelerated by interaction of the inclusion with some external components, such as surface skin (or hair).

In some cases, it is not really an increase in solubility that is seen but simply an increase in dissolution kinetics. This means that, for a guest component which could dissolve slowly in water, its inclusion in a cyclodextrin used in a large amount of water (bath) will accelerate the dissolution of the product, while a great dilution will result in the great dissociation of the inclusion and an almost complete release of the guest.

Such an increase in solubility can allow the preparation of hydrogels with poorly water-soluble active ingredients, such as indomethacin [26] or tretinoin [27].

(e)   Increase in Biological Activity.   That which is an advantage for a dermo-pharmaceutical product can be a drawback in cosmetology. It results from the higher solubility of the inclusion as compared with the guest molecule and from the dissocation of inclusion compound supermolecules. After dissociation, free hydrophobic molecules of the guest compound will be in contact with lipoidic biological membranes. If the absorption rate is high, this mechanism can increase the biological activity of the guest. On the other hand, if the absorption rate is low, reprecipitation of the free guest can occur and its behavior will be almost the same as that of the unincluded guest.

C.   Cyclodextrins and the Skin

Many papers or patents are related to the dermal use of cyclodextrins and their derivatives [28,29]. However, this use requires a good knowledge of cyclodextrins/skin interactions.

  Authors, such as Uekama et al. reported the irritant effects of cyclodextrins on the skin [30] and related these effects to the ability of cyclodextrins to include various biological membrane constituents (such as cholesterol, triglycerides, phospholipids) and, as a consequence, to displace proteins [31–34].

  Despite the high external hydrophilicity of cyclodextrins, which normally should prevent dermal absorption, Arima et al. demonstrated that a few hours after their dermal application, an appreciable amount was capable of penetrating the skin in the following order: β-cyclodextrin < dimethyl β-cyclodextrin < hydroxypropyl β-cyclodextrin [35,36].

  Gerlóczy et al. [37], working on dimethyl β-cyclodextrin, demonstrated its transdermal absorption. However, this occurs in small quantities: the product is hardly detectable in the blood and rapidly excreted by the urine. Furthermore, it is not possible to know if the radioactivity detected in the urine corresponds to the dimethyl β-cyclodextrin or to its metabolites.

  More recently, Vollmer et al. [38], working on radiolabeled hydroxypropyl β-cyclodextrin, demonstrated its penetration and showed its localization in high amounts at the level of compact stratum corneum and in smaller amounts at the level of sebaceous glanas and blood vessels in the dermis and muscular layer. No accumulation was noticed in the hair roots. Thus, hydroxypropyl β-cyclodextrin seems to be able to spread out in lipophilic zones of the stratum corneum as well as in hydrophilic parts of the epidermis. It must be pointed out that morphological examination of the skin 3 h after 4.9% hydroxypropyl β-cyclodextrin solution application did not reveal any skin alteration.

  The absence of irritating effect of the three natural cyclodextrins as well as of the dimethyl and hydroxypropyl β-cyclodextrins was demonstrated by applying, on 1 cm^2 of the skin of healthy volunteers, amounts equivalent to 2 mg of cyclodextrin dissolved or dispersed in water or petroleum jelly, respectively. The measurement of a possible increase in vasodilatation induced by the possible irritating power of cyclodextrins, assessed by laser Doppler velocimetry, revealed no significant difference between skin treated by water or petroleum jelly and skin treated by any of the cyclodextrin preparations [39].

III. CYCLODEXTRINS IN COSMETICS

Cyclodextrins are used in the preparation of cosmetics not only in the form of an inclusion compound with one (or more) of the active components of the preparation but also as free (or "empty") cyclodextrins.

A. Use of "Empty" Cyclodextrins

1. Scrubbing Particles

Water-insoluble cyclodextrin polymer beads can be prepared from β-cyclodextrin treated by epichlorohydrin. These products have been acclaimed for their mild scrubbing effect. Furthermore, associated to oily ingredients, such as moisturizers or circulation promoters, they can form stable inclusion compounds, gradually releasing the active component [40].

2. Entrapment of Unpleasant Odors

By their ability to include gas and other volatile molecules, empty cyclodextrins have been proposed as deodorants in various forms [41]. They can be associated to fragrances [42,43] and antiseptics or antimicrobial preservatives [43]. For this purpose, almost any kind of cyclodextrin may be used.

(a) *Mouth Odors.* Cyclodextrin associated with plants extracts, propolis, marine algae extracts, iron, or copper chlorophyl sodium derivatives is endowed with deodorant effects resulting from the control of the sulfur-compound mouth bacteria metabolites [44]. Products concerned can be either dentifrices [45] or patches [46].

(b) *Body Odors.* The product can be presented in the form of an underarm adhesive tape [47,48], constituted, for example, of stretchable polyurethane film and an adhesive acrylic layer containing 3% of a mixture of α-, β-, and γ-cyclodextrins (6:3:1) [48]. Another product is proposed to control underarm or foot odors [49] and contains hydroxypropyl β-cyclodextrin 5, ethanol 20, 1,3-butylene glycol 3, citric acid 0.03, sodium citrate 0.07, sodium hexametaphosphate 0.01, fragrance 0.3, methyl paraben 0.1, and water to 100 (by weight). Antiperspirant powder-type aerosols are also proposed [50].

(c)   Hair Odors.   The association of either α-, β-, γ-, or δ-cyclodextrin (or their cationic derivatives) to cationic surfactants is claimed as presenting a synergistic deodorant activity for a prolonged period [51]. It is surprising to see the possible presence in the formulation of δ-cyclodextrin which, due its low stability, is more a laboratory curiosity than an industrial product. However, the scalp deodorant contains stearylmethylammonium chloride 0.5, α-cyclodextrin 0.5, ethanol 30.0, and water to 100 (by weight).

3.   Absorption of Fatty Components

Fatty components can be included in cyclodextrins, which, in consequence, can be used as cleansing agents capable of removing either makeup or sebum.

(a)   Spray-Dried Particles.   A product consisting of a spray-dried slurry containing calcium phosphate, squalane as moisturizing agent, and β-cyclodextrin (in excess with respect to squalane) is presented as particles showing high absorption ability for sebum [52].

(b)   Cosmetics Packs.   A peel-off type of pack constituted of polyvinyl alcohol 12, carboxymethyl cellulose 3, methyl β-cyclodextrin 20, ethanol 10, perfume 0.05, paraben 0.05 and water to 100%, is claimed as having excellent cleansing and moisturizing effects without irritating the skin [53]. Another cosmetic pack with very similar components contains hydroxypropylated β-cyclodextrin [54].

(c)   Shampoos.   Powder and aerosol shampoos containing microporous silica, β-cyclodextrin, modified starch, clay, and plant extracts [55] have been patented. The two shampoos are already marketed in France by Laboratories Klorane (Pierre Fabre); the high sebum-absorbing power of the cyclodextrin is highlighted.

4.   Additives to Formulation

In these formulations, cyclodextrins are not used to improve some physicochemical characteristics of a guest active component but as an additive necessary for the physical state of the formulation.

(a)   Powder-Like Preparations.   Unlike in the preceding shampoos, where the powder form of the preparation was due mainly to the presence of silica, starch, and clay, in the following formulation cyclodextrins are used as adsorbents of the liquid products, leading to the powder form.

    Cosmetics of powder (or paste) form are prepared by mixing 5–90% of a mixture of cyclodextrins or derivatives to other cosmetic base materials in order to obtain a powder or paste form showing good storage stability that is easily restorable to the liquid state [56].

(b)   Emulsions Stabilized by Cyclodextrins.   Emulsions without surfactants can be stabilized by cyclodextrins [57,58]. Cyclodextrins can include constituents from the oil phase, forming a rigid film at the surface of the dispersed phase [59].

Cosmetic emulsions have been described [60,61]. An example of a composition is liquid paraffin 15, β-cyclodextrin 5, carboxymethyl cellulose 1.5, glycerol 5, preservative q.s., perfume q.s., and distilled water to 100% [60]. A viscous cold cream contains petroleum jelly 3, beeswax 2, cetanol 2.5, lanolin 5, squalane 20, paraffin oil 10, fragrance 0.5, preservative q.s., antioxidant q.s., hydroxypropylated β-cyclodextrin 3.9, propylene glycol 5, and water to 100% [62].

B.   Inclusions in Cyclodextrins

1.   Stabilization

(a)   Dyes.   Hair dyes are very often susceptible to oxidation. Inclusion of the dye in a cyclodextrin can result in a product with a longer shelf-life and better color-fastness than those of conventional hair dyes. Hair-dye powders [63] or hair-dye couplers and developers for creams [64] have been stabilized by inclusion in α-, β-, or γ-cyclodextrin.

(b)   Kojic Acid.   Kojic acid is effective for the inhibition of melanin formation but can color itself during the aging of cosmetic products. Its inclusion in β-cyclodextrin can prevent this phenomenon [65–70]. An example of a cream is kojic acid 1.0, β-cyclodextrin 4.0, beeswax 6.0, cetanol 5.8, reduced lanolin 8.0, squalane 30.0, glycerides 4.0, hydrophobic glycerol monostearate 2.0, polyoxyethylene sorbitan monolaurate 2.0, distilled water 37.20%, and perfumes, preservatives, and antioxidants [65].

(c)   Royal Jelly.   Royal jelly can be presented in stable powder form by mixing with β-cyclodextrin in the proportions 7:3, 8:2 or 9:1 (wt/wt) and freeze-dried [71]. Another possibility is to dry a mixture of royal jelly with cyclodextrin (45–60% of royal jelly) containing a catalytic amount of ethanol [72], the physiological active compound; 10-hydroxy-2-decenoic acid is stabilized by this treatment.

2.   Decrease in Undesirable Side Effects

(a)   Permanent-Wave Odor.   Most permanent-wave setting products have a very unpleasant odor due to reducing thioglycolate compounds. Their inclusion in a cyclodextrin [73], which can be not only β-cyclodextrin [74] but also α-, γ-, or δ-cyclodextrin [75] or a dimethyl and trimethyl derivative [76], results in odor

inhibition. A synergistic effect can be obtained by association of the cyclodextrin to a cationic polymer [75].

(b)   Odor from Other Cosmetic Products.   Mineral, animal, or vegetal tars incorporated in skin preparations give them an unpleasant odor. Their inclusion in cyclodextrins, especially in the case of soybean tar, inhibits the odor, and also the color without decreasing their activity [76].

The unpleasant odors of enzymes contained in bath preparations can be masked by inclusion in β-cyclodextrin [77].

A different case is presented by skin-tanning compositions, which develop a characteristic odor when the tanning agents such as dihydroxyacetone react with the skin. This effect can be prevented by incorporation of cyclodextrins, especially γ-cyclodextrin, in the preparation [78]. Here, there is no malodorous product from the preparation that is included but rather a product formed during the use of the preparation. The principle of such odor prevention is very similar to that of the cyclodextrins associated to antiperspirant preparations seen above.

(c)   Masking of Unpleasant Odor and Taste of Menthol.   Menthol is very often added to various cosmetic products for its refreshing activity. However, it has an unpleasant odor and a bitter taste. Its inclusion in β-cyclodextrin [79,80] or hydroxypropyl β-cyclodextrin [81] can overcome this drawback. This technique is applied to mouthwashes [80,82], shampoos [82], or skin cosmetics with cooling activity [81].

3.   Inclusion of Perfumes

(a)   Interest of Inclusions.   The inclusion of perfumes in cyclodextrins is very often studied in order to obtain dissolution in water without using irritating surfactants. From this standpoint, despite the fact that various cyclodextrins can be used [83], hydroxypropyl β-cyclodextrin has been intensively studied and is among the best solubilizers [84], especially when its substitution degree is between 3.4 and 4.6, an increase to 8.1 resulting in a decrease in solubilization [85,86]. Furthermore, the percutaneous absorption of paraben preservatives is decreased [87], probably because of the possible interaction between the cyclodextrin and the preservative [88–90]. However, such an interaction does not systematically result in a decrease in the preservative effect because, depending on the exact nature of the preservative and of the cyclodextrin employed, the active group can be outside the cyclodextrin cavity and thus still efficient [88,90]. Another advantage of the inclusion of a fragrance in a cyclodextrin is that it can result in prolonged release of the perfume [91].

(b)   Hair or Skin Lotion.   When methylated β-cyclodextrin is used to solubilize aromatic substances, it is possible to decrease the amount of ethanol and surfactants used in conventional lotions. The product obtained does not produce any

skin irritation and gives more sustained release of the fragrance than in the absence of the cyclodextrin [92].

(c)   Bath Preparations.   Aromatic essential oils included in cyclodextrins can be stabilized [93] and can dissolve easily in bath water. For example, the preparation of the inclusion compound consists of mixing an essential oil (containing hinokition) 1–10 g and cyclodextrins 50 g with a small amount of water for 1–3 h, followed by drying. Then, the bath preparation consists of sodium sulfate 45–48, sodium acid carbonate 39–42, borax 2, coloring agent trace, and the inclusion compound 1–8% [94].

Perfumed tablets to dissolve in the bath can be prepared with solid inclusion of the fragrance. Bath tablets of 10 g are prepared from jasmine oil partially methylated cyclodextrin inclusion compound 10, sodium chloride 40, borax 47, and boric acid three parts. The tablets dissolve completely in 1 L of warm water at 42°C [95].

(d)   Toothpastes.   The inclusion of a fragrance in a cyclodextrin for toothpaste is worthwhile simply for better stability, but it can be also a good indicator of duration of tooth brushing [96]. For example, a toothpaste contains limonene-β-cyclodextrin inclusion compound, another flavor such as isoamyl acetate, and β-cyclodextrin. When this toothpaste is used, the flavor of the paste changes to that of limonene in about 30 s, indicating the length of time for adequate brushing.

(e)   Soaps.   Fragrance compounds stabilized by inclusion in cyclodextrins can be coated with oils and incorporated into soap [97]. The preparation of the inclusion requires 100 g of water, 100 g of β-cyclodextrin, and 25 g of lemon oil. After mixing, the product is dried, pulverized, and mixed with 100 g of liquid paraffin for coating. This product is used for the soap preparation.

(f)   Talc.   To prepare perfumed talcs, it is possible to adsorb the perfume onto starch and to incorporate this powder into the talc. However, as starch is a suitable medium for the growth of microorganisms, the use of a cyclodextrin can prevent this drawback and improve the antimicrobial efficacy of talc powders. The starch can be partially or totally replaced by β-cyclodextrin or β-cyclodextrin–fragrance complex [98]. The normal moisture of the skin is enough to progressively dissolve the inclusion and provide a prolonged release of the aroma.

(g)   Sustained Release.   In the previous example, the inclusion of a fragrance in a cyclodextrin resulted in prolonged release. This effect can be examined for different applications [99,100], especially for underarm deodorants [101].

(h)   Bathtub.   Why not perfume the bathtub itself rather than the bath water? It seems to be possible by manufacturing the bathtub from an unsaturated polyester, glass fibers, and a hinoki perfume–cyclodextrin inclusion compound [102].

4. Solubilization

As seen above, the inclusion of an aromatic substance in a cyclodextrin leads to a powder form, stabilizes the aroma, and solubilizes it. Many other products are included in cyclodextrins in order to obtain their aqueous solubilization.

(a) Liposoluble Vitamins. Vitamin A (retinol) and acid vitamin A (retinoic acid) are used against skin aging, but their low water-solubility prevents them from being formulated in hydrogel form and can require the use of fatty or hydro-alcoholic carriers; the latter, however, increases skin irritation. These products can be solubilized by inclusion in cyclodextrins.

Retinoic acid [103,104] or its derivatives, such as palmitate [105], can be included in β-cyclodextrin. It can increase the efficiency of the product which can be incorporated into a cream base to be used in the treatment of acne [103] or to improve skin thickness and elasticity [105].

Vitamin E, tocopherol, which also has low water solubility, is already marketed under the form of an inclusion compound in β-cyclodextrin incorporated in a cream: Luminys from Roan S.p.a., Italy.

(b) Fatty Components. Slightly water-soluble components—such as oils, fats, and higher fatty acids—can be included in hydroxypropylated γ-cyclodextrin in order to prevent skin roughness [106]. The product is constituted of hydropxypro-pylated γ-cyclodextrin 1.0, benzophenone 0.05, composed perfumes 0.1, and de-ionized water three parts, mixed with a second preparation containing deionized water 82.0948, glycerol 1.0, 1,3-butylene glycol 2.0, lactic acid 0.005, sodium lactate 0.2, monoammonium glycyrrhizinate 0.05, aloe extract 0.5, 95% ethanol 10, and dye 0.0002.

Dispersibility of oily higher fatty acids or their esters is increased by adding α-cyclodextrin [107]. For example, 1 g of medicinal triglycerides, consisting of palmitic acid 30, stearic acid 5, oleic acid 45, linolic acid 10, and γ-linolenic acid 7, is mixed with 10 g of α-cyclodextrin. This mixture is well dispersed in 100 mL of water.

(c) Bath Products. Squalane can be included in cyclodextrins and used in bath preparations [108]. Its slow release will result in a better moisturizing effect of the skin. An aqueous solution of a mixture of α-, β-, and γ-cyclodextrin (6:3: 1) is homogenized with squalane and spray-dried. The powder obtained, 10 parts, is mixed with sodium sulfate 40 and sodium acid carbonate 50 parts, and flavoring materials, leading to the bath preparation.

It could be very fashionable to incorporate milk in the bath water. Several formulations are proposed in which milk components are included in cyclodextrin [109,110]. An example of a bath preparation contains sodium sulfate 35, sodium acid carbonate 45, cyclodextrin-milk component inclusion compound 20% (by weight), and milk-like fragrance [110].

Another pleasant type of bath could be an effervescent bath. It is possible by including carbon dioxide in cyclodextrins [111]. Thus, 1 kg of α-cyclodextrin is treated with carbon dioxide at 8 kg/cm^2 pressure to absorb 40.4 g of gas. A 50-g tablet contains this product, 80 parts, sodium sulfate 10, sodium succinate 5, dextrin 4.5, a perfume 0.5 parts, and a coloring agent.

(d) Miscellaneous. Minoxidil, a hair-care compound stimulating keratinocyte growth and promoting hair regrowth, has a low water-solubility that can be increased by α-cyclodextrin [112]. The hair preparation contains minoxidil 1, α-cyclodextrin 0.5, propylene glycol 1, 95% ethanol 42, and purified water q.s. to 100 mL.

Iodine is a powerful microbicidal agent, but it is water-insoluble and has a characteristic unpleasant taste. These drawbacks can be prevented by inclusion of iodine in maltocyclodextrin. A mouthwash [113] can be prepared by dissolving potassium iodide 0.6 parts in 1 part of water mixing with iodine 0.75 and isoleeat (maltosylcyclodextrin) 7.5 parts, and kneaded with water to prepare a paste that is added with water to 100 parts and stirred for at least 4 h.

An inclusion compound of L-carnitine in β-cyclodextrin, called Cyclosome, is presented in Cellutex cream marketed by Regena Ney Cosmetic in Germany. It is presented as new way of infiltrating L-carnitine, which increases its effect in reducing the embedded fats.

IV. CONCLUSION

There are many ways to use cyclodextrins in cosmetic formulations: for their inclusion ability toward external compounds or for the inclusion of components presenting drawbacks, such as volatility, poor stability, poor water-solubility, irritating effects, bad taste or odor, etc. They can be added to almost any kind of cosmetic preparation, in which they can exert several roles simultaneously: solubilization, stabilization, a decrease in side effects, etc. Owing to their absence of irritating power, they seem to be promising adjuvants for improving the quality of cosmetic products.

REFERENCES

1. Villiers A. Sur la fermentation de la fécule par l'action du ferment butyrique. C R Acad Sci 1891; 112:536.
2. Schardinger F. Über thermophile Bakterien aus verschiedenen Speisen und Milch, sowie über einige Umsetzungsprodukte derselben in kohlenhydrathalien Nährolö-

sungen, darunter krystallisierte Polysaccharide (Dextrine) aus Stärke. Z Unters Nahrungs-Genußmittel Gebrauchsgegenstände 1903; 6:865.

3. Schardinger F. Mitteilung aus der staatlichen Untersuchungsanstalt für Lebensmittel in Wien, Azetongärung. Wien Klin Wochenschr 1991; 17:207.

4. Schardinger F. Bildung kristallisierter Polysaccharide (Dextrine) aus Stärkekleister durch Mikrobien. Zentralbl Bakteriol Parasitenk Infektionskr 1911; II29:188.

5. Vaution C, Hutin M, Glomot F, Duchêne D. The use of cyclodextrins in various industries. In: Duchêne D, ed. Cyclodextrins and Their Industrial Uses. Paris: Editions de Santé, 1987:299–350.

6. Duchêne D, Debruères B, Brétillon A. Les cyclodextrines. Nature, origine et intérêt en pharmacie galénique. Labo-Pharma Probl Tech 1984; 32:842.

7. Duchêne D, Wouessidjewe D. Physicochemical characteristics and pharmaceutical uses of cyclodextrin derivatives. Pharm Techn Int 1990; 2(2):21.

8. Schmid G. Preparation and application of γ-cyclodextrin. In: Duchêne D, ed. New Trends in Cyclodextrins and Derivatives. Paris: Editions de Santé, 1991:188.

9. Uekama K. Pharmaceutical applications of methylated cyclodextrins. Pharm Int 1985; 6:61.

10. Uekama K, Irie T. Pharmaceutical applications of methylated cyclodextrin derivatives. In: Duchêne D, ed. Cyclodextrins and Their Industrial Uses Paris: Editions de Santé 1987:393.

11. Yoshida A, Yamamoto M, Irie T, Hirayama F, Uekama K. Some pharmaceutical properties of 3-hydroxypropyl and 2,3-dihydroxypropyl β-cyclodextrins and their solubilizing and stabilizing abilities. Chem Pharm Bull 1989; 37:1059.

12. Koizumi K, Okada Y, Kubota Y, Utamara T. Inclusion complexes of poorly water-soluble drugs with glucosyl-cyclodextrins. Chem Pharm Bull 1987; 35:3413.

13. Yamamoto M, Yoshida A, Hirayama F, Uekama K. Some physicochemical properties of branched β-cyclodextrins and their inclusion characteristics. Int J Pharm 1989; 49:163.

14. Yoshida A, Arima K, Uekama K, Pitha J. Pharmaceutical evaluation of hydroxyalkyl ethers of β-cyclodextrins. Int J Pharm 1988; 46:217.

15. Stella RA, Stella VJ. Pharmaceutical applications of cyclodextrins, 2, In vivo drug delivery. J Pharm Sci 1996; 85:1142.

16. Szejtli J. Highly soluble β-cyclodextrin derivatives. Starch/Stärke 1984; 36:429.

17. Duchêne D. New trends in pharmaceutical applications of cyclodextrin inclusion compounds. In: Huber O, Szejtli J, eds. Proceedings of the 4th international Symposium on Cyclodextrins. Dordrecht Boston London: Kluwer, 1988:265–375.

18. Gal-Füzy M, Szente L, Szejtli J, Harangi J. Cyclodextrin-stabilized volatile substances for inhalation therapy. Pharmazie 1984; 39:558.

19. Szejtli J, Szente L, Bánky-Elöd E. Molecular encapsulation of volatile easily oxidizable labile flavour substances by cyclodextrins. Acta Chim Acad Sci Hung 1979; 101:27.

20. Uekama K, Oh K, Irie T, Otagiri M, Nishimiya Y, Nara T. Stabilization of isosorbite 5-mononitrate in solid state by β-cyclodextrin complexation. Int J Pharm 1985; 25:339.

21. Duchêne D, Debruères B, Vaution C. Improvement of drug stability by cyclodextrin inclusion complexation. STP Pharma 1985; 1:37.

22. Szejtli J, Bolla-Pusztai É, Szabó P, Ferenczy T. Enhancement of stability and biological effect of cholecalciferol by β-cyclodextrin complexation. Pharmazie 1980; 35:779.

23. Glomot F, Benkerrour L, Duchêne D, Poelman M-C. Improvement in availability and stability of a dermocorticoid by inclusion in β-cylodextrin. Int J Pharm 1988; 46:49.

24. Szejtli J. Cyclodextrins in foods, cosmetics and toiletries. In: Szejtli J. ed. Proceedings of the first international symposium on cyclodextrins. Budapest: Akadémiai Kiadó, 1982:469.

25. Fujioka K, Kurosaky Y, Sato S, Noguchi Te, Nogichi Ta, Yamahira Y. Biopharmaceutical study on inclusion complexes: I. Pharmaceutical advantages of cyclodextrin complexes of bencyclane fymarate. Chem Pharm Bull 1983; 31:2416.

26. Lin S-Z, Wouessidjewe D, Poelman M-C, Duchêne D. In vivo evaluation of indomethacin/cyclodextrin complexes, Gastrointestinal tolerance and dermal anti-inflammatory activity. Int J Pharm 1994; 106:63.

27. Amdidouche D, Montassier P, Poelman M-C, Duchêne D. Evaluation by laser Doppler velocimetry of the attenuation of tretinoin induced skin irritation by β-cyclodextrin complexation. Int J Pharm 1994; 111:111.

28. Duchêne D, Wouessidjew D, Poelman M-C. Dermal uses of cyclodextrins and derivatives. In: Duchêne D, ed. New Trends in Cyclodextrins and Derivatives Paris: Editions de Santé, 1991:449.

29. Duchêne D, Wouessidjewe D, Poelman M-C. Les cyclodextrines dans les préparations pour usage dermique. In: Seiller M, Martini M-C, eds. Formes Pharmaceutiques pour Application Locale. Paris: Lavoisier Tec Doc, 1996:481.

30. Uekama K, Irie T, Sunada M, Otagiri M, Arimatsu Y, Nomura S. Alleviation of prochlorperazine-induced primary irritation of skin by cyclodextrin complexation. Chem Pharm Bull 1982; 30:3860.

31. Irie T, Otagiri M, Sunada M, Uekama K, Ohtani Y, Yamada Y, Sugiyama Y. Cyclodextrin induced hemolysis and shape changes on human erythrocytes in vitro. J Pharmacobio-Dyn 1982; 5:741.

32. Okamoto H, Komatsu H, Hashida M, Sezaki H. Effects of β-cyclodextrin and di-O-methyl β-cyclodextrin on the percutaneous absorption of butyl paraben, indomethacin and sulfanilic acid. Int J Pharm 1986; 30:35.

33. Othani Y, Irie T, Uekama K, Fukunaga K, Pitha J. Differential effects of α-, β- and γ-cyclodextrins on human erythrocytes. Eur J Biochem 1989; 186:17.

34. Yoshida A, Yamamoto M, Irie T, Hirayama F, Uekama K. Some pharmaceutical properties of 3-hydroxypropyl and 2,3-dihydroxypropyl β-cyclodextrins and their solubilizing and stabilizing abilities. Chem Pharm Bull 1989; 37:1059.

35. Arima H, Adachi H, Irie T, Uekama K. Improved drug delivery through the skin by hydrophilic β-cyclodextrins: enhancement of anti-inflammatory effect of 4-biphenylylacetic acid in rats, Drug Invest. 1990; 2:155.

36. Arima H, Adachi H, Irie T, Uekama K, Pitha J. Enhancement of the anti-inflammatory effect of ethyl 4-biphenylylactic acetate ointment by β-cyclodextrin derivatives: increased absorption and localized activations of the prodrug in rats, Pharm Res 1990; 7:1152.

37. Gerlóczy A, Antal S, Szejtli J. Percutaneous absorption of heptakis-(2,6-di-O-14C -

methyl)-β-cyclodextrin in rats. In: Huber O, Szejtli J, eds. Proceedings of the 4th international Symposium on Cyclodextrins. Dordrecht, Boston, London: Kluwer 1988:415.

38. Vollmer U, Stoppie P, Mesens J, Wilffert B, Peters T. Hydroxypropyl β-cyclodextrin in transdermal absorption in vivo in rats. In: Hedges AR, ed. Minutes of the 6th International Symposium on Cyclodextrins. Paris: Editions de Santé, 1992:535.

39. Montassier P. Inclusion de la trétinoïne dans les cyclodextrines, Thesis. 495, University of Paris XI, pharmaceutical sciences, Dec. 19, 1996.

40. Imamura K, Tsuchama Y, Tsunakawa H, Okamura K, Okamoto R, Harada K. (Merushan Kk, Kogyo Gijutsuib). Cosmetics containing water-insoluble cyclodextrin polymers as scrubbing particles, JP 05,105,619 [93,105,619]. Feb. 14 1991.

41. Trinh T, Phan DV (Procter and Gamble Co.). Articles containing small particle size cyclodextrin for odor control, PCT Int Appl WO 94 22,501. March 31, 1993.

42. Ito K, Matsuda H (Shiseido Co. Ltd.). Aromatic deodorizing compositions, JP 07,241,333 [95,241,333]. March 4, 1994.

43. Trinh T, Cappel JP, Geis PA, McCarty ML, Pilosof D, Zwerdling SS. (Procter and Gamble Co. USA). Cyclodextrin solutions for odor control on surfaces, WO 96 04,937. Aug. 12, 1994.

44. Cho H, Torii M, Kanamori T (Lion Corp.). Deodorant controlling mouth odor, JP 63,264,516 [88,264,516]. April 20, 1987.

45. Imaoka KK, Sakamoto T, Gomi T (Lion Corp.). Dentifrices containing tranexamic acid, triclosan and menthyl acetate, JP 04,139,117 [92,139,117]. May 13, 1992.

46. Kishi T, Iwakawa M (Sekisui Chemical Co. Ltd.). Oral odor-controlling patch compositions, JP 63,280,014 [88,280,014]. May 12, 1987.

47. Kishi T (Sekisui Chemical Co. Ltd.). Adhesve deodorant tape for the control of underarm odor, JP 63,280,011 [88,280,011]. May 12, 1987.

48. Kishi T (Sekisui Chemical Co. Ltd.). Deodorant adhesive tapes for the control of underarm odor. JP 63,280,010 [88,280,010]. May 12, 1987.

49. Matsuda H, Ito K (Shiseido Co. Ltd.). Body deodorants containing hydroxyalkylated cyclodextrins, JP 03,284,616 [91,284,616]. Dec. 16, 1991.

50. Maekawa A (Sunstar Inc.). Antiperspirant aerosol compositions containing cyclodextrin, JP 03,170,415 [91,170,415]. Nov. 30, 1989.

51. Yamagata Y, Yoshibumi S (Lion Corp.). Hair preparations containing cationic surfactants and cyclodextrins, JP 62,267,220 [87,267,220]. May 15, 1986.

52. Saeki T, Morifuji T (Sekisui Plastics; Tanaka Narikazu, Japan). Moisturizer-containing cyclodextrin composite particles for cosmetics, JP 08,151,317 [96,151,317]. June 11, 1996.

53. Yagi Y, Sato M (Sanraku Co. Ltd.). Cosmetic packs containing methylated cyclodextrin, JP 01,193,209 [89,193,209]. Aug. 3, 1989.

54. Matsuda H, Ito K (Shiseido Co. Ltd.). Cosmetic packs containing hydroxyalkylated cyclodextrin, JP 03,287,512 [91,287,512]. March 31, 1991.

55. Jeanjean M, Sénégas N, Fabre B (Pierre Fabre Dermo-Cosmétique). Phytogenic dry shampoo comprising micronized powders, WO 96 00,563. June 28, 1994.

56. Natori Y, Nishida J (Lion Corp.). Cosmetics containing cyclodexrivatives for hair and skin, JP 05,194,153 [93,194,153]. August 3, 1993.

57. Shimada K, Ohe Y, Ohguni T, Kawano K, Ishii J, Nakamura T. Emulsifying proper-

ties of α-, β- and γ-cyclodextrins. Nippon Shokuhin Kogyo Gakkaishi 1991; 38: 16.

58. Hibi T, Kurahashi R, Sugimori K, Kamikama K (Nippon Tobacco Sangyo). Preparation of emulsions without emulsifiers, JP 06,128,149 [94,128,149]. May 10, 1994.
59. Laurent S, Graille J. Unpublished results. CIRAD, Montpellier, 1993.
60. Masuda T, Ishida S, Hashimoto S (Sunstar Inc.). Emulsions containing carboxymethyl cellulose sodium salt, JP 61,133,138 [86,133,138]. June 20, 1986.
61. Ogino S, Kamiya H (Kao Corp.). Manufacture of emulsions with methylated β-cyclodextrin for cosmetics, JP 63,194,726 [88,194,726]. Feb. 6 1987.
62. Matsuda H, Ito K (Shiseido Co. Ltd.). Emulsified cosmetics containing hydroxy-alkylated cyclodextrins, JP 03,284,611 [91,284,611]. Dec. 18, 1991.
63. Oishi T, Nakanishi F, Yamamoto T (Hoyu Co. Ltd.). Hair dye powders containing cyclodextrin-dye inclusion compounds, EP 279,016. Aug. 24, 1988.
64. Rose D, Hoeffkes H (Henkel KGaA). Stabilization of oxidative hair dye precursors with cyclodextrins, DE 4,418,990. Dec. 7, 1995.
65. Hatae S, Nakajima K (Sansei Pharmaceutical Co. Ltd.). Skin whitening cosmetics, JP 61,109,705 [86,109,705]. Nov. 1, 1984.
66. Hatae S (Sansei Pharmaceutical Co. Ltd.). Use of kojic acid or its β-cyclodextrin inclusion complex for the treatment of elastosis, EP 207,499. July 2, 1985.
67. Hatae S Nakajima K (Sansei Pharmaceutical Co. Ltd.). Skin compositions containing cyclodextrin inclusion compound of kojic acid, JP 63,08,311 [88,08,311]. Jan. 14, 1988.
68. Motono M (Sansei Pharmaceutical Co. Ltd.). Discoloration prevention in cosmetics containing kojic acid (derivatives) by UV absorbers, β-cyclodextrin, and EDTA, JP 63,188,609 [88,188,609]. Aug. 4, 1988.
69. Motono M (Sansei Pharmaceutical Co. Ltd.). Skin lightening compositions containing maltosyl cyclodextrins and kojic acid, J. 63,208,510 [88,208,510]. Aug. 30, 1988.
70. Tseng CY, Yu ZR, Li CF. Preparation of royal jelly and property characterization of the products during storage. Zhongguo Nongye Huaxue Huizhi 1994; 32:113.
71. Sanraku-Ocean Co. Ltd. Stable royal jelly powder, JP 60. 37,942 [85 37,]. Aug. 9, 1983.
72. Komiyama T. Oxidizing agents for hair wave-setting, JP 04,356,412 [92,356,412]. Dec. 10, 1992.
73. Kubo S, Nakamura F (Shiseido Co. Ltd.), Low-odor permanent wave preparations, DE 3,416,075. Oct. 31, 1985.
74. Iwao S, Kuwana H (Lion Corp.). Hair-cosmetic composition containing a cationic polymer and a cyclodextrin for inhibition of perm odor, EP 246,090. Nov. 19, 1987.
75. Sato M, Yagi Y, Ishikura T (Sanraku Co. Ltd.). Methylated cyclodextrins in reducing agents for setting hair waves, JP 62,161,714 [87,161,714]. July 18 1987.
76. Takeuchi K, Ichikawa S (Fujinaga Pharmaceutical Co. Ltd.). Odorless and colorless skin preparations containing tar and cyclodextrin, JP 01,287,042 [89,287,042]. May 13, 1988.
77. Yorozu H, Okawa W (Kao Corp.). Bath preparations containing enzymes, sulfite salts, and inclusion compounds. JP 03,271,218 [91,271,218]. March 19, 1990.

78. Lentini PJ, Zecchino JR (Estee Lauder Inc.). Skin tanning compositions containing cyclodextrins, WO 95 22,960. Aug. 31, 1995.

79. Ogino S, Hirota H (Kao Corp.). Liquid shampoos containing menthol and β-cyclodextrin, JP 61,197,509 [86,197,509]. Feb. 27, 1985.

80. Shimada T, Mukogasa K, Gomi T, Yokoo T (Lion Corp.). Oral compositions containing cationic bactericides and cyclodextrin, JP 07,101,842 [95,101,842]. April 18, 1995.

81. Ito K, Nagai I (Shiseido Co. Ltd.). Cosmetic containing menthol derivatives, JP 06,329,528 [94,329,528]. May 20, 1993.

82. Oshino K, Egichi Y (Kao Corp.), Effervescent mouth-refreshing solid compositions containing exothermic agents and l-menthol inclusion compound, JP 07,238,008 [95,238,008]. Jan. 7, 1994.

83. Buschmann HJ, Knittel D, Schollmeyer E. α-Cyclodextrin as a perfume oil complexing agent. Parfuem Kosmet 1991; 72:586.

84. Matsuda H, Yomogida K, Miyazawa K, Uchikawa K, Taniguchi T, Tanaka M, Taki A, Sumiyoshi H, Uekama K. Release control of fragrance materials by 2-hydroxypropyl-β-cyclodextrin. In: Hedges AR, ed. Minutes of the 6th International Symposium on cyclodextrins Chicago, April 21–24, 1992. Paris: Editions de Santé, 1992, p. 516.

85. Matsuda H, Ito K, Tanaka M, Sumiyoshi H. Inclusion complexes of various fragrance materials with 2-hydroxypropyl-β-cyclodextrin. STP Pharma Sci 1991; 1: 211.

86. Matsuda H, Ito K, Fujiwara Y, Tanaka M, Taki A, Uejima O, Sumiyoshi H. Complexation of various fragrance materials with 2-hydroxypropyl-β-cyclodextrin. Chem Pharm Bull 1991; 39:827.

87. Matsuda H, Ito K, Taniguchi K, Tanaka M, Uekama K. Application of 2-hydroxypropyl-β-cyclodextrin to perfumes and cosmetics. Proceedings of the 7th International Cyclodextrins Symposium. Tokyo, Japan, 1994:516–519.

88. Lehner SJ, Müller BW, Seydel JK. Interactions between p-hydroxybenzoic acid esters and hydroxypropyl-β-cyclodextrin and their antimicrobial effect against Candida albicans, Int J Pharm 1993; 93:201.

89. Loftsson T, Stefánnsdóttir Ó, Fridriksdóttir Ö, Gudmundsson Ö. Interactions between preservatives and 2-hydroxypropyl-β-cyclodextrin. Drug Dev Ind Pharm 1992; 18:1477.

90. Matsuda H, Ito K, Sato Y, Yoshizawa D, Tanaka M, Taki A, Sumiyoshi H, Utsuki T, Hirayama F, Uekama K. Inclusion complexation of p-hydroxybenzoic acid esters with 2-hydroxypropyl-β-cyclodextrins: on changes in solubility and antimicrobial activity, Chem Pharm Bull 1993; 41:1448.

91. Citernesi U, Sciacchitano M. Cyclodextrins in functional dermocosmetics, Cosmet Toiletries 1995; 110(3):53.

92. Ogino S, Kamya H (Kao Corp.). Methylated β-cyclodextrin-containing skin or hair lotion preparations with no irritation to the skin, JP 63,192,706 [88,192,706]. Feb. 6, 1987.

93. Ogawa H (Lion Corp.). Bath preparations containing water-soluble sulfides and fragrances, JP 02 11,512 [90 11,512]. June 30, 1988.

94. Hasebe K, Ando Y, Chikamatsu Y, Hayashi K. Bath preparation containing aro-

matic essential oil-cyclodextrin inclusion compounds, JP 63 08,309 [88 08,309]. June 26, 1986.

95. Sato M, Yagi Y, Ishikura T. Bath preparations containing perfume-methylated cyclodextrin inclusion compounds, JP 62,161,720 [87,161,720]. Jan. 13, 1986.

96. Sato H (Sunstar Inc.). Toothpastes containing a flavor that changes to a different flavor during tooth brushing, JP 61,218,513 [86,218,513]. March 23, 1985.

97. Terajima Y, Tokuda K, Nakamura S. (Shiseido Co. Ltd.). Cosmetics containing fragrant compound-cyclodextrin inclusion compounds coated with oils, JP 63 35,517 [88 35,517]. July 31, 1986.

98. Szejtli J, Szente L, Kulcsar G, Kernoczy LZs. β-cyclodextrin complexes in talc powder compositions, Cosmet Toiletries 1986; 101:74.

99. Tanaka M, Matsuda H, Sumiyoshi H, Arima H, Hirayama F, Uekama K, Tsuchiya S. 2-hydroxypropylated cyclodextrins as a sustained-release carrier for fragrance materials. Chem Pharm Bull 1996; 44:416.

100. Ishii K, Yoshioka T, Takada S (Shiseido Co. Ltd.). Perfumes containing saccharides and organic substances with specified inorganic balance (IOB), JP 08,176,587 [96,176,587]. Dec. 27, 1994.

101. Yamamoto K, Hikichi S, Kitaoka K, Hara K, Tomizawa N. Perfumes for underarm applications, JP 08,183,719 [96,183,719]. Dec. 28, 1994.

102. Chikahisa N. Perfume-containing thermosetting resin moldings, JP 01,279,973 [89,279,973]. Oct. 26 1987.

103. Chen Z, Zhuang Z, Li G, Tan Z. Preparation and clinical effect of inclusion of retinoic acid-β-cyclodextrin. Guangdong Yaoxueyuan 1996; 12(2):85.

104. Zhuang Z, Li G, Shi X. Study on the stability of retinoic acid-β-cyclodextrin inclusion. Zhongguo Yiyuan Yaoxue Zazhi 1995; 15:362.

105. Wadstein J, Samuelsson M. Vitamin A–cyclodextrin complex for prevention of skin aging, WO 94 21,225. Sept. 29, 1994.

106. Matsuda H, Ito K, Taki A, Uejima O (Shiseido Co. Ltd.). Cosmetic composition containing inclusion product with hydroyalkylated cyclodextrin for preventing skin roughness, EP 366,154. Oct. 28, 1988.

107. Kawamura K, Takahashi K, Kamiyama F, (Sekusui Cemical Co. Ltd.). Dispersion of higher fatty acids and their esters in water and effect of cyclodextrin on the dispersion, JP 62 96,432 [87 96,432]. May 2, 1987.

108. Kuwabara N, Ohu S (Ensuiko Sugar Refining Co. Ltd.). Bath preparations containing cyclodextrin-squalane inclusion compounds, JP 63,190,819 [88,190,819]. Feb. 4, 1987.

109. Shibauchi I, Nakamura K. Bath preparation containing cyclodextrin inclusioncompounds of milk or milk products, JP 60,181,013 [85,181,013]. Feb. 27, 1985.

110. Shibauchi I. Bath preparation containing cyclodextrin-milk component inclusion compound, JP 61,286,318 [86,286,318]. June 13, 1985.

111. Yorozu H, Eguchi Y. Cyclodextrins containing carbone dioxide for use as bath additive, JP 61,277,610 [86,277,610]. June 3, 1985.

112. Navarro R, Delaunois M (Pierre Fabre Dermo-Cosmétique), Minoxidil-based hair care composition, WO 95 25,500. Sept. 28, 1995.

113. Kawakami M, Sakamoto M, Kumazawa T (Kyushin Seiyaku Co. Ltd.), Mouthwashes containing iodine-maltocyclodextrin inclusion compound as a microbicide with high water-solubility, JP 63,150,217 [88,150,217]. Dec. 16, 1986.

15

Microcapsules in Cosmetics

Yelena Vinetsky and Shlomo Magdassi
The Hebrew University of Jerusalem, Jerusalem, Israel

I. INTRODUCTION

Today more than ever, the skin-care and cosmetic formulator is being challenged to develop efficacious and clearly distinctive topical formulations. New raw materials, sometimes with profound effects on the various skin layers, are used for these preparations. Examples of such materials are vitamins, natural and synthetic lipids and phospholipids that mimic those in human epidermis, and, of course, alpha hydroxy acids (AHAs). Topical delivery systems that were developed primarily for pharmaceutical products are now applied for skin-care and cosmetic products because they facilitate the formulation of unstable compounds and may offer improved efficacy and esthetics.

The objectives of topical formulations can be classified in two major areas [1]: to modulate or assist the barrier function of the skin and to administer an active ingredient to one or more skin layers or compartments while minimizing systemic involvement.

The topical formulations used in cosmetics and personal-protection products that contain encapsulated active ingredient usually are one of the following delivery systems, or a combination of them: (1) microcapsules, (2) nanoparticles and microparticles, and (3) porous polymeric systems.

II. MICROCAPSULES IN TOPICAL FORMULATIONS

Microcapsules are small particles that contain an active agent or core material surrounded by a coating layer or shell. Sometimes each microcapsule may contain several cores, either of the same component or various components. At present, there is no universally accepted size of microcapsules, and their diameter may vary from 1 to 1000 μm. Capsules smaller than 1 μm are often called *nanocapsules*, and capsules larger than 1000 μm are called *macrocapsules* or beads. Commercial microcapsules typically have a diameter between 3 and 800 μm [2]. The wall thickness of microcapsules ranges from 0.5–150 μm, and the content of active ingredient (encapsulated component) may vary from 25–90% of the total weight.

Microcapsules may have a variety of structures (Fig. 1), ranging from spherical and irregular shapes with one core to multicores and even multilayer coatings. Obviously, the structure of the microcapsules depends on many parameters, such as the rigidity of the coating layer, the type of core material (solid or liquid), the preparation method, etc.

Simple
Microcapsule

Microcapsule with
Irregular Form

Multi-Nuclear
Microcapsule

Multi-Layer
Microcapsule

Figure 1   Types of microcapsule's structures. (From Ref. 1.)

III. METHODS OF MICROENCAPSULATION

The most important parameter in microencapsulation in general and in cosmetic products in particular is the release of the microencapsulated ingredient during storage and after its application. The microencapsulated components can be released in two main routes [3]: mechanical release by applying pressure during squeezing or rubbing or release by diffusion of the encapsulated component through the coating. These two principal release mechanisms influence both the shelf life of the product and the method of application to the skin or hair.

Several methods may be used for microencapsulation of cosmetic products, of which the in situ polymerization and phase separation methods are probably the most popular.

A typical process of microencapsulation usually includes the following stages (Fig. 2):

1. Emulsification stage (a): if the core material is liquid, two main types of systems exists [4]. If the encapsulated material is an oil, an oil-in-water (O/W) emulsion is prepared at the first step. If the encapsulated material is in aqueous phase, a water-in -oil (W/O) emulsion is prepared with the hydrophilic material in the internal phase. In both cases, emulsification parameters—such as the type and concentration of

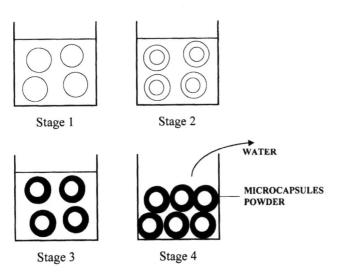

Stage 1   Stage 2

WATER

MICROCAPSULES POWDER

Stage 3   Stage 4

Figure 2 A general scheme of microencapsulation process: stage 1, emulsification; stage 2, microcapsule formation; stage 3, hardening of microcapsule walls (if necessary); stage 4, removal of liquid phase (to obtain a powder).

emulsifiers, the type of homogenization equipment, etc., are critical in controlling the size distribution of the core materials within the micro-capsules. Dispersion stage (b): if the core material is solid, the solid particles should first be dispersed in a medium from which the coating process will take place.

2. Formation of an insoluble film around each droplet or particle of the core material (microcapsule wall formation).

3. Hardening of microcapsule walls (if necessary) in order to obtain the proper physical properties, such as rigidity, porosity etc.

4. Removal of the external liquid phase if it is necessary to obtain the microcapsules in a dry powder form.

The efficiency of an active component depends largely on its bioavailability—that is, the way in which this component is actually distributed and delivered to the organism. It is indeed essential that the active substance reach its site of action and is released for a prolonged period of time at the site of action. This property is important for pharmaceutical applications, which require controlled release of an active compound in order to achieve a constant concentration of the active ingredient over prolonged period, thus avoiding the need for several applications. An example of a release profile for a free active ingredient and a "slow-release ingredient" (in polymeric matrix) is presented in Fig. 3 [5]. So far, long-acting cosmetic formulations are not widely used, and the authors believe that the use of "long-lasting" or "slow-release formulations" will become more common in the cosmetic market.

The following section will describe some of the most common microencapsulation processes.

A. Complex Coacervation

This method of microencapsulation is based on the ability of water-soluble, oppositely charged polymers to interact while forming a new liquid polymer-rich phase, which is called a *complex coacervate*. Usually the interaction between various type of positively charged gelatin (at pH below 8) and a negatively charged polymer is used to form a complex coacervate which, if present on a core material, may lead to formation of a microcapsule's wall (coating layer) (Fig. 4). Although gelatin is the most widely used protein for microencapsulation, the method is universal, and various oppositely charged polymer pairs can be used (Table 1).

Since the active groups of the polymer that interact with gelatin may vary significantly, the coacervates may vary in the physicochemical properties: Some coacervates are fluid while others are very viscous, some contain about equal amounts of gelatin and polyanion while others contain predominately gelatin.

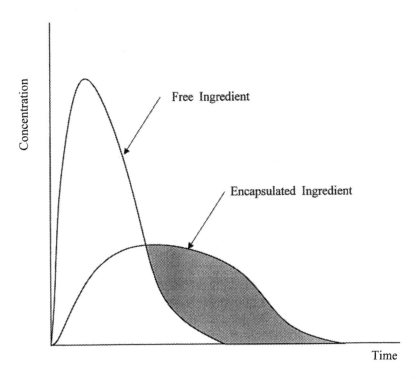

Figure 3   Evolution of the concentration of an active ingredient. Curve 1, free ingredient;
Curve 2, encapsulated ingredient. (From Ref. 5.)

Successful microencapsulation with a specific gelatin/polyanion combination is
dependent on various factors, such as pH, gelatin/polymer molar ratio, presence
of electrolytes in solution, etc. If the encapsulated core material is liquid, the
additional emulsification stage adds other parameters, such as interfacial tension,
adsorption onto the interface, and homogenization equipment. The use of a liquid
core material may also lead to manufacturing problems at the final stages of the
process, during the removal of the continuous liquid phase, as in the case of
encapsulation of volatile oils (such as orange oil), which are widely used as fragrance and flavor additives.

Microcapsules with complex coacervate walls containing protein are usu-

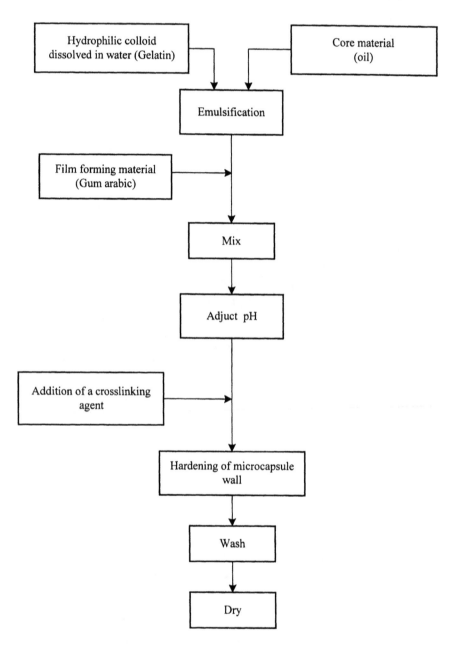

Figure 4   Microencapsulation by complex coacervation.

Table 1 Examples for the Use of Various Polymers in Complex Coacervation Method for Cosmetic Applications

Positively charged polymer	Negatively charged polymer
Acid- or alkaline-processed gelatin	Gum arabic [6]
	Poly(acrylic acid) [7]
	Carboxy-methyl cellulose (CMC) [8] or ethyl cellulose [9]
	Sodium alginate [10]
Albumin	Poly(acrylic acid) [8]

ally further treated with cross-linkers, such as formaldehyde or glutaraldehyde, in order to impart suitable rigidity and porosity of the microcapsules [11].

B. Microcapsule Formation by Complex Precipitation

Another way of forming an insoluble shell around core material is based on interaction between polymers or proteins with oppositely charged surfactants. Sovilj et al. [12] obtained paraffin oil microcapsules by the formation of an insoluble complex between a protein (gelatin) and an ionic surfactant (sodium dodecylbenzene sulfonate).

In our recent reports, we demonstrated the possibility for microcapsule formation by the creation of an insoluble complex between a similar protein, gelatin, and another anionic surfactant, sodium dodecyl sulfate (SDS) [13–15]. It was found that positively charged gelatin has the ability to bind anionic surfactants, such as SDS, and various soap molecules via electrostatic and hydrophobic interactions. At a certain molar ratio of surfactant to protein, gelatin became uncharged and therefore was precipitated by the surfactant. It was also shown that under suitable conditions, in which another protein (β-lactoglobulin) is positively or negatively charged, it may form an insoluble complex with both anionic (SDS) and cationic (dodecyltrimethylammonium bromide, DTAB) surfactants [16]. Based on these investigations, β-lactoglobulin microcapsules were obtained by the formation of insoluble complexes between the protein and SDS or DTAB.

Like the complex coacervation method, the method of microencapsulation by complex precipitation may also include an emulsification stage (Fig. 5) while the surfactant is added at a proper concentration, which is defined by the pH and protein concentration in the first step.

From the description of this process, it is clear that the formation of microcapsules can be achieved by various protein-surfactant or polymer-surfactant

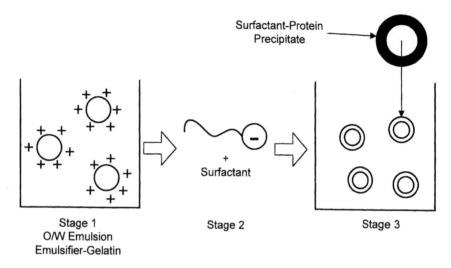

Figure 5   Microencapsulation based on formation of an insoluble complex between protein and surfactant.

combinations provided that either the protein, polymer, or surfactant has some emulsifying activity and that the interaction between the two can lead to precipitation of a suitable coating layer. The investigations carried out in our laboratory demonstrated that it is possible to encapsulate various fragrances such as orange oil by interaction between positively charged egg albumin (below its isoelectric point) and oppositely charged commercial surfactant, lecithin [17]. The orange oil microcapsules are presented in Fig. 6. The same methodology was applied with other proteins to obtain either positively or negatively charged microcapsules that have a core material such as a UV screening agent [18]. These microcapsules can be incorporated into various cosmetic formulations.

C.   In Situ Polymerization

In situ polymerization is a process whereby the shell material is directly polymerized onto the core. Microcapsules by in situ polymerization are mainly prepared by the reaction of urea-formaldehyde and/or melamine-formaldehyde resin condensates with acrylamine-acrylic acid copolymers [19]. Polymerization occurs exclusively in the continuous phase and on the continuous-phase side of the interface formed by the dispersed core material and continuous phase. Polymerization of reagents located there produces a relatively low-molecular-weight prepolymer. In situ polymerization by the reaction of urea with formaldehyde at low pH is suitable for the microencapsulation of various water-miscible liquids [20], micro-

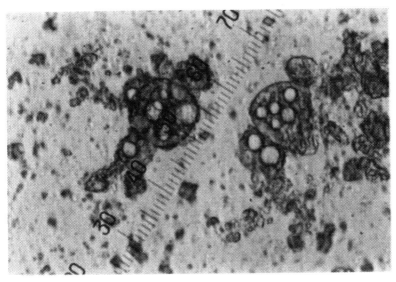

Figure 6 Microcapsules obtained by formation of insoluble egg albumin/lecithin complex as coating layer around orange oil droplets.

encapsulation by melamine-formaldehyde interaction is usually conducted in aqueous media. This process does not require special equipment, but the presence of a reactive component limits the types of liquids that can be encapsulated. Representative examples of wall materials are polyamide, polyurea, polyurethane, polyphenylester, and epoxy resin [21]. The resulting microcapsule's walls are highly cross-linked and do not swell greatly when immersed in water or stored under high-humidity conditions. Such microcapsules undergo brittle fracture under impact. As in the case of complex coacervation and complex precipitation, the first step of an in situ encapsulation procedure is to form an O/W emulsion (Fig. 7) when the core material is a liquid.

As was noted by Thies, the nature and amount of emulsifier in the aqueous phase have a significant effect on the microcapsule production process [11].

Coating with this method results in a uniform shell deposition and thickness ranging from 0.2–75 μm [22]. In situ polymerization is used to produce small (3–6 μm in diameter) microcapsules loaded with perfume oils [23] or larger capsules for cosmetic applications usually loaded with mineral oils [24].

IV. NANOCAPSULES

A special group of systems that contain encapsulated core material comprises the nanocapsules. Nanocapsules are reservoir systems made up of a continuous

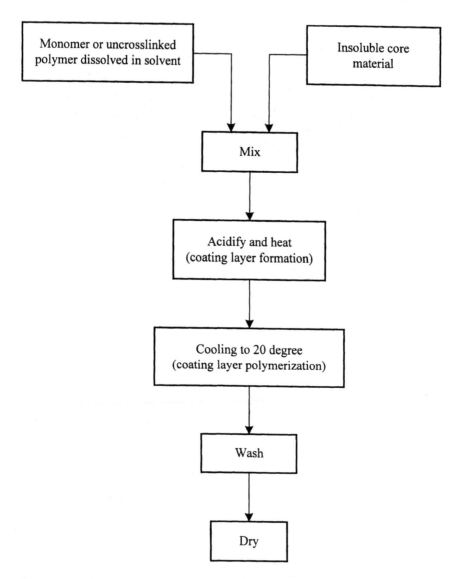

Figure 7   In situ polymerization method for microencapsulation of fragrances. (From Ref. 23.)

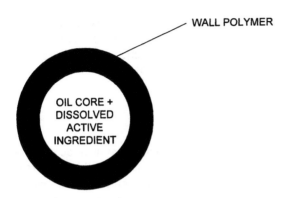

WALL POLYMER

OIL CORE +
DISSOLVED
ACTIVE
INGREDIENT

Figure 8   Schematic description of a nanocapsule. (From Ref. 25.)

polymerized envelope surrounding a liquid or gelified core (Fig. 8) [25]. The substances that are encapsulated are most often lipophilic and may be composed of a dispersion or oily mixture. As noted by Seiller et al. [25], ''certain nanoparticles are hybrid nanocapsules—nanospheres insofar as the active substances are intimately integrated into the gelified core of a nanocapsule.'' (The nanospheres are matrix systems of polymer composed of a solid core with a discontinuous envelope, or even without a coating layer.)

Typically, nanocapsules are composed of acrylic derivatives of polyalkylcyanocrylate or derivatives of copolymers such as styrene, lactic and glycolic acids, cross-linked polysiloxanes, or biological macromolecules (gelatin, dextrin, or albumin).

The procedure for the preparation of nanocapsules is basically similar to that for preparation of microcapsules, i.e., it is based on formation of a coating layer around a nanodroplet or nanoparticle. The first method is based on interfacial polymerization of a monomer, which is dissolved in the dispersed phase. For example, a monomer (water-insoluble) is dissolved in an oil phase and then emulsified by using proper surfactant, usually at high shear forces. At this stage, a submicron emulsion is formed and the monomer is polymerized at the interface of the O/W emulsion. This process can be applied, in principle, for hydrophilic or hydrophobic substances and sometimes requires the addition of a solvent.

The second method of obtaining nanocapsules is based on converting a dissolved polymer into a solid material. In principle, the polymer is dissolved with the active substance in an organic solvent, which is brought into contact with another solvent, which does not dissolve the polymer. If the contact is made in such conditions, leading to the formation of nanodroplets, the polymer is precipitated while encapsulating the nanodroplets of the active substance.

By these methods, nanoparticles are formed (150–800 nm) having a membrane thickness of 3–10 nm.

The incorporation of active ingredients in nanocapsules for cosmetics seems very promising; however, their use in the cosmetic market at present is very limited. Recently, Lancome has introduced a cosmetic product called "Primordiale" that contains nanocapsules of vitamin E [26].

V. POROUS POLYMER SYSTEMS

In contrast to the structure of microcapsules, porous polymer systems do not have a continuous shell or membrane: these are particulate systems in which the external and internal phases are in contact through small channels present in the matrix.

The most widely used systems (Polytrap from Dow Corning and Microsponge from Advanced Polymer Systems) in this class are the materials that have ability to cross-link polymers (substituted acrylates) and to entrap active components, usually by sorption mechanisms [1]. Both of these systems are made by a suspension-polymerization process that lead to very high internal–surface area production. Both systems are considerably lipophilic and have a significant oil absorption capacity, which can be used for oily-skin formulations. This type of particulate system seems to be very promising for the cosmetic industry. For example, Boots recently introduced a new lipstick, "Moisture-Stay Lipocolor," that contains an active ingredient, entrapped in microsponges, that is released while the lips are pressed together [27].

VI. COSMETIC APPLICATIONS

As noted by Greff [3], the use of microcapsules in cosmetic applications may help in solving problems such as incompatibility of substances, in conversion of odorous substances into materials with little or no odor (for example, vitamin F [26]), in protection of substances liable to oxidation or affected by atmospheric moisture (enzymes and vitamins), in the conversion of volatile substances into nonvolatile substances, or even conversion of liquids into dry powder forms. The following section contains a description of the possible applications of various encapsulated active ingredients in cosmetics: this information, which is taken from published patents, is presented according to the final cosmetic application.

A. Skin-Care Products

The use of microcapsules in skin-care products is gaining more interest recently owing to the requirement of combining various active agents within one formulation. The patent literature on this subject indicates various possible applications of microcapsules in such products, as is demonstrated in the following patents:

1. UV Protective Creams

The cosmetic composition with microcapsules obtained by Lahmani and Simoneau-Agopian [28] contains at least one sun filter present in microcapsules 1–1250 μm in size whereby the sun filters do not come in contact with the skin. The walls of the microcapsules contain gelatin, which has been hardened with glutaraldehyde. The sun filters used by the authors are selected from molecules such as octyl methoxycinnamate, camphorous methylbenzylidene, isoamylmethoxy-cinnamate, PABA, and octyl salicylate.

An interesting UV protective product, introduced by Eurand Company [26], uses microencapsulated solid antisunburn micromirrors ("Star Dusts") which are incorporated into the cosmetic vehicle.

2. Depilatory Creams

Boelcke et al. [29] suggested enzyme-containing cosmetics for application as depilation pastes, which are prepared by using microencapsulated enzymes (e.g., proteases or keratinases) for protection against inactivation caused by the presence of surface-active agents such as sodium stearate and sodium lauryl sulphate.

3. Skin Tanning

Suares and Dobkowski recently [30] proposed a cosmetic product and method for imparting a natural-appearing tan to skin, the product being in the form of a multicompartment dispenser. A first and second substance are stored in separate compartments of the dispenser and at least one of the substances includes a silicone. Additionally, the first substance contains an alpha-hydroxy aldehyde such as dihydroxyacetone and the second substance contains at least one amino acid such as glycine. Both the first and second components are W/O emulsions encapsulated within a microcapsule.

4. Cleansing Creams

Norbury et al. [31] proposed cleansing creams containing frangible microcapsules. As described, these are stable compositions that can be applied as skin cleansers and contain brittle microcapsules with encapsulated oils such as mineral oil, castor oil, jojoba oil, vegetable oil, octyl hydroxystearate, benzoates, isopropyl palmitate, and isopropyl myristate. The larger microcapsules also provide an abrasive action, therefore increasing the efficiency of the cleansing cream.

5. Skin Depigmentation

The inhibition of lipid peroxidation in the skin can be achieved, as suggested by Kono et al. [32], by using encapsulated antioxidants such as tocopherols.

6. Emollients

Noda et al. [33] proposed the use of gelatin microcapsules in which the core substances are composed of liquid paraffin, cetylisooleate, and vitamin A palmitate in a W/O type of emollient lotion. It was found that the formulation had better vitamin A palmitate stability at high temperatures as compared with a non-encapsulated product.

Soper et al. describe the microencapsulation of various oils by gelatin and carboxy-methyl cellulose (CMC) [8], which leads to the formation of a free-flowing product, such as encapsulated citrus oil. This powder can be used in a variety of creams and lotions.

7. Deodorants

Charle et al. [34] prepared a sprayable aerosol composition by encapsulating a liquid cosmetic material in microcapsules and introducing into microcapsules a fluid under a pressure essentially equal to the pressure in the aerosol container from which the composition is dispensed.

8. Bath Powders Containing Microencapsulated Oils

A cosmetic powder discussed by Tararuj et al. [35] is a thin, dry layer of powder particles that are fragrance-enhanced by fragrance-containing microcapsules. Upon application of a light finger pressure, the microcapsules break to release the oils, resulting in the consistency and scent of a fragrance powder.

A method of microencapsulating an agent to form a microencapsulated powder was recently presented by Tice and Gilley [36]. Here an effective amount of the agent is dispersed in a solvent containing a dissolved wall-forming material to form a dispersed. An extraction of the solvent from the microdroplets forms the microencapsulated product in a powder form. The microencapsulation process can be performed by using a various mediums such as water, organic solvent, or oil with or without addition of surfactants. The wall forming material [poly(lactide), poly(glycolide), poly(caprolactone), or poly(hydroxybutyrate)] enable encapsulation of various components such as fragrances, dyes, oils, or pigments.

B. Hair Products

1. Hair Creams and Ointments

Shroot et al. [37] propose a composition for the treatment of acne, physiologically dry skin, seborrhea, and hair loss, and for the promotion of the regrowth of hair. The composition comprises one or more additives such as hydrating agents, antiseborrheic agents, antiacne agents, antibiotics, and agents promoting the regrowth of hair. The product is in the form of gelatin capsules consisting of starch, talc, gum arabic, a solvent, and a thickener.

Another invention discussed by Shroot et al. [38] consist of a dispersion of ionic or nonionic lipidic microcapsules including an antiseborrheaic agent, an antibiotic, an anti-inflammatory agent, an antipsoriasis agent, an agent for promoting hair growth, a carotenoid, an A2 phospholipase inhibitor, and an antifungal agent. The composition can be prepared in the form of a cream, ointment, or gel.

2. Shampoos

Aqueous compositions proposed by Mausner [39] contain capsules of hair-conditioning agents. Typically, a shampoo–hair conditioning composition combines an effective amount of at least one surfactant for shampooing (triethanolamine lauryl sulfate), and discrete visible gelatin capsules of an encapsulated mineral oil in an amount effective to condition the hair.

A recent composition for use as a shampoo and hair conditioner was presented by Behan et al. [40]. The process enables hydrophobic materials to be microencapsulated in structures that have a high loading of the material in the presence of other materials such as surfactants and mineral oils. The process is based on forming an emulsion containing the hydrophobic material in the presence of silica nanoparticles and gelling this system. The resulting composition is a hydrophobic material encapsulated by a silica coating.

3. UV Screening Lotions

The invention discussed by Forestier et al. [41] relates to a cosmetic product containing humectants, surfactants, antifoams, fragrances, oils, waxes, lanolin, colorings, and pigments. It is intended to protect the human epidermis and contains up to 8% by weight of UV screening agents. These cosmetic formulations, which contain hard gelatin capsules, can be used as shampoos, gels, or lotions.

C. Nail Varnish and Makeup Removers

1. Nail Varnish Removers

Charle et al. [42,43] presented a cosmetic composition for removing nail enamel in a form of a multiplicity of rupturable microcapsules, a portion of which contain a first solvent for the enamel and the remainder a second solvent exhibiting a solvent action on the enamel greater than the first as well as a perfume to mask the odor of the solvents. The walls of the microcapsules are obviously inert to the solvents and the perfume; they are composed of a material such as polystyrene/maleic acid, polyethylene, gelatin, casein, natural wax, paraffin wax, polyamide, urea formaldehyde, and polyethylene/ethyl cellulose mixtures.

2. Makeup Removers

A cosmetic makeup remover cream composition for the skin is described in a patent by Charle [44]. It comprises microcapsules, the walls of which are formed from a polymer such as polyurethane, cellulose acetate, urea formaldehyde, polystyrene, styrene-acrylonitile, polyvinyl chloride, polyethylene, and epoxy resin. The cosmetic composition can be used in the form of a cosmetic cotton or a thin sheet.

   Another cosmetic formulation that can be applied as makeup remover [45] contains microcapsules homogeneously dispersed in a paper towel. This patent describes a process for the preparation of a cosmetic blotting-paper towel containing a multiplicity of microcapsules. The microcapsules (which are made of polypropylene) release the makeup remover upon the application of pressure, which ruptures the microcapsules.

D. Cosmetic Colorant Compositions

1. Lipsticks

Murphy and Lieberman [46] proposed a lipstick formulation comprising oils encapsulated in a water-soluble shell dispersed within an anhydrous base—a mixture of various waxes, oils, and coloring agents. When the lips are remoistened, the encapsulated oils are released, imparting a shiny, wet look to the lips over a prolonged period of time.

2. Blushes and Eye Shadows

Pahlck et al. [47] describe a "cosmetic formulations having activatable dormant pigments dispersed in an anhydrous base or vehicle. Shear forces applied to the cosmetic formulation following application to the skin cause activation of the dormant pigment, thereby releasing the pigment and giving the original color of the anhydrous base or vehicle renewed intensity or a 'long wearing' effect." Each microcapsule of this cosmetic formulation contains a colorant that is released into the base phase when the microcapsules are fractured by mechanical action applied to produce an intense shade. This formulation may be used as a blush and eye shadow.

   The same authors also suggest a method of making porous hydrophilic-lipophilic copolymeric powders for application as blush and eye shadow [48] and a cosmetic formulation having renewable dormant pigments made up of microcapsules dispersed in base phase, while each microcapsule has a core comprising up to 70% pigment by weight.

3. Face Powders

J. W. Charbonneau, and D. B. Pendergrass [49] recently proposed a fragrance-delivery composition composed of cake-forming components and microcapsules containing fragrance oil. The coherent cake is substantially free of encapsulated fragrance oil and contains a colorant such that the cake presents a uniform color as applied to the skin and the fragrance is released upon rupture of the microcapsules. The cake contains 20–40% microencapsulated fragrance oil, 20–60% talc, 10–40% pearlescent pigment, and 0–2% preservative.

Table 2 summarizes several cosmetic applications that make use of microcapsules, and information related to the preparation of microcapsules.

Table 2 Microcapsules in Various Cosmetic Applications

Shell material	Encapsulated material	Method of encapsulation	Application	Ref.
Gelatin	Oil-phase perfume component	Coacervation	Shampoo and cleansing compositions	50
Urea/melamine	Colorant	In situ polymerization	Body powder	51
Gelatin/CMC[a]	Citrus oil	Complex coacervation	Creams and lotions	8
Gelatin/CMC	Oily core material	Complex coacervation	Body powder	52
Urea/formaldehyde	Cosmetic oils	Polymerization	Cleansing cream	31
Gelatin/gum arabic	Pigments, mineral and vegetable oils	Complex coacervation	Lipsticks	53
Gelatin/carrageenan	Oily components	Complex coacervation	Lipsticks	54
Gelatin/gum arabic	Vitamin A palmitate, oils	Complex coacervation	Emollient lotion	33
Starch	Colorants	Phase separation	Lipsticks	46
Gum arabic	Antioxidants	Solvent evaporation	Skin depigmentation cream	32

[a] Carboxy-methyl cellulose.

In conclusion, there are many ways of obtaining microcapsules by simple methods. These methods can be applied for various cosmetic components; however, at present the use of cosmetic products that contain microcapsules is very limited.

We expect that the need for sophisticated and active cosmetic products will lead to the extensive use of microcapsules in the near future.

REFERENCES

1. Nacht S. Cosmet Toilet 1995; 110(9):25.
2. Thies C. In: Benita S. ed. Microencapsulation: Methods and Industrial Application. New York: Marcel Dekker, 1996:1.
3. Greff D. Soap Perf Cosmet 1977; 50(12):501.
4. Tadros TF, Adv Coll Interface Sci 1993; 46:1.
5. Huc A, Buffevant C, Perrier E. Seifen Öle Fette Wachse 1993; 118:1147.
6. Arneodo C, Benoit JP, Thies C. STP Pharma 1986; 2(15):303.
7. Kobayashi T. French Patent 1568500 690523. 1969.
8. Soper JC. U.S. Patent 89-401189. 1991.
9. Rawlings RM, Procter D. Canadian Patent 1086127 800923. 1980.
10. Sivik B, Gruvmark J, Jakobsson M. In: Lambertsen, G. ed. Proceedings of the 16th Scandinavian Symposium on Lipids. Bergen: Lipidforum, 1991:147.
11. Thies C. Polym Mater Sci Eng 1990; 63:243.
12. Sovilj V, Djakovic L, Dokic P. J Colloid Interface Sci 1993; 158(2):483.
13. Magdassi S, Vinetsky Y. J Microencaps, 1995; 12:537.
14. Vinetsky Y, Magdassi S. Colloid Surface A, 1997; 122:227.
15. Magdassi S, Vinetsky Y. In: Benita S. ed. Microencapsulation: Methods and Industrial Application. New York: Marcel Dekker, 1996: 21.
16. Magdassi S, Vinetsky Y, Relkin P. Colloid Surface B, 1996; 6:353.
17. Vinetsky Y, Magdassi S. Unpublished research.
18. Magdassi S, Vinetsky Y. Unpublished research.
19. Finch CA. Microencapsulation and controlled Release. Spec Publ R Soc Chem 1993; 138:1
20. Cakhshaee M, Pethrick RA, Rashid H, Sherrington DC. Polymer Commun 1985; 26:185.
21. Watanabe Y, Hayashi M. In: Nixon JR, ed. Microencapsulation, vol 3. New York and Basel: Marcel Dekker, 1978:14.
22. Sparks RE. In: Kirk RE, Othmer DE, eds. Kirk-Othmer Encyclopedia of Chemical Technology, vol 15. New York: InterScience, 1991:470.
23. Greenblatt HC, Dombroski M, Klishevich W, et al. Microencapsulation and controlled release, Spec Publ R Soc Chem 1993; 138:148.
24. Pahlck HE, Martin SR, Squires ME. U.S. Patent 89-426204. 1990.
25. Seiller M, Martini MC, Benita S. In: Benita S. ed. Microencapsulation: Methods and Industrial Application. New York: Marcel Dekker, 1996: 587.

26. Product of the Eurand Company: French agents for the cosmetic industry: SEDERMA.
27. European Cosmetic Market 1997; 14 (4).
28. Lahmani P, Simoneau-Agopian L. U.S. Patent 5 455 048. 1995.
29. Boelcke U, Wagner HR. German Patent 1 940 105. 1971.
30. Suares AJ, and Dobkowski BJ. U.S. Patent 5 612 044. 1997.
31. Norbury RJ, Chang RW, Zeller LC. U.S. Patent 5 013 473. 1991.
32. Kono Y, Hatate Y, Mitani H Japan Patent 08 259 422. 1996.
33. Noda A, Yamaguchi M, Aizawa M, Kumano Y. EP 316, 054. 1989.
34. Charle R, Kalopissis G, Zviak C. U.S. Patent 3 679 102. 1972.
35. Tararuj C, Schaab CK. U.S. Patent 4 952 400. 1990.
36. Tice TR, Gilley RM. U.S. Patent 5407609. 1995.
37. Shroot B, Lang G, Maignan J, Colin M. U.S. Patent, 4 654 354. 1987.
38. Shroot B, Hensby C, Maignan J, et al. U.S. Patent 5268494. 1993.
39. Mausner JJ. U.S. Patent 4 126 674. 1978.
40. Behan JM, Ness JN. Perring KD. U.S. Patent 5 500 223. 1996.
41. Forestier S, La Grange A, Lang G, et al. U.S. Patent 5 223 533. 1933.
42. Charle R, Zviak C, Kalopissis G. U.S. Patent 3 686 701. 1972.
43. Charle R, Zviak C, Kalopissis G. U.S. Patent 3729569. 1973.
44. Charle R, Zviak C, Kalopissis G. U.S. Patent 3 691 270. 1972.
45. Charle R, Zviak C, Kalopissis G. U.S. Patent 3 978 204. 1976.
46. Murphy JH, Lieberman G. U.S. Patent 3 947 571. 1976.
47. Pahlck HE, Martin SR. Squires ME. U.S. Patent 5 320 835. 1994.
48. Pahlck HE, Martin SR, Squires ME. U.S. Patent 5 382 433. 1995.
49. Charbonneau JW, Pendergrass DB. U.S. Patent 5 391 374. 1995.
50. Tanner AK, Dubois ZG. WO 9 516432, 1995.
51. Davis RA, Nichols SM, Buttery HJ, WO 9 511 661. 1995.
52. Scarpelli JA, Soper JC. U.S. Patent 5 043 161. 1991.
53. Pahlck HE, Martin SR, Squires ME. WO 9 106 277. 1991.
54. Fellows CT, Brown G, Haines RC. U.S. Patent 86-867199. 1986.

16

Applications of Polyvinyl Alcohol Microcapsules

Kentaro Kiyama
Lion Corporation, Tokyo, Japan

I. INTRODUCTION

Nowadays, new functions are required in the cosmetic field in order to respond to the diversification of consumer needs. It has become an indispensable technology to be able to incorporate new base materials [1–3]. As these new base materials are generally difficult to incorporate stably, the use of a microcapsule having a protective function attracts the public eye. However, a widely used gelatin-wall microcapsule had the disadvantage that it did not function fully because a core substance elute gradually in a product containing a surface-active agent.

The following properties are considered necessary for impermeable-wall microcapsules for cosmetics:

1. To offer stable protection to the core substance in final products over a long period of time
2. Not to be broken down at production, but to be easily breakable while in use
3. To control the size of microcapsule's particles with ease
4. To be prepared easily
5. To be harmless to the human body and to have minimal environmental impact

Table 1 shows various microencapsulation methods and their characteristics [4]. The safety of an interfacial polymerization method and in situ polymer-

Table 1 Preparation Methods to Satisfy the Necessary Requirements for Microcapsules

Microencapsulation methods	Stability in products	Easy breaking	Controlling of particle size	Easy preparation	Innocuousness
Interfacial polymerization	△	×	○	△	△ Monomer
In situ polymerization	○	×	△	△	△ Monomer
Spray-drying	△	○	△	○	○
Solidifying in liquid (orifice)	○	×	△	○	○
Phase separation from aqueous solution	○	○	○	○	○

Key: △ = fair; × = poor; ○ = good.

ization method is suspected, because monomers remain. A spray-drying method provides less stability, caused by porous wall formation and difficulty in controlling particle sizes. Using a solidifying-in-liquid method (orifice method), it is hard to make a microcapsule of small particle diameter (not more than 500 μm) in terms of production methods, and this creates difficulty in breaking down microcapsules while in use because of its thicker wall. In contrast to these methods, a phase-separation method from aqueous solution is likely to fulfill each condition. Therefore the method was selected.

Models of microcapsules in final products are shown in Fig. 1. The low stability of gelatin microcapsules is considered to be caused by the fact that surface-active agents permeate swollen gelatin walls, dissolving the core substance and making the core substance elute from the microcapsules into bulk phase.

On the other hand, the use of polymers having crystalline parts as a wall material, which prevents a molecule from permeating and diffusing, is considered to block the permeation of surface-active agents. Polyvinyl alcohol (PVA) was selected because it is crystalline, is prepared by means of phase separation from an aqueous solution, and is harmless to the human body [5].

The present study made it possible to develop microcapsules with the PVA wall having crystalline parts, to compare the microcapsules with widely used gelatin microcapsules, and to enhance the stability of core substances greatly.

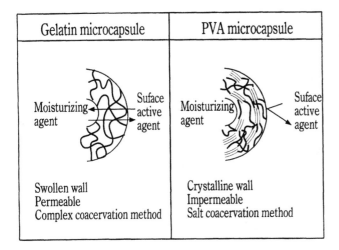

Gelatin microcapsule	PVA microcapsule
Moisturizing agent → Suface active agent	Moisturizing agent → Suface active agent
Swollen wall Permeable Complex coacervation method	Crystalline wall Impermeable Salt coacervation method

Figure 1 Models of PVA and gelatin microcapsules.

II. EXPERIMENTAL

A. Materials

Polyvinyl alcohol (PVA) was purchased from Nippon Synthetic Chemical Industry Co., Ltd. (Osaka, Japan). Pectin was supplied by Herbstreith & Fox (Neuenbürg, Germany). Gelatin (Type A, 300 Bloom) was purchased from Nitta Gelatine Co., Ltd. (Osaka, Japan) and sodium carboxymethyl cellulose (CMC; degree of etherification 60%, Mn = 31500) was from Daicel Chemical Industries, Ltd. (Osaka, Japan). All chemicals used were of reagent grade unless otherwise stated.

B. Methods

1. Preparations of PVA- and Gelatin-Wall Microcapsules

The preparation procedure of PVA-wall microcapsules by means of a phase-separation method, an improved salt coacervation method from aqueous solution, is shown in Fig. 2. First, 100 g of an aqueous solution containing 5% by weight of PVA, 0.6% by weight of pectin, and 40 g of squalane, a hydrophobic moisturizing agent, were added to a 300 cm^3 beaker. The mixture was stirred by a stirrer having three blades (12 mm long and 45-degree inclined angle) at 303 K to disperse and prepare particles with a predetermined particle size. Subsequently, 100 g of an aqueous solution of 25% sodium chloride was added gradually, keep-

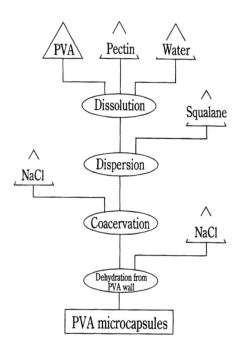

Figure 2 Preparation of PVA microcapsules.

ing the temperature at 303 K. By addition of sodium chloride, PVA coacervated droplets were generated and the surface of oil droplets of squalane was covered. In this preparation, pectin was served to induce phase separation and to prevent coagulation during the process for preparing microcapsules. Then the dispersion was subjected to a dehydration process, that is, 15 g of sodium chloride was added additionally to reduce the water content of PVA for strong PVA wall, thus, a water dispersion of PVA microcapsules was obtained.

On the other hand, as a widely used method, a water dispersion of gelatin microcapsules was prepared by the complex coacervation method from an aqueous solution as shown in Fig. 3. First, 100 g of an aqueous solution containing 3% by weight of gelatin and 1.5% by weight of CMC and 40 g of squalane were added to a 300 cm³ beaker, the mixture was stirred by the stirrer mentioned above at 303 K to disperse and prepare particles with a predetermined particle size. Next, by adjusting pH to 4.0 using acetic acid, coacervated droplets of gelatin and CMC were generated, and the surface of oil droplets of squalane was covered. The dispersion was cooled to 283 K for gelation to solidify the microcapsule wall, then 6 g of 30% by weight of glutaraldehyde was added, and the dispersion

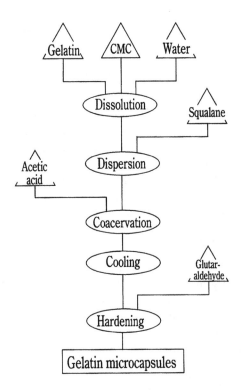

Figure 3 Preparation of gelatin microcapsules.

was stirred for 12 h to harden the gelatin wall; thus, a water dispersion of gelatin microcapsules was obtained.

2. Particle Size and Wall Thickness Evaluation

A Model BH-2 optical microscope (Olympus Optical Co., Ltd., Tokyo, Japan) and a S-520 scanning electron microscope (Hitachi, Ltd., Tokyo, Japan) were used on various samples for the evaluation of size, shape, and wall thickness of the microcapsules.

3. Measurement of Microcapsule's Hardness

Microcapsules with various wall thicknesses were prepared. One microcapsule was taken after sieving them into a predetermined particle size of 250–297 μm with a sieve, and the hardness of the microcapsule was measured by a Rehometer (Fudo Kogyo Co., Ltd., Tokyo, Japan).

4. Measurement of Release Rate of the Moisturizing Agent

The release rate of the moisturizing agent from microcapsules in product, or a facial cleanser containing surface-active agents synthesized from an amino acid, was measured by means of an accelerating test, which preserves samples at 323 K. The moisturizing agent released into the facial cleanser was measured with a Model 655A-12 liquid chromatograph (Hitachi, Ltd., Tokyo, Japan), and the release rate of each time was obtained. To analyze the diffusion mechanism of the moisturizing agent in microcapsule wall, in experiment was carried out using microcapsule sample sifted to particle size of 250–297 μm by a sieve.

5. Measurement of Crystallinity of PVA Wall

The crystallinity of the PVA wall was obtained with differential scanning calorimetry (DSC). The heat of fusion H_1 (J/g) of the PVA wall was measured with a TAS-200 Differential Scanning Calorimeter (Rigaku Co., Ltd., Tokyo, Japan) after extracting and cleansing the moisturizing agent of microcapsules. The PVA samples were sealed in aluminium sample pans and heated from 293–513 K at a rate of 5 K/min. The crystallinity was calculated in accordance with the following equation [6].

$$\text{Crystallinity } [\%] = (H_1/H_2) \times 100 \qquad (1)$$

H_1 is the heat of fusion of PVA wall and H_2 is the heat of fusion per repeating unit of PVAs crystalline part [—CH_2—$C(OH)H$—] = 53.1368 (J/g).

III. RESULTS AND DISCUSSION

A. Preparations of PVA-Wall Microcapsules

Figure 4 shows microphotographs of preparation processes for PVA microcapsules. First, squalane, which is a moisturizing agent, was dispersed in the aqueous solution of PVA and pectin. In the phase-separation process, it is observed that the surfaces of oil droplets of squalane were thinly coated by coacervate of PVA. In the dehydration process, PVA walls could not be discerned by microscopic observation, as they shrank due to their lowered water content caused by a salting-out effect from the further addition of sodium chloride.

Adjusting the particle size of microcapsules becomes vital from the points of appearance and usability. The mean particle size of microcapsules, 300 μm, is required in terms of product appearance and the feeling of touch while in use. The relationship between the particle size distribution of microcapsules and stirring rate is shown in Fig. 5. The more the stirring rate increases, the smaller the particle size becomes and the sharper the particle size distribution becomes. Therefore, the stirring rate enables us to control the mean particle size of microcapsules.

Microcapsules having a mean particle size of 300 μm and various average wall thicknesses were prepared by varying the volume rate of core and wall materials. The results of measuring the hardness of microcapsules are shown in Fig. 6, which indicate that thicker walls require stronger force to break and that the thickness of walls is able to control the hardness of microcapsules. Microcapsules having a mean wall thickness of 3 μm and a hardness of 22 g were selected because they satisfy the product design of breaking while in use yet not breaking down during production.

Figure 7 shows SEM photographs of surfaces and cross sections of prepared microcapsules of PVA wall as well as those of gelatin wall. For both microcapsules, the surface is dense and there are no micropores. The wall thickness is even over the entire cross-sectional area.

B. Release Behavior of the Moisturizing Agent from Microcapsules

Using the above-mentioned preparation procedures for microcapsules, three kinds of PVA microcapsules having a mean diameter of about 300 μm and a mean wall thickness of 3 μm were obtained with variations in preparation conditions of temperature and time. In a similar manner, gelatin microcapsules having a mean diameter of about 300 μm and a mean wall thickness of 30 μm were obtained by the above-mentioned microcapsule preparation method.

Figure 8 shows the results of measuring the residual fraction of the moisturizing agent, squalane, in microcapsules. PVA microcapsules increased the residual fraction significantly, while gelatin microcapsules made almost all core substances elute for 1 month. It became apparent that by varying preparation conditions of PVA microcapsules, the release rates could be controlled.

C. Prediction of Diffusion Coefficients of the Moisturizing Agent in Microcapsule Walls

The diffusion coefficients in microcapsule walls were calculated by applying Fick's diffusion equation, represented by the following equation, to the above-mentioned measured results of release curves and by a curve-fitting method.

$$\frac{dQ}{dt} = -D_m \cdot A \cdot \frac{dC_m}{dx} \qquad (2)$$

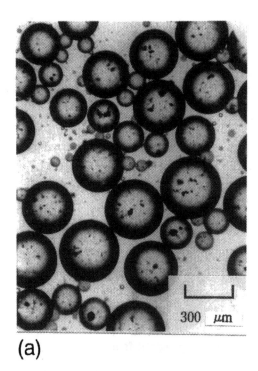

(a)

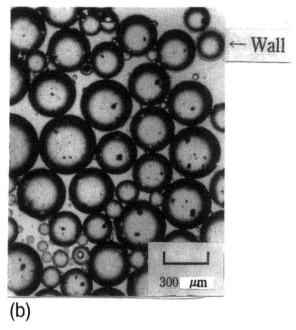

← Wall

(b)

Figure 4 Optical microscopic photographs of preparation processes of PVA microcapsules: (a) dispersion, (b) phase separation, and (c) dehydration.

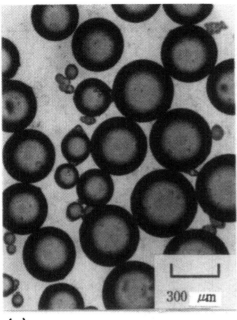

(c)

where

d = differential
Q = amount of the moisturizing agent (mol)
t = release time (s)
D_m = diffusion coefficient of the moisturizing agent in microcapsule wall (m²/s)
A = cross-sectional area for diffusion (m²)
C_m = concentration of the moisturizing agent in microcapsule wall (mol/m³)
x = wall thickness (m)

The calculation was based on the following assumptions:

1. The concentration of squalane in microcapsules is considered to be lowered because water penetrates in the reverse direction, because squalane is released from microcapsules into the bulk phase.

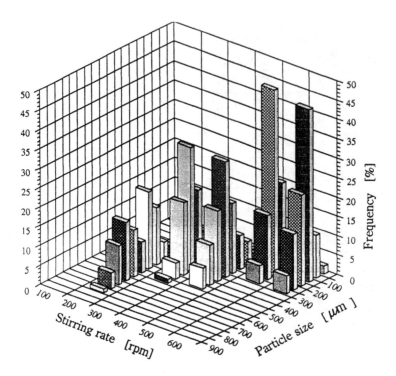

Figure 5 Correlation between the particle size distribution of PVA microcapsules and stirring rate. (From Ref. 3.)

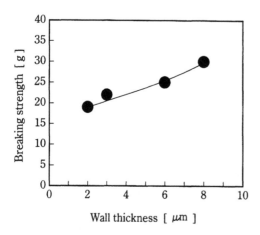

Figure 6 Correlation between breaking strength and wall thickness. (From Ref. 3.)

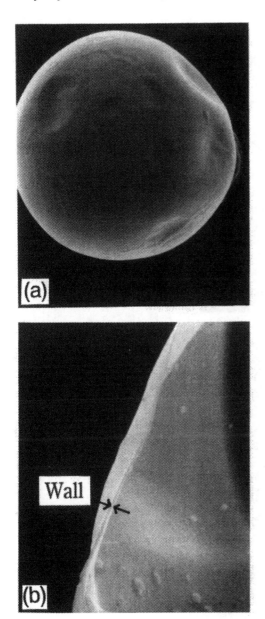

Figure 7   SEM photographs of microcapsules: (a) PVA microcapsules, surface; (b) PVA microcapsules, cross section; (c) gelatin microcapsules, surface; and (d) gelatin microcapsules, cross section.

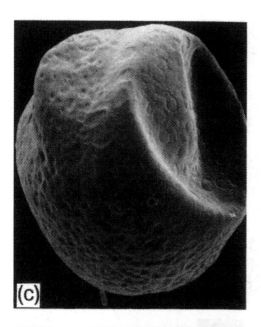

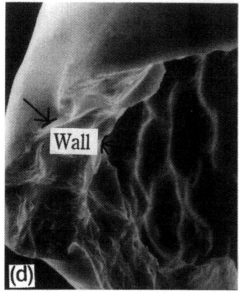

Figure 7   Continued

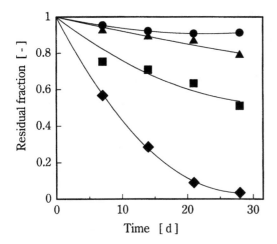

Figure 8 Stability of microcapsules in facial cleanser at 323 K, PVA microcapsules; sample 1 (●), sample 2 (▲), sample 3 (■), gelatin microcapsules (◆). (From Ref. 3.)

2. Concentrations of squalane are linearly equilibrated between the wall component phase of microcapsules and the liquid phase.
3. The quantity of squalane entrapped in the wall of microcapsules shall be negligible when the mass balance is computed.

Table 2 shows experimental data and calculated results of the diffusion coefficients of squalane in the walls of microcapsules. The obtained diffusion coefficient gives the apparent diffusion coefficient on the assumption that the wall is treated as a uniform medium. Using the preparation conditions of the PVA microcapsules, the diffusion coefficient of sample 1 was reduced to one-seventh compared with sample 3. The value was also about 1/250 comparing with the gelatin microcapsule.

D. Correlation Between the Diffusion Coefficient in Microcapsule Wall and the Crystallinity

Figure 9 shows the results of measuring the crystallinity of the PVA wall. An endothermic peak of PVA was found around 453 K, and crystallinity was calculated in accordance with the previously noted formula. Based on the result, it is found that the crystallinity of PVA is controllable in a range from 14–33% by varying preparation conditions of PVA microcapsules.

As shown in Fig. 10, a favorable correlation was found between the diffusion coefficient calculated by the measured results of the release curves and the

Table 2 Experimental Data and Calculated Diffusion Coefficients in Microcapsule Walls

		PVA microcapsules		Gelatin microcapsules
	Sample 1	Sample 2	Sample 3	
Mean diameter (μm)	276.6	277.7	268.1	273.9
Mean wall thickness (μm)	3	3	3	30
Residual 7 (d)	0.954	0.924	0.753	0.567
fraction of 14 (d)	0.922	0.890	0.709	0.286
squalane 21 (d)	0.908	0.867	0.634	0.090
(at 323 K) 28 (d)	0.913	0.789	0.510	0.035
Diffusion coefficients in microcapsule wall (m²/s)	6.0×10^{-18}	25.0×10^{-18}	40.0×10^{-18}	1.5×10^{-15}

(d) = days.

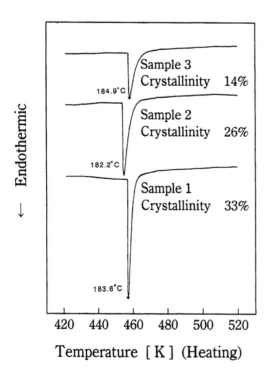

Figure 9 DSC analysis of PVA wall. (From Ref. 3.)

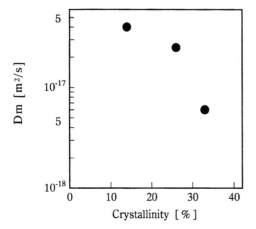

Figure 10 Correlation between the diffusion coefficient in the microcapsule wall and crystallinity. (From Ref. 3.)

crystallinity of PVA wall obtained with the DSC method. Based on the result, the improved impermeability of PVA microcapsules is probably attributed to the maze effect due to rigid crystalline parts.

E. Evaluation of Microcapsules in Products

Figure 11 shows evaluation results when PVA microcapsules were incorporated into facial cleansers. Foaming in use was significantly lowered by direct incorporating squalane, but favorable foaming was maintained in microcapsules. While the direct incorporation of squalane which has an antifoaming effect, damages the foaming, microcapsules have a mechanism for eluting squalane through the destruction of the microcapsules during facial cleansing, which does not damage the original foaming.

While the direct incorporation of squalane lowers the moisturizing effect greatly, the use of microcapsules improves the moisturizing effect. In microcapsules, squalane is considered to be effectively absorbed because the microcapsules are broken down on the skin surface. On the other hand, in direct incorporation, squalane would not be absorbed on the skin, because the squalane would be finely emulsified.

IV. CONCLUSIONS

An improved salt coacervation method from aqueous solution for the preparation of impermeable microcapsules and PVA having crystalline parts for the wall

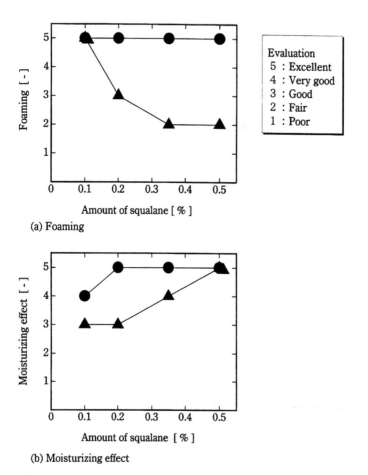

(a) Foaming

(b) Moisturizing effect

Figure 11   Evaluation of PVA microcapsules in facial cleanser. Sensory test: microcapsules (●), blank (▲). (From Ref. 3.)

material was selected. The mean particle size of 300 μm is required for good product appearance and feeling while in use. Microcapsules, which have a mean wall thickness of 3 μm and a hardness of 22 g, were selected because they satisfy the product requirement of not breaking during production but breaking easily while in use. The particle size of microcapsules is adjustable by stirring rate, and the mean wall thickness is also adjustable by the volume rate of core and wall materials.

The release rates of the moisturizing agent from microcapsules were measured. In gelatin-wall microcapsules, most core substances eluted. However, in

PVA-wall microcapsules, the residual fraction was increased greatly. It is found that the release rate is controllable by varying preparation conditions (temperature and time) of PVA microcapsules. Good correlations were found between the diffusion coefficient of the moisturizing agent in microcapsule walls, led by the release curves, and the crystallinity of PVA walls, led by the DSC method. This indicates that the existence of rigid crystalline parts improves the impermeability of materials. As a result of evaluating microcapsule combined facial cleansers, both the foaming and the moisturizing effect while in use were preferable to the case where the moisturizing agent was directly incorporated into facial cleansers.

REFERENCES

1. Takizawa M, Kiyama K. Microcapsules. Kinousei Kesyouhin. Tokyo:CMC Co, 1990: 136.
2. Kinoshita M, Kiyama K. Development and application of microcapsules in cosmetics and toiletries. Fragr J 1991; 3:33
3. Kiyama K, Yamamoto H, Nakano H, Kiyomiya A. Development and release behavior of impermeable PVA-wall microcapsules in cosmetics. J Soc Cosmet Chem Jpn 1995; 29:258.
4. Kondo T. Microcapsules: their preparation and properties, Surf Coll Sci 1978; 10:1.
5. Nagano K, Yamane S, Toyoshima K. Polyvinyl Alcohol, Tokyo, Kobunshi Kankokai Co, 1970.
6. Tubbs RK, Inskip HK, Subramanian PM. Relationships between structure and properties of polyvinyl alcohol and vinyl alcohol copolymers. Monograph No. 30, Soc Chem Ind Lond 1968; 30:88.

17
Delivery by Nylon Particles

V. Parison
Elf Atochem, Paris, France

I. INTRODUCTION

Today's consumers expect high-performance as well as multifunctional and safe products. Cosmetic products must deliver more attributes today than ever before. They must moisturize, protect skin from sun damage, provide the best possible emolliency, and have appealing tactile properties.

Nylon microspheres (orgasol) have been used for many years in makeup and skin-care products because of their spheroid structure. Owing to the fineness of their particle size and the very narrow particle size distribution, a nice smooth feeling and a suitable adherence are obtained.

Apart from this very interesting improvement in product texture, nylon microspheres offer another advantage resulting from their porous structure.

There are two basic ways to benefit from this porous structure:

1.   Using nylon microspheres directly in a formulation, it is possible to reach a matte aspect, which is directly related to the absorption of one part of the oil content on the skin. It can avoid all kind of exudation observed in lipstick or foundation formulations. It also provides a nice matte film on the skin, which gives a more natural and a long-lasting effect to the final cosmetic product. For skin care, nylon microspheres are very interesting in solving all problems of oil control by absorbing oil-secretion excess in masks, lotions, and creams.

2.   Nylon microspheres may also be used as a vector for active ingredients. Like several other polymers, nylon microspheres have chemical characteristics that allow them to be compatible with most active cosmetic materials. Depending on the grade, these particles are able to hold lipophilic or hydrophilic ingredients up to 50% of their weight, including vitamins, fragrances, sun filters, moisturizers, etc.

When the finished cosmetic product is then applied to the skin, the physical desorption will lead to a delivery of the active cosmetic material for a prolonged period, thus improving the absorption by the skin.

II. NYLON PARTICLES: PHYSICOCHEMICAL PROPERTIES

A. Chemical Structure and Properties

1. Principle of Polyamide Synthesis

Polyamide is synthesized from lactame by polymerization.
The lactame formula is as follows:

$$
\begin{array}{cc}
O & H \\
\| & | \\
C \!\!-\!\!-\!\!-\!\! N \\
\diagdown \quad \diagup \\
(CH_2)_n
\end{array}
$$

The general polyamide formula is

$$[-HN-(CH_2)_y-NH-CO-(CH_2)_x-CO-]^n$$

Various polyamides can be obtained by the selection of the proper lactame. For example, capro lactame leads to formation of polyamide 6 (nylon 6):

$$[-NH-(CH_2)_5-CO-]^n$$

Lauryl lactame leads to formation of polyamide 12 (nylon 12):

$$[-NH-(CH_2)_{11}-CO-]^n$$

Amino undecanoic acid lactame leads to formation of polyamide 11 (nylon 11):

$$[-NH-(CH_2)_{10}-CO-]^n$$

Lactame polymerization, by opening the cyclic compound, leads to polyamide synthesis. This polymerization is different from monomer polycondensation and tends to produce higher-molecular-weight polymers.

Polymerization by opening cycle consists in a transamidation (transacylation) in which cyclic amides are converted into linear chains:

The amphoteric character of the secondary amide makes the electrophilic and nucleophilic attacks possible. These are followed by cleavages and condensations. This process is followed by the chain's elongation.

Lactame can be polymerized by various ways: anionic polymerization, cationic polymerization, hydrolytic polymerization, and acidolytic polymerization.

2. Synthesis of Polyamide Powders

Polyamide powders are obtained by various methods:

Directly (elaboration and precipitation of the polymer in a solvent)
Indirectly
Polymer solubilization in a solvent at high temperature followed by powder
  precipitation while cooling
Pulverization of a cold polymer solution
Polymer grinding

The nylon powders currently used in cosmetic formulations are based on three main processes:

(a) Direct Synthesis
1. Elf Atochem (Orgasol): this process produces powders with particle size from 5 to 60 μm. The physical structure obtained is a porous spheroidal powder, as presented Figs. 1 and 5.
2. Toray (SP 500) develops an original process that produces spherical empty particles ranging in size from 7–10 μm, as shown in Fig. 2.

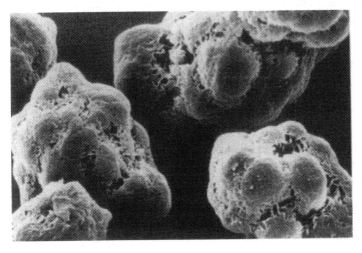

Figure 1   Orgasol EXD N COS: extra D natural cosmetic (10 μm), ELf Atochem.

Figure 2   SP 500 7% μm, Toray.

(b)   Indirect Synthesis by Dissolution/Precipitation.   This type of process is easier than the previous one and leads to a larger range of particle size, from 6–200 μm. This technique may also provide a hybrid powder.

(c)   Indirect Synthesis by Grinding/Sieving.   The method is based on a simple principle: the polymer is ground in its granulated form, which is previously obtained by extrusion. Then the various particles are sieved and classified according to their size as different grades.

The main problem with these products is that their touch is not very smooth and pleasant (''skin feel'') and for this reason they are mainly used in exfoliation formulations. Such particles are presented in Figs. 3 and 4, for the commercial products Huls (Vestosint) and Elf Atochem (Rilsan).

3.   Chemical Properties

Polyamides have a good resistance to many solvents, such as chlorinated solvents, aliphatic hydrocarbons, aromatic solvents, alkali, and acid. In addition, in principle, the polyamides are very inert and therefore suitable for use with a large variety of active ingredients.

B.   Physical Structure and Properties

1.   Powder Aspect

The different processes of manufacturing provide particles with different morphologies, which can be observed by electron microscopy.

Figure 3 Vestosint 20 μm, Huls.

(a) Spherical Particles.

1. Prepared by the direct method: the surface aspect is smooth and the granulometry is homogeneous.
2. Prepared by dissolution/precipitation: the surface aspect is rough and the granulometry is homogeneous.

Figure 4 Rilsan 30 μm, Elf Atochem.

(b) *Pseudospherical Powder.*

   1. Prepared by direct way: the granulometry is homogeneous (Figs. 1, 2, and 5).
   2. Prepared by dissolution/precipitation: the granulometry is homogeneous.

(c) *Ground Powder.* Grinding/sieving: the granulometry is heterogeneous (Figs. 3 and 4).

2. Powder Granulometry

The particle size distributions were measured by a Coulter multisizer. The Coulter multisizer determines the number and size of particles suspended in a conductive liquid by monitoring the electrical current between two electrodes immersed in the conductive liquid on either side of a small orifice, through which a suspension of particles is forced to flow.

While each particle passes through the orifice, it changes the impedance between the electrodes and produces an electrical pulse of short duration having a magnitude essentially proportional to the particle volume.

For a spheroidal particle, the main equation describing the particle's volume is as follows:

$$\Delta R = \frac{\rho_e V_p}{S^2}\left(1 - \frac{\rho_e}{\rho_p}\right)$$

Figure 5   Orgasol EXD N COS: extra D natural cosmetic (10 μm), Elf Atochem.

where

ΔR = electrolyte resistance variation through the aperture
S = Aperture area of the right section
ρ_e = Electrolyte resistivity
ρ_p = Particle resistivity
V_p = Particle volume

A large series of pulses is electronically scaled, counted, and accumulated in a number of size-related channels, which, when their contents are displayed on the integral visual display, produces a size distribution curve.

This instrument is designed to provide accurate particle size distribution curves, within a 30:1 dynamic range by diameter, or a 27,000:1 range by volume, from any orifice used.

3. Orgasol Particle Size Distribution and Its Influence in Cosmetic Applications

One of the main characteristics of Orgasol is its narrow particle size distribution. Such examples are presented in the curves for the 2002 UD N COS (ultra S natural cosmetic 5 μm) and the 2002 EXD N COS (extra D natural cosmetic 10 μm) products (Figs. 6 and 7).

This narrow particle size distribution explains a part of the extremely

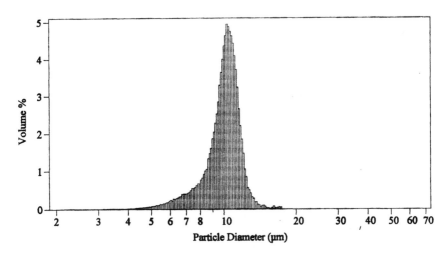

Figure 6 Orgasol EXD N COS: extra D natural cosmetic (10 μm), particle size distribution.

smooth texture obtained when these particles are introduced in makeup products or creams.

The film formed after the application of the finished product on the skin is thinner, due to a *good spreadability* of the powder. This also leads to an important property of these nylon particles: a *transparent aspect* of the makeup.

An additional advantage of this regular and narrow particle size distribution is its good flow and dispersion properties during the processing of the final product.

4. Powder Porosity

As we have seen previously in the microscope photographs, polyamide powders and especially Orgasol, have a porous structure which is very profitable for the formulator (Fig. 8), which seeks a delivery system for active ingredients.

The question is how to optimize this property in order to enable its use in the right application. It can interfere with the oil content within the formulation (by absorption), for example, or help to introduce a liquid component in a powder form.

First, we will define the adsorption/absorption phenomenon. Adsorption consists in the accumulation of a substance between two phases—e.g., at the interface (gas-solid, gas-liquid, liquid-solid, liquid-liquid, solid-solid). This phenomenon finds its origin in the intermolecular attraction strength which is responsible for the condensed phase cohesion. A molecule attracted by other molecules from two different phases will find a favorable energetic position at the surface

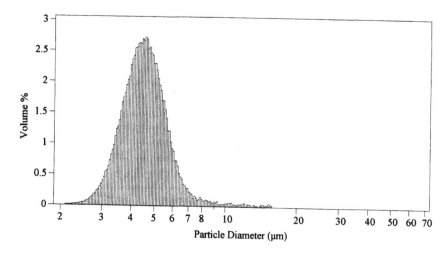

Figure 7   Orgasol UD N COS: ultra S natural cosmetic (5 μm), particle size distribution.

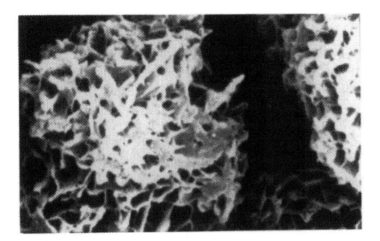

Figure 8   Orgasol UD N COS: ultra S natural cosmetic (5 μm), porosity.

of the most attractive phase. This phase is named the *adsorbant* and the molecule adsorbed is the adsorbat. If the energetic and kinetic conditions allow the molecule to enter into the adsorbant bulk phase, then it becomes an *absorption*.

The evaluation of adsorption/desorption isotherms provides several criteria which classify adsorption in two categories according to interactions which maintain the adsorbat on the surface. Physical adsorption (van der waals and electrostatic forces) or physiosorption. This type of adsorption is reversible and characterized by low energy—i.e., less than 50 KJ/mole.

Chemical interactions: *chemiosorption*. This type of adsorption is partially reversible and characterized by high energy—i.e., more than 100 KJ/mole.

5.   Solid-Gas Adsorption Isotherm

These isotherms define the amount of gas adsorbed per gram of solid as a function of the equilibrium pressure at constant temperature: $V = f(P/P_0)$

where
V = Pore volume (cm^3/g)
P = 760 mmHg
P_0 = adsorbat condensation saturating pressure
T = 273.15 K (0°C) (Fig. 9)

From these isotherms, some other parameters are deduced: S = specific surface area (m^2/g). Pore volume or surface repartition in function of their diameter, D (nm or μm).

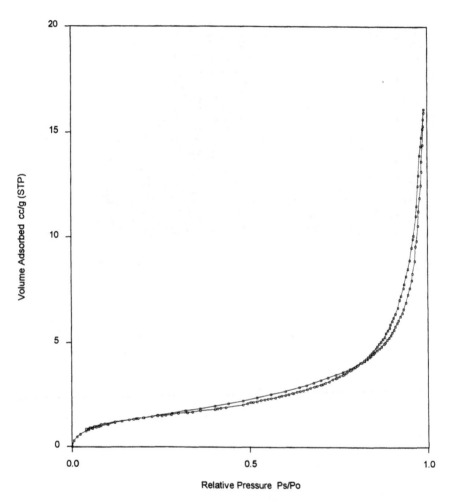

Figure 9 Isotherm, volume adsorbed in function of relative pressure.

In a porous solid, the specific surface area depends on the pore distribution. Two methods are used to access this specific surface.

(a) Physical Adsorption Isotherms.   The physical adsorption of a gas on solid takes place in multilayers. The main theory which describes the adsorption is the Brunauer–Emmett–Teller (BET) theory. The specific surface measurement (S) consists in determining, from the physical adsorption isotherm, the adsorbed gas in a volume which corresponds to the monomolecular covering (V_m), while the various parameters are related as shown in the following equation:

$$S\left(\frac{m^2}{g}\right) = 26.86 \times V_m\left(\frac{cm^3}{g}\right) \times A_m\ (nm^2) \times \frac{1}{M(g)}$$

M = sample weight
A_m = area occupied by an adsorbant molecule for azote,
      A_m = 0.1627 nm² per molecule.

An example of such an isotherm is presented in Fig. 10.

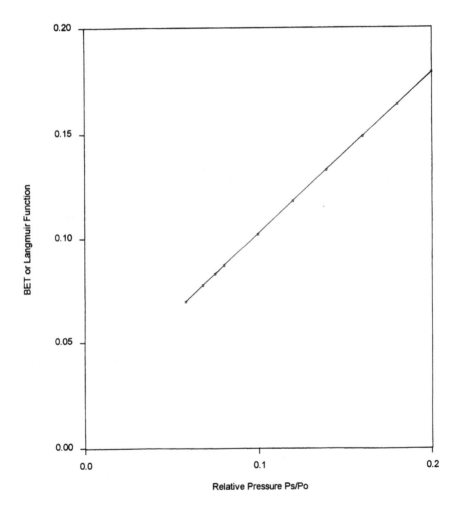

Figure 10 Surface area, BET in function of relative pressure.

(b)   Porosity Measurements.   Various types of porosity exist: we have to distinguish intragranular porosity from intergranular porosity. The pore size, *D*, is very different for various powders:

> Microporosity: *D* is lower than 3 nm
> Mesoporosity: *D* is in between 3 and 50 nm
> Macroporosity: *D* is in between 50 nm and 150 μm

Mercury porosimetry is the main method used for determination of the porosity, since it is simple and quick. It also has the advantage of being accurate for a wide range of pore sizes. With this method, several useful physical parameters can be obtained such as:

> Porous volume
> Porous volume distribution in function of diameter
> Pore surface distribution as a function of diameter (Fig. 11)
> Specific surface, which is based upon the assumption of cylindric pores
>   density

The principle of this measurement consists in bringing into contact the mercury and the powder to be analyzed. Since mercury does not have an affinity for the polyamide, it is necessary to use pressure to push it to enter the pores of the powder. The smaller the pores the higher the required pressure. To find the porous volume, the volume of mercury that has penetrated needs to be determined. At equilibrium, the radius of wet pores at pressure *P* is described by *the Washburn equation*:

$$ r = \frac{-2Y \cos \sigma}{P} $$

where

Y = surface tension of mercury (474 dynes/cm at 25°C)
σ = contact angle of mercury with the material

On the mercury porosimetry Orgasol curves, two porous volume populations appear. One corresponds to intraparticular porosity (pores' distribution inside the polyamide particle) and the other to interparticular space distribution. The intragranular porosity corresponds to a pore distribution with an average pore diameter between 75 and 8000 Å. The intragranular porosity corresponds to a pore distribution with a pore diameter between 1 and 10 μm.

With respect to Orgasol, specific surface area and porosity are directly related to the particle size. The finer the product, the better the specific surface area and porosity. This can be demonstrated for three grades of Orgasol products:

> 1.   Grade UD N COS:

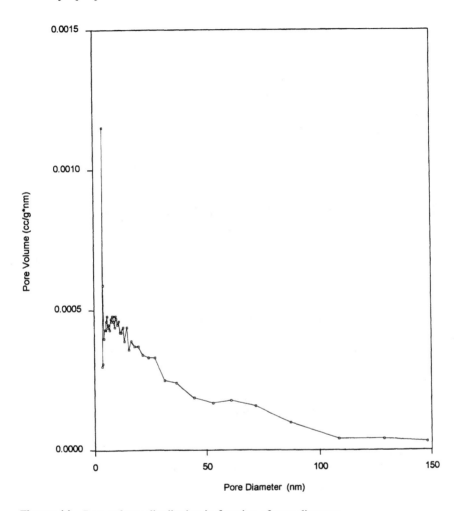

Figure 11 Pore volume distribution in function of pore diameter.

$D = 5 \ \mu m$
SSA $= 10 \ m^2/g$
Porosity $= 0.27 \ cm^3/g$
Oil absorption $= 1.15 \ g/g$

2. Grade EXD N COS:
$D = 10 \ \mu m$
SSA $= 3.5 \ m^2/g$
Porosity $= 0.15 \ cm^3/g$
Oil absorption $= 0.8 \ g/g$

3. Grade D N COS (S natural cosmetic):
   D = 20 μm
   SSA = 1.2 m²/g
   Porosity = 0.09 cm³/g
   Oil absorption = 0.7 g/g

These data on porosity and specific surface area are very useful in selecting the proper Orgasol grade for impregnation of various components within the pores of the nylon particles. The selection is based on screening according to the quantity and the type of active ingredient desired in the final product.

III. ORGASOL AS A DELIVERY SYSTEM FOR COSMETICS AND DERMATOLOGICAL PRODUCTS

Three main applications are targeted by the use of Orgasol as a delivery system in cosmetic products:

1. Carrier for liquid cosmetic ingredients in a powder form, for insoluble cosmetic ingredients in creams, lotions, lipsticks, etc.
2. Protection of sensitive cosmetic ingredients. A posttreatment can be added, in order to cover the external surface of the loaded particles and thus avoid interactions between intraparticular and extraparticular pore environment.
3. Slow delivery and long lasting effect on skin, which increases the contact time of the active component with the skin, thus improving the effectiveness of the cosmetic product.

A. Impregnation Process: Theory and Reality

Impregnation is a combination of the two previously described phenomena: adsorption and absorption. In fact, as we work with a porous powder, one part of the active ingredient is located inside the pores (intragranular porosity) and the other portion, which corresponds to the saturation level, is adsorbed on the particle's surface. It is difficult to exactly quantify the external part (intergranular porosity) but it has to be minimal.

The objective is to load a maximum level of the active ingredient inside the particles, but depending on the process conditions, different results are obtained. Variations come from the evaporating solvent speed, contact time between the powder and the active ingredient in solution, quality of the stirring, affinity between nylon and molecule to absorb, and so forth.

The general process of impregnation is simple and is based on physical absorption by the concentration gradient. If the active ingredient is not liquid or too viscous, it has to be dissolved in a solvent. The powder and solution are brought into contact under slow stirring in a reactor adapted to powder mixing. (Before introducing an active ingredient it is advisable to operate a vacuum and eliminate the intraporosity gas.) For the dissolution of the active ingredient, it is best to choose a volatile solvent, because after the first absorption step, it will be easy to evaporate. It should also be taken under consideration that in some cases, the solvent remains within the particles. For example if the active ingredient tends to crystallize inside the pores, with crystal size bigger than the pore size.

A different case occurs when we try to achieve a moisturizing effect: the solvent, water, is introduced as part of the formulation. However, in most cases, the solvent is evaporated under vacuum. The evaporation kinetics has to be optimized since if it is too fast, large crystals of the active ingredient will be formed. The powder would become rigid and the active ingredient would be mainly present at the external part of the particle, so the main objective, entrapment of the active component—would not be obtained.

B. Impregnated Powder Aspect and Dosages

1. Impregnation Level

Before incorporating the impregnated powder in the formulation and evaluating the delivery kinetics, it is necessary to evaluate the quality of the impregnation. Evaluation by electron microscopy gives a general idea of the impregnation quality. This means *Homogeneity* of the impregnation and proper active ingredient concentration.

If the powder is oversaturated, bridges between particles appear. The main consequence is that the product becomes tacky,with a wet feeling. Its dispersion in the final product or its flow during processing is more difficult, and it is also less realistic to claim a long lasting effect.

If the impregnation is conducted properly, the particles' pores are clogged and a thin film around the particles is visible; each particle is well separated from the others and the particle size is not significantly increased and neither is the size distribution modified, as shown in Figure 12. The resulting optimal impregnated powders have a smooth feeling, a good flow behavior, and an easy dispersion ability in various liquids.

2. Active Cosmetic Ingredient Dosage

Apart from the appearance of the nylon particle, the level of impregnation is measured quantitatively. Typically, the active ingredient is extracted by a solvent, (the same as the one used for the impregnation) and the concentration of the active cosmetic ingredient is determined by conventional methods such as UV,

Figure 12 Orgasol 2002 N10 VE COS: treated with vitamin E (10 μm), ELF Atochem.

fluorescence spectroscopy, titration, and so on. Our experience shows that the impregnation efficiency is high between 90 and 95%. The extraction efficiency is also high, between 95 and 98%. This information confirms that the impregnation process is indeed based on physical interactions, and is not a result of chemiosorption. Thus, the active ingredient remains chemically intact.

C. Evaluation of Slow Delivery Through a Membrane: Franz Cell

1. General Principle

The aim of this method is to evaluate the percutaneous absorption of an active ingredient dispersed in an excipient. The active ingredient diffusion is quantitatively estimated by the total product diffusion (amount which is diffused) through the skin. We have chosen to work with Franz cells which are static cells (Fig. 13).

These cells are composed of two compartments:

1. Donor: a glass cylinder, open to the air whose internal surface is 0.64 cm². This cylinder is under the upper face of the membrane (artificial or skin).
2. Receptor: the lower part of the system consists of a reservoir of 4 cm³ with a cylindrical aperture and a lateral sampling port which allows to balance the liquid volume during the test

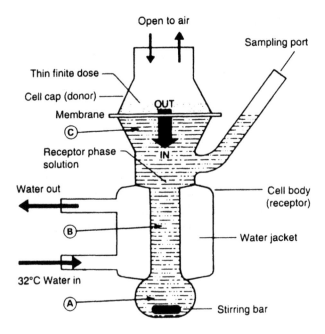

Open to air

Sampling port

Thin finite dose

Cell cap (donor)

OUT

Membrane

Ⓒ

Receptor phase
solution

IN

Water out

Cell body
(receptor)

Ⓑ

Water jacket

32°C Water in

Ⓐ

Stirring bar

Figure 13 Franz cell.

The diffusion cell, connecting lines, and collection chamber are made from an
inert nonreactive material. The membrane or the skin is placed between the two
glass parts of the cell. The stratum corneum face is in contact with the preparation
which has to be tested and the dermic face is in contact with an isotonic solution
(sodium chloride solution 9G/l + Albumine 15G/l). A circulating water bath is
placed around the receptor (32°C) and the diffusion is followed for 24 to 48 h.
In all experiments, the thermodynamic activity of the active ingredient in the
receptor fluid should not exceed 10% of its thermodynamic activity in the donor
medium so as to maintain a favorable driving force for permeation and assure
reasonable and efficient collection of product.

2. Example of a Diffusion Test with Impregnated Orgasol

(a) Tocopherol Acetate: Orgasol 2002 N10 VE COS.   After screening, we selected the 10 micron nylon particles. The optimum impregnation level is around
20% of tocopherol acetate, while the impregnation was performed, with the active
ingredient dissolved in a volatile solvent. The quality and homogeneity of impregnation was evaluated first. Pictures showed that the product conformed to the

proper specifications (particles are separated from each other and pores are clogged), (Fig. 12).

This product was introduced in an aqueous gel (concentration of tocopherol acetate is 0.5%); and 50 µL of the formulation was applied onto the membrane. The tocopherol acetate was tritiated and the detection method used was scintillation in a liquid medium. The membrane used for testing was hairless rat skin. The contact period was 48 h with sampling at 2, 4, 6, 8, 24, 26, 28, 30, 32, and 48 h. The volume sampled was replaced by new liquid samples. The experiments were conducted in six parallel cells.

(b)   Results.   As tocopherol is a large hydrophobic molecule (431 molecular weight), it's absorption level conversely is low, less than 2%. During the first 8 h, the diffusion kinetics were equivalent for the reference and the impregnated tocopherol acetate. After that period, the kinetics were different: the gel containing the Orgasol 2002 N10 VE COS had a significantly slower rate of absorption than the reference gel.

After 48 h, the cumulated absorption amount, expressed as a percentage of the deposited dose, was around 2% for the reference and 1% for the gel containing treated Orgasol.

3.   Tentative Mechanism Explanation

Diffusion through the stratum corneum is a passive phenomenon. It occurs in two steps:

1. The latency period, when the stratum corneum absorbs the tocopherol acetate. At this stage the flux increases progressively.
2. The equilibrium period is maintained as long as the concentration of the product at the skin surface is sufficient. At this stage, the flux remains constant. The absorption level during this period is a function of:

> Product concentration applied on skin.
> Thickness of the stratum corneum.
> Relative solubility of the product in the stratum corneum and the solvent.
> Diffusion of the product in the stratum corneum.

During the first 8 h, diffusion kinetics is similar for the reference and the impregnated tocopherol acetate. The concentration gradient between the skin surface and the inside layers is similar. When the superficial part of impregnated tocopherol acetate in the nylon spheres has diffused out, another concentration gradient exists between the intraporosity and the surface of spheres, which explains the flux delay.

IV. CONCLUSION

Impregnated Orgasol polymeric particles are not only an appealing concept; they are now used in commercial products for various applications, such as:

Moisturizer (hyaluronic acid, fruit acids) in various types of makeup formulations (powders, lipsticks, etc.), since it is a simple way to deliver hydrophilic substances in a dry or lipophilic medium.

Sunfilters, in which lipophilic substances have to be dispersed in aqueous gels, for example, or while it is required to obtain synergy with other materials.

Vitamins (tocopherol acetate, retinol palmitate, etc.) for skin protection while maintaining slow delivery of the vitamin to the skin.

RECOMMENDED READING

Baudet G. Granulometrie par compteur coulter et comparaison avec d'autres methodes. Bull Soc Francaise ceramique Avril–Juin 1972; 95:D4225.

Barry BW. Methods for studying percutaneous absorption. In: Swarbrick J, ed. Dermatological Formulations: Percutaneous Absorption. New York: Marcel Dekker, 1983: 234.

Bronaugh RL. Diffusion cell design. In Shah VP, Maibach HI, eds. Topical Drug Bioavailability Bioequivalence and Penetration. New York and London: Plenum Press, 1993:117.

Edmundson IC. Particles size analysis. In: Advances in Pharmaceutical Sciences, vol 2. London and New York: Academic Press, 1967:153.

Raffinot P. Caracterisation, granulometrie, compteur automatique Coulter. Techniques de l'ingenieur 1978; 7:556.

Sebenda Y. Lactam polymerization, J. Macromol SCI Chem 1972; A6:1145–1199.

Shah VP and Skelly JP. Practical considerations in developing a quality control (in vitro release) procedure for topical drug products. In: Shah VP, Maibach HI, eds. Topical Drug Bioavailability Bioequivalence and Penetration. Plenum Press, New York and London: Plenum Press, 1993:107.

Wester RC, Maibach HI. In vitro testing of topical pharmaceutical formulations. In: Bronaugh RL, Maibach HI, eds. Percutaneous Absorption. New York: Marcel Dekker, 1985:653.

Index